Rotary Kilns

Rotary Kilns

Transport Phenomena and Transport Processes

Second Edition

A. A. Boateng

AMSTERDAM • BOSTON • HEIDELBERG • LONDON
NEW YORK • OXFORD • PARIS • SAN DIEGO
SAN FRANCISCO • SINGAPORE • SYDNEY • TOKYO
Butterworth-Heinemann is an imprint of Elsevier

Butterworth-Heinemann is an imprint of Elsevier
The Boulevard, Langford Lane, Kidlington, Oxford OX5 1GB, UK
225 Wyman Street, Waltham, MA 02451, USA

Copyright © 2016, 2008 Elsevier Inc. All rights reserved.

No part of this publication may be reproduced or transmitted in any form or by any means, electronic or mechanical, including photocopying, recording, or any information storage and retrieval system, without permission in writing from the publisher. Details on how to seek permission, further information about the Publisher's permissions policies and our arrangements with organizations such as the Copyright Clearance Center and the Copyright Licensing Agency, can be found at our website: www.elsevier.com/permissions.

This book and the individual contributions contained in it are protected under copyright by the Publisher (other than as may be noted herein).

Notices

Knowledge and best practice in this field are constantly changing. As new research and experience broaden our understanding, changes in research methods, professional practices, or medical treatment may become necessary.

Practitioners and researchers may always rely on their own experience and knowledge in evaluating and using any information, methods, compounds, or experiments described herein. In using such information or methods they should be mindful of their own safety and the safety of others, including parties for whom they have a professional responsibility.

To the fullest extent of the law, neither the Publisher nor the authors, contributors, or editors, assume any liability for any injury and/or damage to persons or property as a matter of products liability, negligence or otherwise, or from any use or operation of any methods, products, instructions, or ideas contained in the material herein.

ISBN 978-0-12-803780-5

British Library Cataloguing-in-Publication Data
A catalogue record for this book is available from the British Library

Library of Congress Cataloging-in-Publication Data
A catalog record for this book is available from the Library of Congress

For information on all Butterworth-Heinemann publications
visit our website at http://store.elsevier.com/

Publisher: Joe Hayton
Acquisitions Editor: Brian Guerin
Editorial Project Manager: Carrie Bolger
Production Project Manager: Lisa Jones
Designer: Greg Harris

Typeset by TNQ Books and Journals
www.tnq.co.in

Printed and bound in the United States of America

Dedication

To my wife, Sita, my number one cheerleader, to the children, and to all the children of the underdeveloped world, particularly in Africa, who, although faced with little or no technological resources, are still motivated and determined to fulfill the dreams of becoming the scientists and engineers of tomorrow just as their counterparts in the developed world.

> *...In loving memories of two frontiers we lost who inspired my chemical engineering development and career, Professor L.T. Fan of Kansas State University (2014) and J.R. Ferron emeritus professor of chemical engineering, University of Rochester (2015)...*

Contents

Foreword to the First Edition	xiii
Foreword to the Second Edition	xv
Preface to the First Edition	xvii
Preface to the Second Edition	xxi

1 The Rotary Kiln Evolution and Phenomenon — 1
  1.1 The Rotary Kiln Evolution — 1
    1.1.1 Comparison of the Rotary Kiln with Other Contactors — 3
  1.2 Types of Rotary Kilns — 5
    1.2.1 Wet Kilns — 7
    1.2.2 Long Dry Kilns — 7
    1.2.3 Short Dry Kilns — 8
    1.2.4 Coolers and Dryers — 8
    1.2.5 Indirect-Fired Kilns — 9
  References — 11

2 Basic Description of Rotary Kiln Operation — 13
  2.1 Bed Phenomenon — 14
  2.2 Geometrical Features and Their Transport Effects — 15
  2.3 Transverse Bed Motion — 16
  2.4 Experimental Observations of Transverse Flow Behavior — 20
  2.5 Axial Motion — 21
  2.6 Dimensionless Residence Time — 24
  References — 25

3 Freeboard Aerodynamic Phenomena — 27
  3.1 Fluid Flow in Pipes: General Background — 28
  3.2 Basic Equations of Multicomponent Reacting Flows — 33
  3.3 Development of a Turbulent Jet — 35
  3.4 Confined Jets — 37
  3.5 Swirling Jets — 39
  3.6 Precessing Jets — 40
  3.7 The Particle-Laden Jet — 42
  3.8 Dust Entrainment — 43
  3.9 ID Fan — 46
  References — 46

4 Granular Flows in Rotary Kilns — 49
- 4.1 Flow of Granular Materials (Granular Flows) — 49
- 4.2 The Equations of Motion for Granular Flows — 52
- 4.3 Particulate Flow Behavior in Rotary Kilns — 54
- 4.4 Overview of the Observed Flow Behavior in a Rotary Drum — 55
  - 4.4.1 Modeling the Granular Flow in the Transverse Plane — 58
- 4.5 Particulate Flow Model in Rotary Kilns — 59
  - 4.5.1 Model Description — 59
  - 4.5.2 Simplifying Assumptions — 60
  - 4.5.3 Governing Equations for Momentum Conservation — 61
  - 4.5.4 Integral Equation for Momentum Conservation — 64
  - 4.5.5 Solution of the Momentum Equation in the Active Layer of the Bed — 68
  - 4.5.6 Velocity Profile in the Active Layer — 70
  - 4.5.7 Density and Granular Temperature Profiles — 71
  - 4.5.8 An Analytical Expression for the Thickness of the Active Layer — 72
  - 4.5.9 Numerical Solution Scheme for the Momentum Equation — 74
- 4.6 Model Results and Validation — 74
- 4.7 Application of the Flow Model — 77
- References — 79
- Appendix 4A: Apparent Viscosity — 81
- Appendix 4B: Velocity Profile for Flow in the Active Layer — 82

5 Mixing and Segregation — 85
- 5.1 Modeling of Particle Mixing and Segregation in Rotary Kilns — 87
- 5.2 Bed Segregation Model — 89
- 5.3 The Governing Equations for Segregation — 91
- 5.4 Boundary Conditions — 94
- 5.5 Solution of the Segregation Equation — 95
  - 5.5.1 Strongly Segregating System (Case I) — 95
  - 5.5.2 Radial Mixing (Case II) — 96
  - 5.5.3 Mixing and Segregation (Case III) — 97
- 5.6 Numerical Solution of the Governing Equations — 97
- 5.7 Validation of the Segregation Model — 99
- 5.8 Application of Segregation Model — 100
- References — 102
- Appendix 5A: Relationship between Jetsam Loading and Number Concentration — 103
- Appendix 5B: Analytical Solution for Case III Using Hopf Transformation — 104

6 Combustion and Flame — 107
- 6.1 Combustion — 107
- 6.2 Mole and Mass Fractions — 108
- 6.3 Combustion Chemistry — 110
- 6.4 Practical Stoichiometry — 112
- 6.5 Adiabatic Flame Temperature — 113

	6.6	Types of Fuels Used in Rotary Kilns	114
	6.7	Coal Types, Ranking, and Analysis	115
	6.8	Petroleum Coke Combustion	116
	6.9	Scrap Tire Combustion	117
	6.10	Pulverized Fuel (Coal/Coke) Firing in Kilns	118
	6.11	Pulverized Fuel Delivery and Firing Systems	120
	6.12	Estimation of Combustion Air Requirement	122
	6.13	Reaction Kinetics of Carbon Particles	122
	6.14	Fuel Oil Firing	123
	6.15	Combustion Modeling	126
	6.16	Flow Visualization Modeling (Acid–Alkali Modeling)	128
	6.17	Mathematical Modeling Including CFD	129
	6.18	Gas-Phase Conservation Equations Used in CFD Modeling	131
	6.19	Particle-Phase Conservation Equations Used in CFD Modeling	132
	6.20	Emissions Modeling	133
		6.20.1 Modeling of Nitric Oxide (NO_x)	133
		6.20.2 Modeling of Carbon Monoxide	135
	6.21	CFD Evaluation of a Rotary Kiln Pulverized Fuel Burner	135
	References	141
	Appendix 6A: Calculation of Primary Requirements for 10,000 lb/h Coal Combustion	142
7	**Freeboard Heat Transfer**	145
	7.1	Overview of Heat Transfer Mechanisms	145
	7.2	Conduction Heat Transfer	146
	7.3	Convection Heat Transfer	151
	7.4	Conduction–Convection Problems	152
	7.5	Shell Losses	154
	7.6	Refractory Lining Materials	155
	7.7	Heat Conduction in Rotary Kiln Wall	155
	7.8	Radiation Heat Transfer	159
		7.8.1 The Concept of Blackbody	159
	7.9	Radiation Shape Factors	161
	7.10	Radiation Exchange between Multiple Gray Surfaces	163
	7.11	Radiative Effect of Combustion Gases	164
	7.12	Heat Transfer Coefficients for Radiation in the Freeboard of a Rotary Kiln	165
	7.13	Radiative Exchange from the Freeboard Gas to Exposed Bed and Wall Surfaces	166
	7.14	Radiative Heat Transfer among Exposed Freeboard Surfaces	167
	References	170
8	**Heat Transfer Processes in the Rotary Kiln Bed**	173
	8.1	Heat Transfer between the Covered Wall and the Bed	174
	8.2	Modified Penetration Model for Rotary Kiln Wall-to-Bed Heat Transfer	175

8.3	Effective Thermal Conductivity of Packed Beds	178
8.4	Effective Thermal Conductivity in Rotating Bed Mode	181
8.5	Thermal Modeling of Rotary Kiln Processes	182
8.6	Description of the Thermal Model	182
8.7	One-Dimensional Thermal Model for Bed and Freeboard	184
8.8	Two-Dimensional Thermal Model for the Bed	187
8.9	The Combined Axial and Cross-Sectional Model—The Quasi-Three-Dimensional Model for the Bed	189
8.10	Solution Procedure	189
8.11	Model Results and Application	192
8.12	Single-Particle Heat Transfer Modeling for Expanded Shale Processing	196
	References	200

9 Mass and Energy Balance — 203

9.1	Chemical Thermodynamics	203
9.2	Gibbs Free Energy and Entropy	204
9.3	Global Heat and Material Balance	206
9.4	Thermal Module for Chemically Reactive System	207
9.5	Mass Balance Inputs	208
9.6	Chemical Compositions	208
9.7	Energy Balance Inputs	209
9.8	Site Survey—Measured Variables	209
9.9	Shell Heat Loss Calculations	209
9.10	Calcination Module Calculation	209
9.11	Combustion	213
9.12	Energy Balance Module	213
9.13	Sensible Energy for Output Streams	218
	References	228

10 Rotary Kiln Minerals Process Applications — 231

10.1	Lime Making	231
10.2	Limestone Dissociation (Calcination)	232
10.3	The Rotary Lime Kiln	236
10.4	The Cement-Making Process	238
10.5	The Cement Process Chemistry	238
	10.5.1 Decomposition Zone	240
	10.5.2 Transition Zone	240
	10.5.3 Sintering Zone	241
10.6	Rotary Cement Kiln Energy Usage	242
10.7	Mineral Ore Reduction Processes in Rotary Kilns	243
	10.7.1 The Rotary Kiln SL/RN Process	244
	10.7.2 Roasting of Titaniferous Materials	245
10.8	The Rotary Kiln Lightweight Aggregate-Making Process	247
	10.8.1 Lightweight Aggregate Raw Material Characterization	250

		10.8.2 Lightweight Aggregate Feedstock Mineralogy	250

		10.8.2 Lightweight Aggregate Feedstock Mineralogy	250
		10.8.3 Lightweight Aggregate Thermal History	252
	10.9	The Rotary Kiln Pyrolysis/Carbonization	255
		10.9.1 Pyrolysis	255
		10.9.2 Slow Pyrolysis Products	256
		10.9.3 Fast Pyrolysis Products	257
		10.9.4 The Rotary Kiln Pyrolysis for Carbonization	258
	References		263
11	**Rotary Kiln Petroleum Coke Calcination Process: Some Design Considerations**		**265**
	11.1	Introduction	265
	11.2	The Rotary Calcining Integrated System	266
	11.3	Direct-Fired Kiln Characteristics	266
	11.4	Mass Balance	269
	11.5	Thermal Balance and Energy Use	270
	11.6	Volatile Matter	271
	11.7	Dust Output and Pickup	271
	11.8	Fixed Carbon Recovery and Burnout	271
	11.9	Kinetics and Product Quality	272
	11.10	Direct-Fired Pet-Coke Kiln Temperature Profiles	274
	11.11	Design Structure	278
		11.11.1 Mass Balance Scheme	278
		11.11.2 Comments	280
	11.12	Computational Fluid Dynamic Modeling (Pyroscrubber)	285
	References		289
12	**Rotary Kiln Environmental Applications**		**291**
	12.1	Basic Regulatory Framework for Waste Burning Kilns	291
		12.1.1 Destruction and Removal Efficiency	295
		12.1.2 Flame Turbulence and DRE	299
	12.2	Hazardous Waste Incineration	299
		12.2.1 Risk Assessment	301
	12.3	Dual Use—Combustion Systems	301
		12.3.1 Municipal Solid Waste and Power Generation	301
	12.4	Carbon Emissions, Reduction, and Capture	304
		12.4.1 Oxycombustion Cement Plant	304
		12.4.2 Cement Plant CO_2 Limitations in the United States	305
	References		306

Appendix 309
Index 359

Foreword to the First Edition

In contrast to the situation for other gas—solid reactors such as fluidized beds, the rotary kiln reactor has so far lacked an authoritative treatment in a single book. Process engineers, operators, designers, researchers, and students have struggled with a widespread collection of papers, from many industries, with key articles and classical contributions that are over 40 years old and difficult to access. At the same time, new applications for the kiln have arisen, new fuels are being used, and new tasks—such as waste incineration—are being applied to kilns originally designed for other uses. Much improved understanding of bed processes has been developed in the past decade or so, and tools for design and analysis of existing operations, such as computational fluid dynamics, have evolved from rare to widespread use. There has been a pressing need for these developments to be brought together and assessed such that the kiln engineer, for example, can appreciate what can be achieved from different types of mathematical modeling and so that the fluid particle researcher can understand what has been accomplished by recent research and where further effort is needed. The timely publication of this new book, *Rotary Kilns: Transport Phenomena and Transport Processes,* by Akwasi Boateng, admirably fills this gap.

Written by an engineer gifted in his understanding of the fundamentals of fluid–solid systems and their mathematical modeling, and having extensive experience in rotary kiln practice, the book provides the reader with a unique vantage point from which to analyze rotary kiln processes. The book provides a clear exposition of kiln behavior and of the bases and assumptions of the various mathematical models used to describe it. Examples are provided for the purposes of illustration.

In terms of the content, *Rotary Kilns: Transport Phenomena and Transport Processes* provides chapters devoted to the kiln's basic description and to its numerous applications. Several chapters on freeboard processes, aerodynamics, combustion and flames, and heat transfer are linked to a detailed exposition of bed processes, including particulate behavior, granular flow models, mixing and segregation, and heat transfer and their effects on kiln performance. Overall mass and energy balances, measurements, and site surveys are presented in the penultimate chapter, after the reader has been well prepared by the prior treatment of the individual processes, rather than at the beginning of the book. The final chapter contains practical information on several key rotary kiln processes in the minerals industry. An extensive appendix is included.

Readers will find that this book starts most topics at a very basic level. Some examples of this are the chapters on combustion and on mass and energy balances. Soon, however, the complexities become evident, such as in the chapters on granular flow and segregation.

This book will become a benchmark for the study of rotary kiln phenomena, for the design of new rotary kiln processes, and for the modeling and analysis of existing processes. It will be an indispensable addition to the bookshelves of engineers in the cement, lime, pulp and paper, and mineral industries. It will prove invaluable to researchers and academics studying the complexities of solid−gas flows, mixing, and transport processes.

Paul Watkinson, PhD
Vancouver, British Columbia
October 2007

Foreword to the Second Edition

The first edition of this book has made a significant contribution to the study of rotary kilns, providing all who work with or study these simple, but complex, devices a wealth of knowledge in one volume. This new edition contains a number of additions and improvements, including two additional chapters that reflect the growing importance of petroleum coke and environmental applications for rotary kiln operations. These additions are a significant enhancement to the value of this already excellent book.

Andrew A. Shook, PhD
Perth, Western Australia
2015

Preface to the First Edition

The author was born in Ghana in a little village called Kentikrono, now a suburb of Kumasi, the Ashanti capital. After a brief stint as an engineering cadet in the merchant marine corps of Ghana, he accepted a scholarship to study marine engineering in the former Soviet Union. Changing fields, he pursued studies in mechanical engineering in Moscow specializing in thermodynamics and heat engines where he redesigned a turboprop engine for hydrofoil application. He pursued graduate school in Canada, thereafter, and completed an MS in thermofluids mechanics at the University of New Brunswick with a thesis under J.E.S. Venart on energy conservation in greenhouses using thermal night curtains to prevent low-temperature nighttime infrared radiation heat losses through polyethylene roofs. He accepted a faculty position at the University of Guyana in South America, where his wife hails from, and taught undergraduate thermodynamics and heat transfer. His research in fluidized bed thermochemical conversion of rice hulls to provide bioenergy and utilization of the rice hull ash for cement applications earned him a faculty Fulbright award to the United States, where he spent a year working in Dr. L.T. Fan's laboratory at the department of chemical engineering at Kansas State University. Upon an invitation to Canada, he joined Dr. Brimacombe's group at the Center for Metallurgical Process Engineering at the University of British Columbia, where he pursued a PhD in rotary kilns under the sponsorship of Alcan Canada. His dissertation, *Rotary Kiln Transport Phenomena—Study of the Bed Motion and Heat Transfer,* supervised by P.V. Barr, presented some pioneering works on the application of granular flow theories for the modeling of particle velocity distribution in mineral processing kilns from which heat transfer within the kiln bed could be adequately and sufficiently solved.

After a brief stint as an Assistant Professor at Swarthmore College in Pennsylvania, he joined Solite Corporation, a rotary kiln lightweight aggregate manufacturing company in Virginia founded by Jane and John Roberts (Swarthmore 1939) as a research and production engineer. At Solite, he developed a two-part training manual on rotary kiln transport phenomena for project 10-10-10, an operational campaign promoted by John Roberts to increase production and product quality by 10% and also reduce fuel consumption by 10%. After Solite restructured in 1997, the author returned to Pennsylvania and joined Fuel and Combustion Technology (FCT) founded by colleagues from the United Kingdom, Peter Mullinger and Barrie Jenkins, who, having also completed their PhD works on kiln combustion had developed methods of optimizing turbulent diffusion flame burners to match cement and lime kiln processes. When FCT's owner, Adelaide Brighton Cement, was acquired by Blue Circle Cement, the author joined the process group of Fuller-FL Schmidt (FFE) Minerals, now FLS Minerals in Bethlehem, PA, where he participated in works leading to the design of several large direct-fired mineral processing kilns including limestone calcination, vanadium extraction, and soda ash production. He later worked for Harper International in Lancaster, NY, a lead provider of indirectly heated, high-temperature, rotary kilns employed for niche applications including inorganic materials. After Harper, he became a consultant to the industry providing process expertise including training to the rotary kiln community where he was dubbed "the kiln doctor." He is now a senior Research Scientist with the Agricultural Research Service of the US Department of Agriculture pursuing research in biofuels and bioenergy.

Rotary Kilns: Transport Phenomena and Transport Processes, is a culmination of the author's work in rotary kilns in both academic research and industry. It captures the author's experiences in production, process design, and commissioning, and more importantly, attempts to bridge the classroom and the rotary kiln industry. The focus of *Rotary Kilns: Transport Phenomena and Transport Processes* is to provide the process engineer and the researcher in this field of work some of the quantitative descriptions of the rotary kiln transport phenomenon including freeboard and bed process interactions. The latter combines the transverse bed motion and segregation of granular materials and the resultant effect of these phenomena on the bed heat transfer. Although other bed phenomena, such as axial segregation (sequential banding of small and large particles along the kiln length) and accretion (deposition or growth of material onto the refractory wall forming unwanted dams) are also not well understood, these are only qualitatively described. However, these phenomena can be better explained after careful elucidation of the transverse bed motion, segregation, and heat transfer. The work has been divided into sequential topics beginning with the basic description of the rotary kiln operation followed by fluid flow in rotary kilns where the freeboard phenomenon is presented. Here the similarities of fluid flow in conduits are drawn to describe the characteristics of confined flows that manifest themselves in combustion and flames typical of the rotary kiln environment.

In Chapter 4 the granular flow phenomenon in rotary kilns is presented. In rotary kilns, often the material being processed is composed of granules, hence the underlying theories for such flows are important to the bed motion, gas–solids reactions, and solid–solid reactions that take place in the bed. With the knowledge of these flows, it is only prudent to cover mixing and segregation as they develop in rotary kilns. This is accomplished in Chapter 5. The severity of mixing phenomena impacts greatly on the quality of the product since it influences the thermal treatment of any granular material. Mixing and segregation determines the extent to which the rotary kiln can be classified as a continuous stirred reactor. The flame is the heart of direct-fired kilns, thus combustion and flame is treated in Chapter 6. The types of flames developed in rotary kilns depend on the flow distribution in the freeboard, which, in turn, determines the heat fluxes to the charged material and also emissions. Treatment of heat transfer in freeboard is therefore a logical sequence and this follows in Chapter 7 with a review of the fundamentals of process heat transfer. Many mathematical models have previously been applied to describe freeboard heat transfer in rotary kilns including one-dimensional zone models, and two- and three-dimensional computational fluid dynamics. Some of these are presented, including recent developments. Freeboard treatment is followed by bed heat transfer in Chapter 8. Like fixed-bed heat transfer, rotary kiln bed heat transfer is composed of particle-to-particle conduction, convection, and radiation. However, superimposed on this phenomenon is an *advective* transport component that is generated due to granular flow that sets apart rotary kiln heat transfer from packed-bed heat transfer. Some existing packed-bed models and their extension to rotary kilns are presented here. Following the bed heat transfer, the mass and energy balance is established in Chapter 9 by considering the kiln operation as a thermodynamic system that interacts with the atmosphere. A simple mass and energy balance is presented for a lime kiln. Having established all the above, it is only prudent to present some specific mineral processing applications for which the rotary kiln has been the main workhorse in Chapter 10. Some of the processes discussed include lime making, cement making, carbothermic reduction kilns, and lightweight aggregate kilns.

The author is indebted to the many students both in the colleges he has taught and in the industry where he has lectured. He is grateful to Solite Corporation, which gave him an unprecedented opportunity to test his theories and mathematical models on large rotary kiln processes in the early years. He is also indebted to FCT, FLS Minerals, Harper, and all the many members of the family of rotary kiln operators particularly Utelite Corporation, Graymont, Inc., and others who gave him an unparalleled education beyond the classroom. Finally, the author is indebted to Dr. Gus Nathan, Dr. Phillip Shaw, and Dr. Peter Cooke for the critical feedback they provided on the manuscript for this book.

A. A. Boateng, PhD
Royersford, PA
2008
aboaten1@gmail.com

Preface to the Second Edition

Since the publication of the first edition of this book in 2008, feedback from our numerous readers has been very positive but certainly not without suggestions for improvement. While reviewers of the first edition generally agreed upon the adequacy and sufficiency of the material, they requested that adding a content on pyrolysis processes with low temperatures, adding a new material on the calcination of petroleum coke, and providing environmental compliance information on the use of rotary kilns as incinerators for chlorinated hydrocarbons and carbon dioxide capture would bring the book in line with recent developments. More computational fluid dynamics (CFD) information was also requested.

We have attended to these requests by either expanding on existing chapters or by providing complete new chapters. Specifically, we have expanded on Chapter 10 with a section on rotary kiln pyrolysis/carbonization where we have provided information on pyrolysis processes through a case study carried out at the UK Biochar Research Centre (UKBRC) of the University of Edinburgh on producing "standard biochar" for use in biocarbon applications exemplifying features of a rotary kiln for carbonization.

We have included two new chapters, Chapters 11 and 12. Chapter 11 provides new material and information on the characteristics of petroleum coke calcination in the rotary kiln and the use of this information as a design case for sizing a rotary kiln for the said application, so that the reader, a process engineer or the student in the process engineering class could identify with the design methodologies underlying such application. This chapter is used to highlight new developments in CFD where CFD is used as a tool to optimize a new kiln design or to improve an existing one.

Finally Chapter 12, contributed with the help of Stephen P. Holt, addresses rotary kiln environmental applications and introduces the reader to the technical basics of the US environmental regulations that are applicable when an industrial rotary kiln is permitted to burn hazardous waste for energy and also for incineration of toxic materials. The chapter also introduces new but preliminary developments in carbon emissions and capture.

The author is grateful for the contributions made to this second edition by Dr. Ondrez Masek of the UKBRC of the University of Edinburgh (www.biochar.ac.uk) and to Mr. Stephen P. Holt, P.E., (spholt@aol.com) Director, Environmental Affairs, Giant Cement Holding, Inc., Summerville, South Carolina.

A. A. Boateng, PhD
Royersford, PA
2015
aboaten1@gmail.com

The Rotary Kiln Evolution and Phenomenon

1.1 The Rotary Kiln Evolution

Rotary kilns have been synonymous with cement and lime kilns probably because of the history of their evolution and development. It has been reported that cement deposits characterized by Israeli geologists in the 1960s and the 1970s place cement making at 12,000,000 BC when reactions between limestone and oil shale occurred during spontaneous combustion to form a natural deposit of cement compounds (Blezard, 1998). Between 3000 and 300 BC, cement evolution had continued with the Egyptians who used mud mixed with straw to bind dried bricks to carry out massive projects such as the pyramids. This evolution continued with the Chinese who used cementitious materials for building the Great Wall. Projects such as the building of the Appian Way by the Romans later led to the use of pozzolana cement from Pozzuoli, Italy, near Mt Vesuvius. However, it is reported that the technology that uses the burning of lime and pozzolana to form a cementitious admixture was lost and was only reintroduced in the 1300s. In the United States, projects such as the construction of a system of canals in the first half of the nineteenth century, particularly the Erie Canal in 1818, created the first large-scale demand for cement in this country, which led to various cement production businesses to compete for the market share. By 1824, Portland cement had been invented and developed by Joseph Aspdin of England; this involved the burning of finely ground chalk with finely divided clay in a lime kiln yielding carbon dioxide as an off-gas (Peray, 1986). In these early days, stationary kilns were used and it is said that the sintered product was wastefully allowed to cool after each burning before grinding. The history of cement (Blezard, 1998) has it that in the late 1870s Thomas Millen and his two sons, while experimenting with the manufacture of Portland cement in South Bend, Indiana, burned their first Portland cement in a piece of sewer pipe. This perhaps marked the first experimental rotary kiln use in America. By 1885, an English engineer, F. Ransome, had patented a slightly tilted horizontal kiln that could be rotated so that material could move gradually from one end to the other. The underlying principle of this invention constitutes the rotary kiln transport phenomenon that we know of today.

Because this new type of kiln had much greater capacity and burned more thoroughly and uniformly, it rapidly displaced the older type kilns. It has been further mentioned that the factor that contributed to the tremendous surge of Portland cement between 1880 and 1890, reportedly from about 42,000 to 335,000 barrels, was the development of the rotary kiln (Blezard, 1998). Like most early inventions in the United States, it is said that Thomas A. Edison played a role in furthering the development of the rotary kiln. He is credited for introducing the first long kilns used in the industry at his Edison Portland Cement Works in New Village, NJ, USA, in 1902.

His kilns are believed to have been about 150 ft long in contrast to the customary length at that time of 60–80 ft. Today, some kilns are more than 500 ft long with applications ranging far wider than cement and lime making. By the 1900s, most of the advances in the design and operation of cement and lime kilns had undergone a systematic evolution since the days of the ancient Egyptians. By this time, almost countless variations of patented kilns had been invented and promoted, although some of these never found useful applications. It is fair to say that kilns have evolved from the so-called field or pot kilns that were crudely constructed of stone and often on the side of hills, to vertical shaft and rotary kilns with each evolution step carried out with the improvement of labor intensiveness, productivity, mixing, heat transfer, and product quality in mind.

Following cement, other industries also joined in the bandwagon. For example, the rotary kiln process for making lightweight aggregate (LWA) was developed by Stephen Hayde in the early 1900s in Kansas City, Missouri (Expanded Shale, Clay, and Slate Institute). In the expanded shale industry, natural LWAs had been used to make lightweight concrete since the days of the early Greeks and Romans, but it was not until the discovery of expanded shale, manufactured by the rotary kiln process, that an LWA with sufficient strength and quality became available for use in the more demanding reinforced concrete structural applications.

Currently, rotary kilns are employed by industry to carry out a wide array of material processing operations, for example, reduction of oxide ore, reclamation of hydrated lime, calcining of petroleum coke, and hazardous waste reclamation. This widespread usage can be attributed to such factors as the ability to handle varied feedstock, spanning slurries to granular materials having large variations in particle size, and the ability to maintain distinct environments, for example, reducing conditions within the bed coexisting with an oxidizing freeboard (a unique feature of the rotary kiln that is not easily achieved in other reactors). The nature of the rotary kiln, which allows flame residence times of the order of 2–5 s and temperatures of over 2000 K, makes such kilns a competitive alternative to commercial incinerators of organic wastes and solvents. However, the operation of rotary kilns is not without problems. Dust generation, low thermal efficiency, and nonuniform product quality are some of the difficulties that still plague rotary kiln operations. Although the generally long residence time of the material within the kiln (typically greater than 1 h) aids in achieving an acceptably uniform product as the early users had intended, there is considerable scope for improving this aspect of kiln performance. In order to achieve this improvement a more quantitative understanding of transport phenomena within the bed material is required, specifically of momentum transport, which determines particle motion and energy transport, which, in turn, determines the heating rate for individual particles. This book seeks to present the quantitative understanding of the transport phenomena underlying the rotary kiln.

Fundamentally, rotary kilns are heat exchangers in which energy from a hot gas phase is extracted by the bed material. During its passage along the kiln, the bed material will undergo various heat exchange processes, a typical sequence for long kilns being drying, heating, and chemical reactions that cover a broad range of temperatures. Although noncontact (i.e., externally heated) rotary kilns are employed for specialized work, most kilns allow direct contact between the freeboard gas and bed material, as shown in Figure 1.1. The most common configuration is countercurrent flow whereby

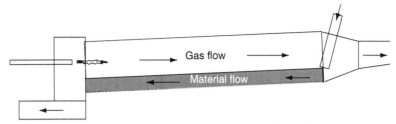

Figure 1.1 Schematic diagram of countercurrent flow rotary kiln configuration.

the bed and gas flows are in opposite directions, although cocurrent flow may be utilized in some instances, for example, rotary driers.

1.1.1 Comparison of the Rotary Kiln with Other Contactors

As can be seen from the history of its evolution, the design and operation of rotary kilns have undergone a systematic evolution since the days of the ancient Egyptians. Improvements include reduced labor, increased productivity, mixing, heat transfer, and product quality. Mineral processing kilns can be classified as vertical, horizontal, or other miscellaneous mixed types (Table 1.1). At one extreme, vertical kilns operate in the packed-bed mode whereby the material being processed (calcined) is charged from a top hopper and contained in a vertical chamber in which the static bed moves, en bloc, downward in plug flow. An example is an annular shaft kiln schematic shown in Figure 1.2.

Here the charge can be either in countercurrent or in parallel flow to the combustion gases that transfer heat to the solids (e.g., limestone) as the gas flows through the

Table 1.1 Typical Features of Rotary and Other Contact Kilns

Vertical Kilns	Horizontal Kilns	Other/Mixed
Traditional shaft-type kilns	Conventional long wet, dry rotary kilns	Fluidized-bed type kilns
Indirect gas-fired	Direct or indirect fired	Gas suspension-type kilns or flash calciners
Large capacity, mixed feed, center burners	Noncontact, externally heated small-capacity kilns used for niche applications	Rotary hearth with traveling grate or calcimatic kilns
Parallel flow regenerative type	Modern with recuperators such as cooler type, preheater kilns, and internals	Horizontal ring type, grate kilns, etc.
Annular or ring type	Cylindrical	Cylindrical, rectangular, etc.

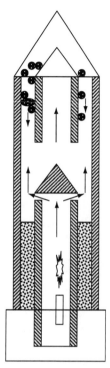

Figure 1.2 Schematic diagram of annular shaft kiln.

particle–gas interstices. To maximize heat and mass transfer in such devices, ample voidage within the particulate charge is necessary. This ensures uniform circulation of hot gases through the packed bed. Feed particle size and distribution must be selected to ensure an optimum voidage. Typically, particle size greater than 50 mm (2 in.) is normal for shaft kilns leading to a typical charge void fraction of about 45%. At the other extreme to packed beds, as encountered in vertical shaft kilns, are fluidized-bed contactors or related kilns whereby the charged particles are suspended by the hot gases in a dilute phase (Figure 1.3).

In fluid-bed suspension kilns, the void fraction can be on the order of 60–90%. The hot gases perform two functions, that is, they fluidize or suspend the particles, and, at the same time, they transfer heat to the particles. Although heat transfer is extremely efficient at the gas–particle level, a tremendous amount of energy is required to keep the bed in suspension and to move the charge. Since fluidization is a function of particle size, feed particles can only be fed as fines. Additionally, because of vigorous mixing associated with fluidization, attrition and dust issues can be overwhelming. In between these two extremes are the horizontal rotary kilns (Figure 1.1) that offer a distinct environment for combustion gases (freeboard) and the charge (bed). Unlike packed-bed vertical kilns, some degree of bed mixing is achieved by kiln rotation and associated phenomena, although not to the extent achieved by fluid-bed suspension kilns. Rotary kilns have evolved as the equipment of choice for most mineral processes

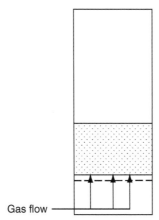
Figure 1.3 Schematic diagram of a fluidized-bed calciner.

because they provide a compromise between the packed- and suspension-bed type modes of operation thereby allowing large-capacity processing with few process challenges.

In spite of the distinctions described herein, most kilns, vertical or horizontal, when used for thermal processing, for example, calcination, oxidation, and reduction, in a continuous operation will have distinct zones along their axial length. These will include a preheating zone where the particles are preheated, a combustion zone that normally coincides with the location along the vessel where the combustion or the flame is situated, and the discharge or cooling zone behind the flame. The extent of the intended reaction and, for that matter, product quality, is most influenced by the conditions in the combustion zone where heat must be supplied to the solids, for example, in limestone calcination, well above the dissociation temperature. For product quality purposes, it is important to ensure that the temperature in the calcining zone is uniform, with no hot or cold spots, and that it can be controlled within a tolerable limit. Of all the furnace types described above, one can say that the rotary kiln offers the best potential to control the temperature profile.

1.2 Types of Rotary Kilns

Several rotary kiln designs have evolved, each specific to the process application it is intended for. They also come in several forms and shapes. Although the majority consist of straight, cylindrical vessels, dumbbell-shaped designs (Figure 1.4) take advantage of the benefits that variable drum sizes can bring to process application. With regard to internal kiln fixtures, most direct-fired kilns are lined with refractory materials for several reasons but the primary purposes are to insulate and protect the outer shell, in high-temperature applications, from thermal damage and to save energy. Kilns may also be equipped with dams to increase the material dwell time or with lifters and tumblers (Figure 1.5) to aid the materials to flow axially and in some cases

Figure 1.4 Schematic diagram of a dumbbell-type rotary kiln.

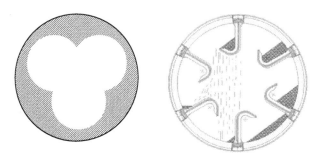

Figure 1.5 Schematic diagram of kiln internal fixtures: trefoil (left) and J-lifter (right).

Table 1.2 Advantages of Using Lifters (Data in Imperial Units)

	Before Lifters Installed	After Lifters Installed	% Change
12 × 250 ft LWA kiln	Added 3 rows of lifters + 3 dams		
Product rate (STPD)	650	970	47
Specific heat consumption (MBTU/t)	2.6–2.8	2	−35
Exit gas temperature (°F)	1200	730	39
Kiln speed (rpm)	1.6	2.7	70
11 × 175 LWA kiln	Added 3 rows of lifters + 1 dam		
Capacity (tpd)	550	625	14
Specific heat consumption (MBTU/t)	2.53	2.24	−12
Exit gas temperature (°F)	1050	850	19
Kiln speed (rpm)	1.75	2.3	31

to improve particle mixing achieved through surface renewal. Table 1.2 presents some of the energy-saving advantages of using lifters in various applications and processes. Some of these savings can be substantial.

Owing to the poor thermal efficiency of earlier long kilns and the need for fuel efficiency, most designs are aimed at maximizing mixing and heat transfer. To accomplish this, kilns are often equipped with heat recuperators, such as preheaters, in which part of the energy in the exhaust gas is recovered to preheat the feed before it enters the kiln. Although coolers are often used to cool the product for safe material handling, they are also used to recuperate the energy, which would otherwise go to waste, as in the earlier-day kilns, to preheat the combustion air and/or to provide other energy needs. Of the modern-day rotary kilns the following can be distinguished: wet kilns, long dry kilns, short dry kilns, coolers and dryers, and indirect-fired kilns. Some of these are discussed in the following sections.

1.2.1 Wet Kilns

Wet kilns are those that are usually fed with slurry materials. Wet kilns are usually long with kiln lengths on the order of 150–180 m (about 500–600 ft). The feed end is usually equipped with chains that serve as a heat "flywheel" by recuperating the heat in the exhaust gas for use in preheating the feed to assist the drying. Chains are also used to break up any lumps that the material might form during the transition phase of changing from slurry to solids upon drying. In the cement industry these kilns are often not efficient and are becoming a thing of the past replaced by long dry kilns. Nevertheless, there are certain applications that are not amenable to the alternative use of long dry kilns, for example, lime mud kilns found in the pulp and paper industry and some food applications.

1.2.2 Long Dry Kilns

These are shorter than wet kilns with lengths on the order of 90–120 m (about 300–400 ft). For long dry kilns, as with wet kilns, the drying, preheating, and calcination all occur in one single vessel (Figure 1.6). However, they work well when the feed particles are large. The reason for the relatively shorter length is that the

Figure 1.6 Wet, long cement kiln.
Courtesy of FLS minerals.

feed is dry with a moisture content resembling that of granular solids rather than that of slurry. Applications include lime kilns and LWA kilns where the mined stones are crushed to about 1.3–5 cm (0.5–1.5 in) before feeding them into the kiln.

1.2.3 Short Dry Kilns

Short dry kilns are usually accompanied by an external preheater or precalciner (Figure 1.7) in which the feed is dried, preheated, or even partially calcined prior to entering the main reactor (kiln). As a result the thermal load on the kiln proper is reduced. Hence kilns equipped with preheaters or precalciners tend to be short, on the order of 15–75 m (about 50–250 ft) depending on the process. The shorter kilns are those in which the entering feed material is almost calcined. Applications include cement and some lime kilns. Because of the large feed particle size encountered in limestone calcination, modern lime kilns are equipped with preheaters that function as a packed bed of stone with a countercurrent flow of kiln exhaust gas rather than the typical cyclone preheaters in cement kiln systems.

1.2.4 Coolers and Dryers

Some coolers and dryers can be in the form of contactors such as the rotary kiln itself, although some are packed-bed contactors such as grate coolers. Rotary coolers can be

Figure 1.7 Cement kiln equipped with cyclone preheaters.

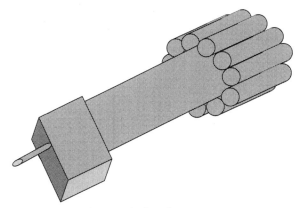

Figure 1.8 Schematic diagram of an attached cooler arrangement.

either in-line or attached (Figure 1.8), the number of which is determined by a simple formula:

$$N = \pi \times (D + d + 2)/(d + 1) \tag{1.1}$$

where D and d are the diameters of the kiln and the cooler, respectively. However, attached coolers place extra mechanical load that must be accounted for in design calculations. They also present maintenance challenges. Rotary coolers and dryers would normally be equipped with tumblers and lifters, which cascade the material well above its angle of repose to take advantage of better solid–gas contact.

1.2.5 Indirect-Fired Kilns

Indirect-fired kilns are those heated externally. They are usually designed for applications where direct contact between the material and the gas providing the heat source is undesirable. In this case, the heat source is external to the kiln (Figure 1.9). Any internally flowing gas that is in the freeboard is used for purging any volatile gas that arises from the bed as a result of chemical/physical reactions. Because of their low thermal efficiency, externally heated kilns are small, typically up to 1.3 m (50 in) in diameter, and are used for niche applications such as calcining of specialty materials.

A unique feature of indirect-fired rotary kilns is multiple and compartmentalized temperature control zones, which can be electrically heated or gas-fired individually. Therefore, they provide the capability of achieving high temperatures. In some cases, for example, graphite furnaces, they can attain temperatures on the order of 2400 °C. The zones can also facilitate tightly defined residence times and controlled atmosphere including flammables. Typical applications include calcination, reduction, controlled oxidation, carburization, solid–state reactions, and purification, including waste remediation on a small scale, which require extremely high temperatures and tight control. Materials processed in indirectly-fired rotary kilns include phosphors, titanates, zinc oxide, and quartz ferrites. These are usually small in quantity but with a high margin of commercial materials that are economical to process in small quantities.

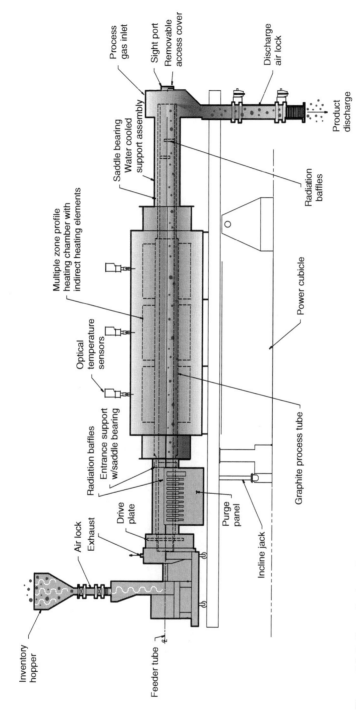

Figure 1.9 Indirect-fired small rotary kiln used for niche applications. Courtesy of Harper International, Lancaster, NY.

References

Blezard, R.G., 1998. Reflections of the History of the Chemistry of Cément. Society of Chemical Industry (SCI) Lecture Series, UK.

Expanded Shale, Clay, and Slate Institute, ESCSI, 2225 Murray Holladay Road, Salt Lake City. http://www.escsi.org/.

Peray, K.E., 1986. The Rotary Cement Kiln. Chemical Publishing Inc., New York.

Basic Description of Rotary Kiln Operation

As seen in Chapter 1, unit operation equipment and other components are added together to form the rotary kiln system for material processing. Perhaps the most important is the rotary reactor, which forms the heart of any process and therefore warrants due attention. The rotary reactor is usually a long horizontal cylinder tilted on its axis. In most rotary kiln process applications, the objective is to drive the specific bed reactions, which, for either kinetic or thermodynamic reasons, often require bed temperatures that, for example, in cement kilns may approach as high as 2000 K. For direct fired kilns, the energy necessary to raise the bed temperature to the level required for the intended reactions, and in some instances, for example, the endothermic calcination of limestone, to drive the reactions themselves, originates with the combustion of hydrocarbon fuels in the freeboard near the heat source or burner. This energy is subsequently transferred by heat exchange between the gas phase (the freeboard) and the bed as is shown in Figure 1.1. Heat transfer between the freeboard and the bed is rather complex and occurs by all the paths established by the geometric view factors in radiation exchange (Figure 2.1). All these manifest themselves into a combined transport phenomenon with the various transport processes coming into play in one application. Often, the analytical tools for handling freeboard transport phenomena have been the subject of considerable research. The ability to simulate the freeboard conditions tends to exceed the ability to accurately determine conditions within the bed. For example, the zone method (Guruz and Bac, 1981) for determining radiative heat transfer and commercial software for calculating fluid flow (and occasionally combustion processes as well) are well established. Thus, the synergy between the freeboard phenomenon and the bed phenomenon, that is, the intended function of the process, gets distorted. One generalized model for estimating bed conditions is the well-mixed phenomenon, synonymous with the continuous stirred tank.

Numerous descriptions of the rotary kiln bed phenomenon have been proposed this way (Wes et al., 1976; Brimacombe and Watkinson, 1979; Tscheng and Watkinson, 1979; Gorog et al., 1981; and others). All these assume that, at each axial position, the bed is well mixed in the transverse plane, that is, the bed material is isothermal over any transverse section of the kiln. However, many kiln operations suffer considerable difficulty in achieving a uniform product, one example being lime kilns, which experience chronic problems in preventing dead burning of larger particles while fully calcining finer particles. Evidence such as this, as well as operator experience, suggests that substantial transverse temperature gradients are generated within the bed. Thus, the well-mixed assumption, although expedient to the modeling of rotary kiln bed transport phenomena, is clearly deficient because one cannot ignore the motion of the bed in the transverse plane or the effect of this motion on the redistribution within

Figure 2.1 Radiation heat exchange in the cross-section.

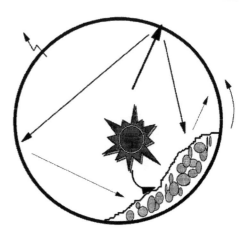

the bed material of energy absorbed at the bed-freeboard interfacial surfaces. To link the freeboard phenomenon to the bed phenomenon, the features of the bed should first be addressed.

2.1 Bed Phenomenon

During the thermal processing of granular materials in rotary kilns, heat transfer within the bed material occurs by the same mechanisms as in any packed bed, such as shaft kilns. Heat transfer paths at play can be particle-to-particle conduction and radiation, as well as interstitial gas-to-particle convection (Figure 2.2). However, the movement of the particles themselves superimposes an advective component for energy transport, which has the potential to dominate heat transfer. Hence, the key feature of a rotary

Figure 2.2 Bed heat transfer paths: (1) Internal conduction. (2) Particle-to-particle conduction. (3) Particle-to-particle radiation. (4) Interparticle convection.

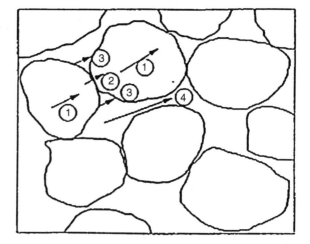

granular bed is the motion in the transverse plane, which sets the axial flow in motion and is dependent upon the rotation rate, degree of fill (volume of the kiln occupied by material), and the rheological properties of the particulate material.

Since the reactor is usually cylindrical and partially filled, it generally possesses two dispersion mechanisms, one in the axial direction that is characterized by an axial mixing coefficient and the other in the transverse direction, often associated with a radial mixing coefficient. The axial mixing component is attributed to the mechanism that results in an overall convection, and causes the bulk of the material to move from the inlet of the cylindrical drum to the outlet with an average velocity equal to the plug flow velocity. The radial mixing component, however, involves mechanisms at a smaller scale that cause local constraints on individual particles and result in velocity components in both the axial direction and the transverse direction. Both axial and transverse mixing coefficients tend to increase with an increase in kiln rotational speed. For low rates of rotation, one expects a spread in the residence time distribution due to the influence of the velocity profile (Wes et al., 1976). The effect of the drum size, particle rheology, and drum internal features are therefore major design considerations. The effect of the drum rotational speed on the transverse flow pattern is illustrated later. For now we present the features of the rotary reactor as a contactor by providing a quantitative description of the dispersion mechanisms, the resultant effects of which are critical to bed heat transfer during material processing.

2.2 Geometrical Features and Their Transport Effects

The key geometrical feature is the vessel size, given in terms of the cylinder diameter and kiln length related by the aspect ratio, that is, the length-to-diameter ratio (L/D) and also the slope. Other pertinent features include the internals, such as constriction dams and lifters, that impact the residence time. The empirical relationship developed by the US Geological Survey in the early 1950s relating the residence time and the kiln geometry has become a design mainstay even to this date (Perry and Green, 1984). It is

$$\bar{\tau} = \frac{0.23L}{sN^{0.9}D} \pm 0.6\frac{BLG}{F}, \tag{2.1}$$

where s is the slope (ft/ft), N is the kiln rotational speed in revolutions per minute (rpm), L and D are the respective kiln length and diameter in ft, G is the freeboard gas velocity defined here in units of lb/h/ft^2, and F is the feed or charge rate in lb-dry material per hour per square foot of cross-sectional area. B is a constant depending on the material and is approximately defined as $B = sd_p^{-0.5}$ where d_p is the particle size.

Because the vessel is partially filled and rotating on its horizontal axis, the freeboard or open space above the bed depends on the bed depth and, for that matter, on the kiln loading (% Fill). The shape of the free surface (the interface between the bed and freeboard) is dependent upon the operational requirements, that is, the feed rate, the drum rotational rate, and the material properties. As a result, the sizing of the rotary kiln depends on the application, typically, the feed rate (capacity) and related transport properties such as temperature, gas flow rates, and bed material velocities that ultimately will determine the residence time. For example, in dry processing applications, cylinder length-to-diameter ratios on the order of 5–12 are typical depending on whether the heat exchange is contact or noncontact. Such *L/D* ratios can result in residence times in the 20- to 120-min range depending upon the kiln rotational speed, the type of internal flights, if any, and the slope in the longitudinal direction, typically in the range of 1°–3°.

The movement of a charge in a rotating cylinder can be resolved into two components mentioned earlier, that is, movement in the axial direction, which determines residence time, and movement in the transverse plane, which influences most of the primary bed processes such as material mixing, heat transfer, and reaction rate (physical or chemical), as well as the axial progress of the charge. Although this linkage between particle motion in the transverse plane and particle velocity in the axial direction was established several decades ago, the literature generally deals with these two types of bed motion as independent phenomena until recent advances in the characterization and application of granular flow theories could be applied to powder processing in such devices (Boateng, 1998).

2.3 Transverse Bed Motion

Depending on the kiln's rotational rate, the bed motion in the transverse plane may be characterized as centrifuging, which occurs at critical and high speeds. This is an extreme condition in which all the bed material rotates with the drum wall. Cascading, which also occurs at relatively high rates of rotation, is a condition in which the height of the leading edge (shear wedge) of the powder rises above the bed surface and particles cascade or shower down on the free surface (Figure 2.3). Although rotary kiln operation in either of these conditions is rare because of attrition and dusting issues, certain food-drying applications take advantage of the high particle-to-heat transfer fluid exposure associated with the cascading mode and the separation effect caused by the centrifugal force component. For example, starting at the other extreme, that is, at very low rates of rotation and moving progressively to higher rates, the bed will typically move from slipping, in which the bulk of the bed material, en masse, slips against the wall; to slumping, whereby a segment of the bulk material at the shear wedge becomes unstable, yields and empties down the incline; to rolling, which involves a steady discharge onto the bed surface. In the slumping mode, the dynamic angle of repose varies in a cyclical manner, while in the rolling mode, the angle of repose remains constant. It has been established (Rutgers, 1965) that the dynamic similarity of the rotary drum behavior, and hence the type of transverse bed motion that occurs during powder processing, is dependent upon the rotational Froude number, Fr, defined as

$$\text{Fr} = \omega^2 R/g, \qquad (2.2)$$

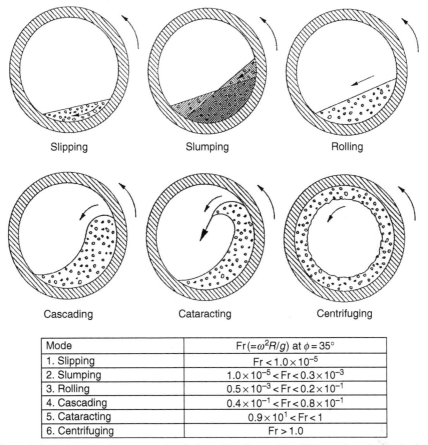

Figure 2.3 Bed motion in cross-sectional plane. Froude numbers (Fr) are given for each of the different modes (Henein, 1980).

where the critical condition for centrifuging implies Fr = 1. The ranges of Froude numbers for the various modes are shown in Figure 2.3.

In the rolling mode (Figure 2.4), where rotary drum mixing is maximized, two distinct regions can be discerned, the shearing region, called the active layer, formed by particles near the free surface, and the passive or plug flow region at the bottom where the shear rate is zero. The particular mode chosen for an operation is dependent upon the intent of the application. A survey of various rotary drum type operations (Rutgers, 1965) has indicated that most operations are in the 0.04–0.2 range of N-critical, which is well below the centrifuging mode and probably the cascading mode as well.

The geometric features of a typical rolling bed are depicted in Figure 2.5. The bed is subtended at the continuous angle of repose ξ. The free surface is subtended at 2θ. Hence, the bed cross-section occupied by material can be defined by this angle. The chord length, L_c, the longest distance traveled by particles on the free surface (path of steepest descent), can also be defined in terms of this angle. The fraction of the

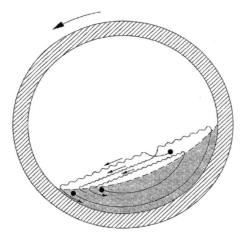

Figure 2.4 Rolling bed.

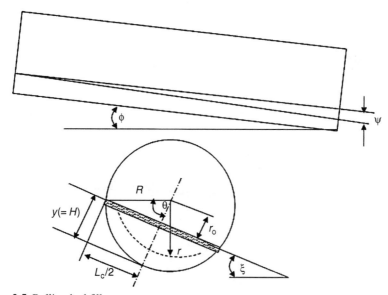

Figure 2.5 Rolling bed fill geometry.

cross-sectional area occupied by material is the kiln loading. This is usually defined as the volume percent occupied by material in the vessel. Granted that the kiln length is constant, the degree of fill is the percent of the cross-sectional area of the cylinder occupied by material (% Fill). The fraction filled defining the bed depth, and based on the geometry, relates the angles at any transverse section as follows:

$$f_c = \frac{1}{2\pi}\left(2\cos^{-1}\left(\frac{R}{R-H}\right) - \sin\left[2\cos^{-1}\left(\frac{R}{R-H}\right)\right]\right) \qquad (2.3)$$

Seaman (1951) developed an approximation for the theoretical residence time of a shallow bed (lightly loaded kiln) and a theoretical relationship for the kiln volumetric flow rate for deep beds (heavily loaded kilns). Nonetheless, no clear definition has ever been given for the range of operation encompassed by the two cases of kiln loadings. Seaman's approximations led to the conclusion that kilns should be considered heavily loaded when the fractional cross-sectional fill of solids exceeds approximately 5%.

As shown in Figure 2.4, two regions of the transverse plane can be discerned: (1) the active region near the top of the bed where surface renewal occurs and (2) the passive region beneath the active region. The active region is usually thinner than the passive region because particles there are not restricted and they move faster. Because the bed is constrained within the cylinder's geometrical domain, the laws of conservation of mass (i.e., particles going into the active layer must come from the plug flow region) require that the depth of the active region be lesser than that of the passive region. This is due to the higher particle velocity there, although there is more to this phenomenon than simple mass conservation. Most of the mixing in the drum cross-section and also dissociation reactions occur in the active region. The deeper the active layer, the better the mixing. To increase the active layer depth, it is essential to increase kiln speed. Tracer studies have shown that particles move forward along the kiln only through the active region (Ferron and Singh, 1991). The kiln's longitudinal slope is usually small (1/2–3/4 in/ft, which is equivalent to $3°-4°$) and far less than the material angle of repose (typically $36°-40°$). Hence, forward motion cannot be sustained by the gravitational component of the stresses alone indicating that the kiln's axial slope assists this forward motion but does not drive it. Therefore, for every kiln revolution, the bed material makes several excursions (possibly three or four) in the cross-section, thereby resulting in an axial advance. Increasing the kiln speed will result in an increase in the number of excursions and ultimately in increased mixing. Kiln speed increase, however, will decrease the residence time of the material, since the bed will move faster axially. It is therefore important for the kiln operator to know, based on the nature of the application, what the critical residence time should be to achieve the desired product quality. Process control by kiln speed is therefore critical if one can provide adequate mixing and maintain sufficient residence time to process the material.

Studies involving rotary kiln bed behavior have resulted in many ways of estimating how the bed will behave at any given operational condition. One such tool is a bed behavior diagram (Henein, 1980), which presents a typical behavior for a sand bed for a 41-cm (1.35-ft) diameter pilot kiln (Figure 2.6). Given the angle of repose, kiln geometry, and speed, users of such diagrams can predict what bed behavior to expect within the kiln cross-section. However, since these delineation curves were generated from room temperature experiments, their industrial use has been limited. Despite this shortcoming, the bed behavior diagram can be helpful for the drying zone and, for the most part, the preheating zone of long kilns. In the calcination zone, however, any softening of the material that increases stickiness or agglomeration of the material will result in an increase in the angle of repose. Because of the importance of mixing on product uniformity, we will elaborate further on rolling bed behavior.

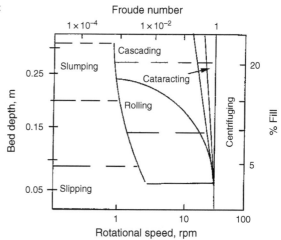

Figure 2.6 Bed behavior diagram: Mapping of bed behavior regimes for different operation conditions (Henein, 1980).

2.4 Experimental Observations of Transverse Flow Behavior

The key step in almost all particulate processing applications is solids mixing. Mixing is primarily used to reduce the nonuniformity in the composition of the bulk to achieve uniform blending or as a first step to improve the convective/advective and diffusion components associated with heat transfer to a particulate bed in thermal processing. Some of the observed flow and transport phenomena in rotary kilns that have provided insights into particulate flow behavior, which lead to accurately stating the mathematical problem or modeling the rotary kiln transport phenomena, are described. Dependent upon the bed depth and the operational conditions, the flow behavior constrained in the transverse plane can be purely stochastic, purely deterministic, or a hybrid of both; hence, mixing can either be modeled by a random walk for very shallow beds (Fan and Too, 1981) or by a well-defined bulk velocity profile estimated by shear flows similar to boundary layer problems (Boateng, 1993). Early workers used tracer particles to observe and characterize mixing (Zablotny, 1965; Ferron and Singh, 1991; and others). Lately, such works have been extended to the use of nonintrusive techniques, such as nuclear magnetic resonance (NMR) (Nakagawa et al., 1993) and positron emission particle tracking (PEPT; Parker et al., 1997).

For bulk processing, fiber-optic probes have been used in the past to establish bulk velocity profiles that have allowed estimates of the parameters necessary for bulk convection heat transport (Boateng, 1993). Boateng conducted experiments on the continuous flow of granular material in the transverse plane of a rotating drum to elucidate the rheological behavior of granular material in rotary kilns. Additionally, for the first time, such work provided data for mathematical modeling of granular flow in the boundaries of a rotary kiln. Using granular materials that varied widely in their physical properties, specifically, polyethylene pellets, long grain rice, and limestone, the mean depth and surface velocities were measured using optical fiber probes (Boateng, 1993; Boateng and Barr, 1997). From these, granular flow behavior including the velocity fluctuations and the linear concentration of particles could be computed. The rotary drum used consisted of a steel cylinder of a 1-m diameter and a 1-m length. It had a glass end piece with a center opening, providing access for flow measurements.

In light of the shape and thickness of the active layer, experiments confined to the observation of the general behavior of material flow in the cross-section could be carried out. It was found that the transition of particles from the plug flow region of the transverse plane, where material moves in rigid motion with the cylinder, to the active layer, where particles are continuously shearing, not only depends on the material's angle of repose but also on the physical properties (e.g., the coefficient of restitution of the material). Because shear stresses generated in the induced flow involve collisional elasticity, particles with a relatively low angle of repose, for example, polyethylene pellets, experience an easy transition from the potential energy position in the plug flow region to kinetic energy in the active layer. Conversely, low coefficient of restitution materials, such as rice and limestone, has a relatively high friction angle, and the energy dissipation is lower than that for polyethylene pellets. As a result, high potential energy is built up during the transition and is accompanied by material buildup prior to release into the active layer. Flow instabilities result, which manifest themselves in the formation of multiple dynamic angles of repose with an unsteady velocity distribution at the exposed bed surface.

The active layer depth and bed flow properties depend on the coefficient of restitution of the material. The flow properties of interest include granular temperature, which is a measure of kinetic energy in random motion of particles, and dilation. Granular temperature was found to be high at regions of low concentration with high mean velocity. These experiments also characterize the shape of the active layer to be parabolic irrespective of the material used. However, its thickness depends on the physical properties of the material and the operational parameters such as rotational rate.

After transferring from the plug flow region into the active layer, particles accelerate rapidly up to around midchord of the surface plane before decelerating with streaming, kinetic, and gravity effects playing various roles in momentum transfer. Increasing the degree of fill provides a longer chord length for material to travel and, for larger kilns, the velocity at the exposed bed surface can become fully developed by midchord. For deep beds, surface velocities can reach as high as 4.5–7.5 times the drum speed. Similar cases have been observed using nonintrusive flow measurement techniques (Parker et al., 1997), reproducing some of the results obtained by Boateng (1993).

2.5 Axial Motion

Axial dispersion is dependent upon transverse dispersion. For each excursion within the cross-section, a particle on the free surface traveling on the chord length makes an axial move either backward or forward. This is due to the random nature of flow within the active layer. The forward advance is based on the cylinder slope in the axial direction as well as the apparent forward angle resulting from the transverse flow pattern. Some of the geometric relationships to the residence time for particles traveling from one end of the kiln to the other have been derived by various investigators since Seaman's work (1951). Although these derivations lack the rigor of a true granular flow, they have combined key empirical relationships to provide design formulas still in use today. Nicholson (1995) assembled some pertinent formulas that have been used to characterize axial bulk movement in kilns that are purely based on geometrical considerations. The major ones are from the early works by Seaman who calculated the material flow properties in the axial direction based purely on geometry of the equilibrium position (Figure 2.5).

For a small kiln slope ϕ, and the dynamic angle of repose ξ, the expression for the axial movement per cascade Z_o, is given as

$$Z_o = \frac{L_c(\phi + \psi \cos \xi)}{\sin \xi} \tag{2.4}$$

where ψ is the bed angle relative to the axial plane (Figure 2.5).

The number of cascades per kiln revolution mentioned earlier, N_c, might be estimated as

$$N_c = \frac{\pi}{\sin^{-1}\left(\frac{L_c}{2r}\right)} \tag{2.5}$$

where r is the radius of the particle path in the bed. Multiplying the axial transport distance per cascade by the number of cascades per revolution and the kiln rotational speed yields the average axial transport velocity, u_{ax}, for a particle at position r anywhere in the radial plane.

$$u_{ax}(r) = nL_c \left(\frac{\phi + \psi \cos \xi}{\sin \theta}\right) \left(\frac{\pi}{\sin^{-1}\left(\frac{L_c}{2r}\right)}\right) \tag{2.6}$$

where n is the kiln rotational speed. For lightly loaded kilns (10–12% fill or less), one can make the following approximation:

$$\sin^{-1}\left(\frac{L_c}{2r}\right) \rightarrow \frac{L_c}{2r} \tag{2.7}$$

The mean axial transport plug flow velocity, u_{ax}, at a point along the kiln length can be calculated by the expression

$$u_{ax} = 2\pi rn \left(\frac{\phi + \psi \cos \xi}{\sin \theta}\right) \tag{2.8}$$

The average residence time can be estimated from Equation (2.8) when knowing the kiln length, L,

$$\bar{\tau} = \frac{L \sin \xi}{2\pi r n(\phi + \psi \cos \xi)} \qquad (2.9)$$

For beds with a low degree of fill and no end constriction dams, one can practically assume a uniform bed depth throughout the entire kiln length ($y = H$) and, $\psi \to 0$. The residence time expression can be reduced to

$$\bar{\tau} = \frac{L \sin \xi}{2\pi r n \phi}, \quad \bar{\tau} = \frac{L\theta \sin \xi}{L_c \omega \tan \xi} \qquad (2.10)$$

It can be easily recognized that Equation (2.10) resembles the empirical formula for the determination of the average residence time (Equation (2.1)), established by the US Geological Survey.

One can also determine the maximum capacity of the kiln by the same geometrical considerations if the flow area can be estimated. From the geometry (Figure 2.5), the chord length can be expressed as

$$L_c = 2(r^2 - r_o^2)^{1/2} \qquad (2.11)$$

$$u dA = 4\pi n \left(\frac{\phi + \psi \cos \xi}{\sin \theta} \right) (r^2 - r_o^2)^{1/2} r dr \qquad (2.12)$$

Integrating Equation (2.12) from $r = r_o$ to $r = R$ gives an expression for the average volumetric axial transport rate as

$$q = \frac{4\pi n}{3} \left(\frac{\phi + \psi \cos \xi}{\sin \theta} \right) (r^2 - r_o^2)^{3/2} \qquad (2.13)$$

For beds with dams, and therefore a nonuniform bed depth along the kiln length, the rate of change of the bed depth can be estimated in terms of the subtended angles as

$$\frac{dr_o}{dz} = \frac{3q \sin \xi}{4\pi n \cos \xi (R^2 - r_o^2)^{3/2}} - \left(\frac{\phi}{\cos \xi} \right) \qquad (2.14)$$

Nicholson (1995) found a good agreement between such theoretical calculations and experimental data for cylinders with and without constrictions at the discharge end.

Several early and recent investigators have derived variations of the throughput, axial velocity, and residence time expressions (Hogg et al., 1974; Perron and Bui, 1990) but Seaman's expressions given here have found most practical applications and are recommended for use in industrial kilns so long as there are no true granular flow models to predict the bed behavior as are available for fluids of isotropic materials.

2.6 Dimensionless Residence Time

Using the same geometric considerations by Seaman, dimensionless residence time and flow rate may be derived (McTait, 1995) as a function of $f(y/R)$. Recognizing that this derivation might be convenient for estimating these parameters for all kilns. Using $L_c = y(2R - y)$ and $\sin \theta = L_c/R$, the dimensionless residence time can be expressed as

$$\bar{\tau}_d = \frac{\bar{\tau} n D \tan \phi}{L \sin \xi} = \frac{\sin^{-1}[(2 - y/R)y/R]^{1/2}}{\pi[(2 - y/R)y/R]^{1/2}} = f(y/R) \qquad (2.15)$$

Similarly, using $A_s = R^2\theta - L_c(R - y)$, a dimensionless volumetric flow can also be expressed in terms of the bed depth, recognizing that $q\bar{\tau} = LA_s$, which is

$$q_d = \frac{q \sin \xi}{nD^3 \tan \phi} = \frac{\pi}{4}[(2 - y/R)]^{1/2} - \frac{\pi}{4} \frac{(2 - y/R)(1 - y/R)y/R}{\sin^{-1}[(2 - y/R)y/R]^{1/2}} \qquad (2.16)$$

and for the condition of a uniform bed with no surface inclination in the axial direction, that is, $dy/dz \to 0$,

$$q_d = \frac{\pi}{6}[(2 - y/R)y/R]^{1/2} \qquad (2.17)$$

Although averaged or plug flow estimations can be deduced from the geometry of the kiln using equilibrium angles, the actual residence time can completely differ from these estimates. This is due to the random nature of solids' motion alluded to earlier. For each cascade or excursion, a particle can move forward in line with the forward geometric projection as discussed, or backward due to several factors including changing material dynamic angle of repose. Evidence of this lies in the situation where large agglomerated particles (notoriously known in the industry as *logs*) that form in high-temperature kilns have been observed to travel back and forth without being discharged. Also, in the cross-sectional plane, particles in the plug flow region can emerge in the active layer in a sequence following the birth–death phenomenon, that is, entirely by random walk (Ferron and Singh, 1991). Hence, residence time is a result of axial dispersion, which in turn depends on transverse dispersion and is truly a distribution function. It is not surprising, therefore, that residence time has been the subject of many tracer experiments. Parker et al. (1997) using the PEPT for particle imaging have shown that the distribution of axial displacement follows a Gaussian distribution. They have observed in their experiments a one-to-one relationship between

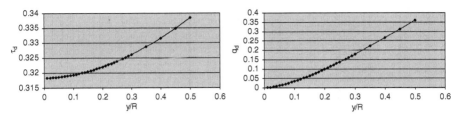

Figure 2.7 Dimensionless residence time and flow rate for a 3.66-m diameter kiln with $L/D = 10$, slope $= 3°$, and material angle of repose $= 40°$.

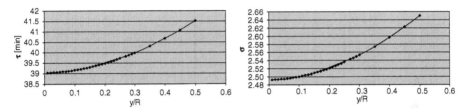

Figure 2.8 Calculated mean residence time and variance for a 3.66-m diameter kiln with $L/D = 10$, slope $= 3°$, and material angle of repose $= 40°$.

the radial entry of a tracer particle and its exit within a bed of industrially processed materials.

This notwithstanding, the variance σ' over the average path traveled by particles on the free surface (some fraction over the chord length, L_c) might be included in the geometric derivations from an operational standpoint (McTait, 1995).

These relationships between residence time and bed depth can be plotted and used in tandem for kiln design. Given the diameter, kiln slope, and desired L/D, the residence time and the feed rate that will result in a desired bed depth at a specified kiln speed can be calculated fairly well using these relationships. Combined with process data, these can be manipulated to achieve optimum kiln size and operating conditions. A plot of the dimensionless residence time and flow rate for a 3.66 m (12 ft) kiln, $L/D = 10$, a kiln slope of $3°$, and $40°$ angle of repose operating at 1 rpm are shown in Figure 2.7. From these, the actual residence time and related variance and also the maximum volumetric flow rate for a typical industrial material with 1600 kg/m^3 (100 lb/ft^3) bulk density can be estimated for design purposes (Figure 2.8).

References

Boateng, A.A., 1998. Boundary layer modeling of granular flow in the transverse plane of a partially filled rotating cylinder. Int. J. Multiphase Flow 24, 499–521.

Boateng, A.A., 1993. Rotary Kiln Transport Phenomena: Study of the Bed Motion and Heat Transfer. The University of British Columbia, Vancouver, Canada.

Boateng, A.A., Barr, P.V., 1997. Granular flow behaviour in the transverse plane of a partially filled rotating cylinder. J. Fluid Mech. 330, 233–249.

Brimacombe, J.K., Watkinson, A.P., 1979. Heat transfer in a direct fired rotary kiln: Part I. Pilot plant and experimentation. Met. Trans. B 9B, 201–208.

Fan, L.T., Too, J.R., 1981. Stochastic analysis and modeling of solids mixing. Proc. Int. Symp. Powder Technol. 81, 697–711.

Ferron, J.R., Singh, D.K., 1991. Rotary kiln transport processes. AIChE J. 37, 747–758.

Gorog, J.P., Brimacombe, J.K., Adams, T.N., 1981. Radiative heat transfer in rotary kilns. Met. Trans. B 12B, 55–70.

Guruz, H.K., Bac, N., 1981. Mathematical modeling of rotary cement kilns by the zone method. Can. J. Chem. Eng. 59, 540–548.

Henein, H., 1980. Bed Behavior in Rotary Cylinders with Applications to Rotary Kilns (Ph.D. dissertation). University of British Columbia, Vancouver.

Hogg, R., Shoji, K., Austin, L.G., 1974. Axial transport of dry powders in horizontal rotating cylinders. Powder Technol. 9, 99.

McTait, G.E., 1995. Particle Dynamics in Rotary Cylinders. Report for Certificate of Postgraduate Study. Cambridge University.

Nakagawa, M., Altobelli, S.A., Caprihan, A., Fukushima, E., Jeong, E.-K., 1993. Non-invasive measurements of granular flows by magnetic resonance imaging. Exp. Fluids 16, 54–60.

Nicholson, T., 1995. Mathematical Modeling of the Ilmenite Reduction Process in Rotary Kilns (Ph.D. thesis). University of Queensland.

Parker, D.J., Dijkstra, A.E., Martin, T.W., Seville, J.P.K., 1997. Positron emission particle tracking studies of spherical particle motion in rotating drums. Chem. Eng. Sci. 52 (13), 2011–2022.

Perron, J., Bui, R.T., 1990. Rotary cylinders: solid transport prediction by dimensional and rheological analysis. Can. J. Chem. Eng. 68, 61–68.

Perry, R.H., Green, D., 1984. Chemical Engineering Handbook. McGraw-Hill, New York.

Rutgers, R., 1965. Longitudinal mixing of granular material flowing through a rotary cylinder: Part I. Description and theoretical. Chem. Eng. Sci. 20, 1079–1087.

Seaman, W.C., 1951. Passage of solids through rotary kilns: factors affecting time of passage. Chem. Eng. Prog. 47 (10), 508–514.

Tscheng, S.H., Watkinson, A.P., 1979. Convective heat transfer in rotary kilns. Can. J. Chem. Eng. 57, 433–443.

Wes, G.W.J., Drinkenburg, A.A.H., Stemerding, S., 1976. Heat transfer in a horizontal rotary drum reactor. Powder Technol. 13, 185–192.

Zablotny, W.W., 1965. The movement of the charge in rotary kilns. Intern. Chem. Eng. 2, 360–366.

Freeboard Aerodynamic Phenomena

3

This chapter presents the rotary kiln freeboard aerodynamic phenomena, drawing parallels with fluid flow in conduits. The goal is to describe the characteristics of confined jets that determine burner aerodynamic mixing and, in turn, combustion efficiency, and flame shape and its character. Having described the flow field, the effect of turbulence on dust pickup from the bed's free surface will also be discussed.

Fluid flow through the kiln freeboard comes from several sources, including combustion air, combustion products, and the air infiltrated into the vessel. In direct-fired kilns, especially those with pulverized fuel combustion systems, the fuel, for example, pulverized coal, is introduced from a pipe. Burner pipe nozzles range from a 25- to 61-cm (10- to 24-in) diameter in size and the kilns (combustion chamber) into which the fuel is discharged are typically 2.4–6.2 m (8–20 ft) in diameter. Hence, with a chamber much larger than the burner pipe, the fuel emerges as a jet. The freeboard flow phenomenon near the combustion zone therefore exhibits the properties of jets, how they are entrained, and how they mix with the surrounding fluid. Therefore, the gross pattern of flow in the region near the burner is determined by the geometry or the physical boundaries surrounding the burner, typically involving a jet confined in a cylindrical vessel and by the manner in which the fuel is discharged. To prevent pulverized fuel from settling in pipes, the conveying air (primary combustion air) is introduced at a high velocity, typically >30 m/s. For direct-fired rotary kilns, the primary air usually comprises about 25% of the combustion air requirements. The secondary air makes up for the remaining 75% of the combustion air. The latter may be introduced through the inlet surrounding the primary air or from discharge coolers that recuperate some of the energy in the discharge product and return it into the kiln to improve combustion and fuel efficiencies. In the region further away from the combustion zone, the flow field is made up of combustion products and any other gases that are released as a result of the bed reactions. The combustion products must be induced into air pollution control devices to be cleaned up before discharge into the atmosphere. Atmospheric discharge is accomplished by an induced draft (ID) fan, which, as its name implies, induces gas flow through the kiln and controls the overall pressure drop across it. Because the ID fan pulls gas through the kiln, the kiln system is under suction and establishes a negative pressure drop across it. The fan's electrical current draw is usually a good indicator of the pressure drop and for that matter the gas flow rate. The higher the pressure (i.e., more negative), the greater the fan amperage. Owing to a slight vacuum, air infiltration into the kiln is almost always evident, originating from the joints between the cylinder and around the fire hood. Infiltration air also comes from several sources including kiln attachments, intended or unintended open access points, such as cleavages around burners, take-off pipes, and holes, with the quantity dependent upon the efficiency of the seals around these curvatures. Kiln seals are therefore a big design and operational challenge, the most prominent being the point between the stationary fire hood and the rotating cylinder. It can be rightfully said that the kiln is like a big conduit and the ID fan drives the flow through it. The flow area (kiln freeboard) depends on the kiln loading or the degree of fill, thereby determining the turbulent intensity of the freeboard flow in the region further away from the combustion zone. Near the feed end, the high turbulent kinetic energy can result in increases in dust generation and discharge through the exhaust. In the near field, however, the interaction of the primary air jet and secondary air in a confined environment introduces intense mixing involving recirculation eddies that return combustion products into the flame region, a phenomenon that underlies mixing, the mainstay of turbulent diffusion flames found in rotary kilns. Such flow properties exhibited by confined jets underlie the freeboard aerodynamic phenomena and will be described further in more detail.

3.1 Fluid Flow in Pipes: General Background

Gas flow downstream from the entry region in kilns follows fundamental principles similar to the generalized pipe flow laws. However, it is more complex in the near field involving entrained jets. These principles are based on the ideal gas laws that demand that $PV/T =$ constant for a nonreacting gas. Therefore, changes in any of the variables P, V, or T will result in a change in the others and will affect the demand on the ID fan. Most theoretical considerations in fluid dynamics are based on the concept of perfect fluids thereby requiring that the fluid is frictionless and incompressible. The no-slip condition in an incompressible fluid means that two contacting layers acting on each other exert only normal or pressure forces but not tangential forces or shear stresses. However, this assumption falls short in real fluids since they offer internal resistance to a change in shape. This results in the concept of viscosity whereby the existence of intermolecular interactions causes the fluid to adhere to the solid walls containing them. Because shear stresses are small for fluids of practical importance such as that encountered in rotary kilns their coefficients of viscosity are small and agree with perfect fluids. For more on the outline of fluid motion with friction, the reader is referred to Schlichting (1979).

The nature of viscosity can be best illustrated by the velocity distribution between two plates, that is, Couette flow (Figure 3.1). The fluid motion is supported by applying a tangential force to the upper plate. Here, the fluid velocity is proportional to the distance, y, between the plates such that

$$u(y) = \frac{y}{h} U \qquad (3.1)$$

The frictional force per unit area or the frictional shear stress, τ, is directly proportional to velocity, U, and inversely proportional to the spacing h yielding

$$\tau = \mu \frac{du}{dy'} \qquad (3.2)$$

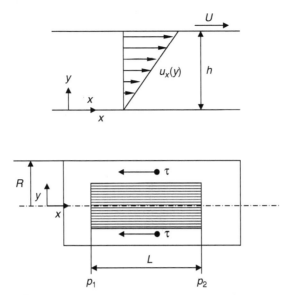

Figure 3.1 Couette flow.

where the constant of proportionality μ is the dynamic viscosity in units of kilograms per meter per second, a property of the fluid. In all fluid motion where frictional and inertia forces interact, such as gas flows in rotary kilns, kinematic viscosity, v, is defined as the ratio between the dynamic viscosity and fluid density:

$$v = \frac{\mu}{\rho} \tag{3.3}$$

These frictional laws that define the viscosity can be applied to flow conditions in cylindrical pipes, in part, similar to conditions in rotary kilns. Fluid flow in conduits was first described by Reynolds' early experiments that identified the flow regimes as "laminar" and "turbulent." For pipes, rotary kilns included, these are characterized by the relationships between pressure drop and flow rate. Laminar flows are characterized by the first Reynolds' principle, which states that the pressure drop of a flowing fluid is proportional to the flow rate at low velocities. We explore this principle in kilns in several ways. For example, a common practice of the industry is the use of a pitot tube for flow measurements during kiln audits. At high velocities, however, the pressure drop of a flowing fluid is proportional to roughly the square of the flow rate, that is, for turbulent flows. The transition regime between laminar and turbulent flows is defined by the Reynolds number, $Re = uD/\mu$. In laminar flow, the fluid moves under the influence of an axial pressure gradient and the individual fluid layers act on each other with shear stress proportional to the velocity gradient. Because there is no pressure gradient in the lateral direction, the flow is streamlined, which by definition means that for each streamline, the velocity is constant. The flow is accelerated by pressure and retarded by shear stress as shown in the classical Couette flow situation (Figure 3.1).

Under these conditions, the equilibrium condition in the flow direction, x, requires that the pressure forces acting on the cylinder cross-section, that is, $(p_1 - p_2)\pi y^2$ must be balanced by the shear stress acting on circumference $2\pi y l \cdot \tau$ resulting in

$$\tau = \frac{p_1 - p_2}{l} \cdot \frac{y}{2} \tag{3.4}$$

Invoking the friction laws (Equation (3.2)) and noting that the u decreases with y results in

$$\frac{du}{dy} = -\frac{p_1 - p_2}{\mu l} \cdot \frac{y}{2} \tag{3.5}$$

This, upon integration, leads to the velocity distribution

$$u(y) = \frac{p_1 - p_2}{\mu l} \left(C - \frac{y^2}{4} \right) \tag{3.6}$$

The integration constant can be obtained by invoking the no-slip condition at the wall, that is, $u = 0$ at $y = R$ for $C = R^2/4$, thereby resulting in a parabolic velocity distribution over the radius as

$$u(y) = \frac{p_1 - p_2}{4\mu l} \left(R^2 - y^2 \right) \tag{3.7}$$

The maximum velocity is at $y = 0$, the centerline, that is,

$$u_m = \frac{p_1 - p_2}{4\mu l} R^2 \tag{3.8}$$

The volume flow rate can be evaluated recognizing that the volume of the paraboloid of revolution is 1/2 times the base times the height, which leads to the Hagen–Poiseuille equation of laminar flow through pipes (Schlichting, 1979) as follows:

$$Q = \frac{\pi}{2} R^2 u_m = \frac{\pi R^4}{8\mu l} (p_1 - p_2) \tag{3.9}$$

Equation (3.9) restates the first Reynolds' experimental characterization indicating that for laminar flows in conduits and pipes, the flow rate is proportional to the first power of the pressure drop per unit length. Equation (3.9) also state that the flow rate is proportional to the fourth power of the radius of the pipe.

For two flows of different fluids, different velocities, and different linear dimensions to be dynamically similar, it is necessary that the forces acting on the fluid particles at all geometrically similar points have a constant ratio at any given time. If we assume a simplified fluid flow, such as incompressible flow that has no elastic and gravitational forces, then the similarity of flows may satisfy Reynolds' criterion that the ratio of the inertia and frictional forces is the same. Reynolds' analogy can be derived by resolving the forces on a control volume in a steady parallel flow in the x-direction, where the magnitude of the inertia forces per unit volume is $\rho u \partial u/\partial x$. The resultant shear force in a control volume ($dx \cdot dy \cdot dz$) is $\partial \tau/\partial y (dx \cdot dy \cdot dz)$, which is, on a per unit volume basis, equal to $\partial \tau/\partial y$. Invoking Equation (3.2) into it gives $\mu \partial^2 u/\partial y^2$. Consequently, for a condition of similarity, the ratio of the inertia forces to the frictional (viscous) forces, identified as the Reynolds number, must be the same.

$$\text{Inertia Forces/Friction Force} = \frac{\rho u \partial u/\partial x}{\mu \partial^2 u/\partial y^2} = \frac{\rho U^2/d}{\mu U/d^2} \qquad (3.10)$$

$$\text{Re} = \frac{\rho U d}{\mu} = \frac{U d}{v} \qquad (3.11)$$

thereby establishing Reynolds' principle of similarity as was earlier derived.

Most practical flows are turbulent. The type of flow for which Equation (3.9) applies exists in reality only for small radius cylinders and for very slow flows with Re < 2300. The Reynolds number of the flow in large cylinders and vessels encountered in practical engineering applications such as rotary kilns are higher. With pipe diameters in the 2- to 5-m range, the Reynolds number after uniform velocity is attained in rotary kilns is typically on the order of 1.5×10^5 (Field et a!., 1967), which falls well into the turbulent regime. For turbulent flows, the pressure drop is no longer directly proportional to the mean velocity but approximately proportional to the second power of the velocity. During transition from laminar into the turbulent regime, the velocity distribution becomes unstable in the presence of small disturbances. After the transition, the laminar or the streamline structure disappears and the velocity at any point in the stream varies with time in both magnitude and direction as turbulent eddies form and decay. At this point, the ability of the fluid to transport momentum, energy, and mass in the mean flow direction is greatly enhanced (Figure 3.2). However, a considerable pressure difference is required to move the fluid through the kiln. This is in part due to the fact that the phenomenon of turbulent mixing dissipates large quantities of energy, which, in turn, causes an increase in the resistance to flow.

Theoretical relationships between fluid flow rate and pressure drop for turbulent flow in pipes of interest similar to that established for laminar flows are not readily available. However, empirical relations based on experiments have been fitted long ago by pioneers such as Blasius and others to provide a quantitative assessment of frictional resistances in terms of Re (Schlichting, 1979). The velocity profile for turbulent flow in large pipes can be characterized as

$$\frac{u}{U} = \left(\frac{y}{R}\right)^{\frac{1}{n}} \qquad (3.12)$$

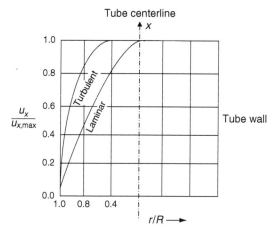

Figure 3.2 Qualitative comparison of laminar and turbulent velocity distribution.

where n is derived from experiment. The correlation between the shear stress at the wall and the flow velocity has been described as

$$\tau_0 = 0.03325 \rho \bar{u}^{7/4} v^{1/4} R^{-1/4} = \rho u_*^2 \tag{3.13}$$

where

$$\frac{u}{u_*} = 8.74 \left(\frac{yu_*}{v}\right)^{\frac{1}{7}} \tag{3.14}$$

and $\frac{yu_*}{v} > 70$ is a condition for purely turbulent friction. The expression for friction velocity can be derived as

$$u_* = 0.150 u^{\frac{7}{8}} \left(\frac{v}{y}\right)^{1/8} \tag{3.15}$$

From which the local skin friction coefficient can be expressed as

$$C'_f = \frac{\tau_0}{\frac{1}{2}\rho U^2} = 0.045 \left(\frac{UR}{v}\right)^{-1/4} \tag{3.16}$$

Although these relations have no direct application to determining flow in the turbulent regime of the rotary kiln, they can be of help in estimating dust loss and wall heat transfer coefficients.

3.2 Basic Equations of Multicomponent Reacting Flows

Consistent with the objective of this chapter, it is important to return to the type of flow encountered in the freeboard of the rotary kiln and address reacting flows. The freeboard flow of interest involves the reacting flow type, which is almost always multicomponent, composed of fuel, oxidizer, combustion products, particulates, and so forth. The thermodynamic and transport properties of multicomponent reacting fluids are functions, not only of temperature and pressure but also of species concentration. The basic equations that describe the simplest case of reacting turbulent flow include conservation equations for mass, concentration, momentum, and enthalpy equations as well as the associated reaction and equations of state for the system (Zhou, 1993),

Continuity Equation:
$$\frac{\partial \rho}{\partial t} + \frac{\partial}{\partial x_j}(\rho u_j) = 0 \tag{3.17}$$

Momentum Equation:
$$\frac{\partial}{\partial t}(\rho u_i) + \frac{\partial}{\partial x_j}(\rho u_j u_i)$$
$$= -\frac{\partial P}{\partial x_i} + \frac{\partial}{\partial x_j}\left[\mu\left(\frac{\partial u_j}{\partial x_i} + \frac{\partial u_i}{\partial x_j}\right)\right] + \rho g_i \tag{3.18}$$

Concentration Transport Equation:
$$\frac{\partial}{\partial t}(\rho Y_S) + \frac{\partial}{\partial x_j}(\rho u_j Y_S)$$
$$= \frac{\partial}{\partial x_j}\left(D\rho \frac{\partial Y_S}{\partial x_j}\right) - w_s \tag{3.19}$$

Enthalpy Transport Equation:
$$\frac{\partial}{\partial t}(\rho c_p T) + \frac{\partial}{\partial x_j}(\rho u_j c_p T) = \frac{\partial}{\partial x_j}\left(\lambda \frac{\partial T}{\partial x_j}\right) + w_s Q_s \tag{3.20}$$

Arrhenius Equation: $w_s = B\rho^2 Y_F Y_{OX} \exp(-E/RT)$ \hfill (3.21)

Equation of State: $P = \rho RT \sum Y_S/M_S$ \hfill (3.22)

We will examine how some commercially available computational tools are used to solve such a system of equations, but for the kiln engineer, perhaps the physical interpretation underlying these complex equations are most important. The similarity parameters inferred from these equations might allow the engineer to make some judgments of what aspects of the flow matter in burner design and operation.

Using the momentum and the energy equations, similarity parameters for reacting flows can be obtained as follows:

1. For fluid flow, the Euler number might be defined as the ratio of the pressure head and the velocity head and is expressed as

$$\mathrm{Eu} = \frac{P_\infty}{\rho_\infty U_\infty^2} \tag{3.23}$$

2. As we saw earlier, the Reynolds number represents the ratio of the inertial force and the viscous force, that is,

$$\mathrm{Re} = \frac{U_\infty L}{v_\infty^2} \tag{3.24}$$

3. The Mach number is a similarity parameter defined as the ratio between kinetic energy and thermal energy:

$$\mathrm{Ma} = \sqrt{\frac{U_\infty^2/2}{c_p T_\infty}} \approx \frac{U_\infty}{a} \tag{3.25}$$

4. For heat transfer within the fluid, the similarity parameter of importance is the Peclet number, which is the ratio of heat convection and heat conduction. We will revisit heat transfer modes later, but for the purposes of flow characterization, the Peclet number is defined in Equation (3.26), where it is also given as the product of the Reynolds number and the Prandtl number, Pr (kinematic viscosity ≠ thermal diffusivity):

$$\mathrm{Pe} = \frac{\rho_\infty U_\infty c_p / L}{\lambda_\infty T_\infty / L} = \frac{U_\infty L}{\lambda_\infty / (c_p \rho_\infty)} = \mathrm{Re} \cdot \mathrm{Pr} \tag{3.26}$$

5. There are two dimensionless parameters for freeboard combustion, and they are the Damkohler I and II, which compare, respectively, the time it will take the reacting fluid to travel the combustion chamber to the reaction time; and the time it takes the reacting molecules to diffuse to the reaction front and the reaction time.

   a. Damkohler I—(heat of reaction)/(heat of convection) = (flow time)/(reaction time)

$$D_\mathrm{I} = \frac{w_{S\infty} Q_S}{\rho_\infty U_\infty c_p / L} = \frac{\tau_\mathrm{f}}{\tau_\mathrm{c}} \tag{3.27}$$

   b. Damkohler II—(heat of reaction)/(heat of conduction) = (diffusion time)/(reaction time)

$$D_\mathrm{II} = \frac{w_{S\infty} Q_S}{\lambda_\infty T_\infty / L^2} = \frac{\tau_\mathrm{d}}{\tau_\mathrm{c}} \tag{3.28}$$

We will treat freeboard combustion later on but suffice it to say that the flow patterns encountered in the combustion zone of a rotary kiln play the most important role in efficient combustion, particularly with pulverized fuel combustion. As the Damkohler numbers indicate, the timescale of combustion involves the diffusion time and the chemical reaction times. Since the reaction will occur when the reactants (air and fuel) are brought together, diffusion is the rate-controlling time step in the combustion process. In turbulent diffusion flames, the controlling measure is the mixing timescale induced by the recirculation eddies of the primary air jet. The conventional wisdom in combusting fluid is that "if it is mixed, it is burnt." As a result, much of the combustion characteristics can be inferred and characterized by the mixing strengths without even resorting to the solution of the full systems of equations presented in Equations (3.17)–(3.22). We will begin with the characterization of the properties of jets as introduced from a primary air pipe into the larger kiln surroundings and discuss the mixing of the flow in the confined region of the kiln as it induces the secondary air from within.

3.3 Development of a Turbulent Jet

It was mentioned earlier that the sources of fluid flow in rotary kilns are the primary air, secondary air, combustion products, and infiltration air. Usually the combustion system is such that the primary air issues from a burner nozzle as a jet into an open tube or into a tube surrounded with secondary air prior to combustion. The primary air nozzle tip velocities can be anywhere between 20 and 100 m/s depending on the firing rate. The secondary air into which the primary air is discharged is at a relatively lower velocity, typically in the range of 5–15 m/s (Table 3.1).

The aerodynamic structure of the jet is therefore similar to that of a submerged free jet (Figure 3.3) except that the structure is distorted due to the boundary constraints of the kiln tube. Unlike a free jet, which has an infinite source of air to entrain, the confined jet (Figure 3.4) has a limited source of entrained fluid. Therefore, when the jet fluid (primary air) emerges from the nozzle, a surface discontinuity is formed between it and the surrounding fluid (secondary air). This discontinuity results in an unstable surface

Table 3.1 **Typical Values of Primary and Secondary Air Velocities and Temperatures**

	Cement Kiln	Lime Kiln	Aggregate Kiln
Primary Air			
Percent of total flow	25–30		15–20
Temperature, K	375		
Velocity, m/s	60–100	20–45	20–30
Secondary Air			
Percent of total flow	70–75		
Temperature, K	750		
Velocity, m/s	5–15	5–10	5–10

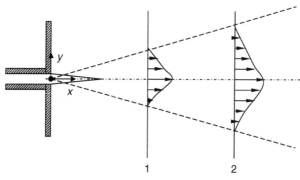

Figure 3.3 Turbulent free jet.

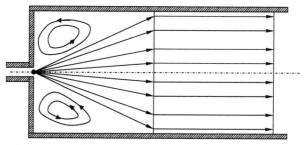

Figure 3.4 Turbulent confined jet.

with waves developing on it that enlarge as they move downstream. They eventually grow into eddies that extend inward to the boundaries of the kiln roughly halfway between the horizontal axis and the outermost points. The boundary becomes distorted such that at any point within the volume occupied by eddies, there will be encountered at some time unmixed fluid from outside the jet, and at other times, fluid that is a mixture of jet fluid and the entrained surrounding fluid. For submerged jets, the unmixed surrounding fluid can be nonturbulent while the jet fluid is turbulent. The turbulence in the mixing region of the jet results from velocity gradients that are introduced by the large-scale eddies in the shear flow. If one considers a time-mean velocity component in the direction parallel to the axis of the jet, one will find in the turbulent-free jet a characteristic sequence of development within the tip of the nozzle. By about eight nozzle diameters downstream the mean velocity profile is fully developed (Field et al., 1967). The stages of development include the potential core, which progressively diminishes in cross-sectional area up to a second region, the transition region, which extends about four to five nozzle diameters before the fully turbulent region (Figure 3.3).

The theory of jets is not the subject of this work. The flow field can be fairly accurately predicted using computational fluid dynamic (CFD) software to mathematically model the conservation equations presented for reacting flows. However, successful analytical predictions of the behavior of jets have been made elsewhere by using certain assumptions concerning mass and momentum transfer, for example, Prandtl's mixing length theory. Such theories can provide useful expressions for the boundary layer velocity profile of free jets that can be applied without the rigor of CFD. However, it might be said that the general characteristics of jets that have applications in rotary kiln combustors are turbulent diffusion jets for which analytical solutions can be challenging and might require added experience. They can be distinguished from laminar jets by the random fluctuations of the velocity that are superimposed on the mean velocity field and also by the fact that energy dissipation occurs through eddies of various sizes including up to and below the 0.1 cm level and in which dissipation is predominantly by the process of molecular diffusion. Some of the velocity and concentration measurements carried out by early investigators, such as van der Hegge Zijnen (1958) and Ricou and Spalding (1961), are still applicable today for establishing velocity profiles for free jets. For example, the transverse velocity and concentration profiles can be established by the relationship for plain and round jet, respectively, as

$$\frac{\bar{u}}{\bar{u}_m} = 0.5\left(1 + \cos\left\{\frac{\pi y}{2x \tan \varphi}\right\}\right) \quad (3.29)$$

$$\frac{\bar{u}}{\bar{u}_m} = 0.5\left(1 + \cos\left\{\frac{\pi r}{2x \tan \varphi}\right\}\right) \quad (3.30)$$

where φ is the jet half-angle, typically 5.5° for free plane (Equation (3.29)) and round (Equation (3.30)) jets. In any plane perpendicular to a round, free jet axis, a relative measure of the entrained ambient fluid was established by Ricou and Spalding (1961) as the ratio of mass flow at x, m_x, to the mass at source, m_0, that is,

$$\frac{\dot{m}_x}{\dot{m}_0} = 0.508\left(\frac{\rho_a}{\rho_0}\right)^{1/2}\left(\frac{x}{h}\right)^{1/2} \quad (3.31)$$

For round jets, particularly pertinent to primary air discharge in rotary kilns, Ricou and Spalding derived the following equation:

$$\frac{\dot{m}_x}{\dot{m}_0} = 0.32\left(\frac{\rho_a}{\rho_0}\right)^{1/2}\frac{x}{d_0} \quad (3.32)$$

Equation (3.32) is applicable to all values of Reynolds numbers at the nozzle $>2.5 \times 10^4$, and holds only for $x/d_0 \geq 6$. At shorter distances from the nozzle, Ricou and Spalding expect that entrainment ($\dot{m}_x/\dot{m}_x - 1$) proceeds at a lower rate of travel along the jet axis and increases progressively until it stabilizes at a constant value.

3.4 Confined Jets

Perhaps the most applicable aerodynamic phenomenon to the rotary kiln combustion system is the confined jet. Unlike the free jet with its unlimited supply of entrainable fluid, when a jet is confined in an enclosed environment, the free supply of surrounding fluid available for entrainment is cut off. The pressure within the jet increases with distance from the nozzle instead of remaining constant as it does in free jets, and the gas entrained in the region near the nozzle originates from the edge of the jet further downstream before the points where the jet impinges on the enclosing wall. The flow of gas back toward the nozzle and close to the surrounding walls is known as recirculation (Figure 3.4).

Recirculation plays a very important role in the shaping of turbulent diffusion flames encountered in rotary kilns. By the entrainment action of the confined jet, the reactants are mixed with further secondary air and with recirculated combustion products. Within the jet, the entrained fluid is mixed with the jet fluid by turbulent diffusion as described earlier. For example, in pulverized fuel combustion, the return of the combustion products by entrainment to the nozzle increases mixing and also the particle residence time in the reaction zone, thereby improving the heterogeneous gas−solid combustion reactions. Granted that such processes are characterized by the Damkohler numbers (Equations (3.27) and (3.28)), the increased diffusional mixing will result in more complete combustion. Several studies including recent works at the International Flame Research Foundation (IFRF) in Ijmuiden, the Netherlands (Haas et al., 1998), have established a direct correlation between flame shape, intensity, and heat flux with the extent of recirculation eddies in direct-fired pulverized fuel rotary kiln combustors. Because of its importance in tube combustion including rotary cement kilns, and the lack of CFD capabilities at the time, early flame researchers proposed a simple theoretical treatment of the confined jet problem based on analytical and empirical relations of free jet entrainment. Some of these early works at IFRF dating as far back as 1953 have been well documented and are still in use today. The confined flame theory established by Thring and Newby (1953), Craya and Curtet (1955), and those that appeared later from the work of Becker (1961) have been used in establishing the optimum aerodynamic characteristics of confined flames. In the rotary cement kiln where the secondary air is introduced at lower velocities compared with the primary air, one can assume that the secondary air is introduced as a stream with negligible momentum and that it is wholly entrained by the primary air jet before any recirculating gas is entrained. Based on this assumption, Craya and Curtet developed a relationship between the momentum ratios known as Craya−Curtet parameter using the similarity laws between free and confined jets. The parameter has found useful application in rotary kilns where it is used as a qualitative measure of the extent of recirculation of entrained gases. It serves as a surrogate to characterizing rotary kiln flame shapes and intensity without the rigor and the complexity of using CFD. The Craya−Curtet parameter is defined as (Jenkins and Moles, 1981):

$$M = \frac{3}{2} R^2 + R + \frac{KR^2}{(r_0/L)^2} \tag{3.33}$$

where R is the discharge ratio, q/Q; $q = (u_0 - u_a)r_0^2$; $Q = u_a(L - \delta^*)^2 + q$; K is a factor relating to the shape of the velocity profile; u_0 is the jet velocity and u_a is the surrounding fluid velocity, δ^* is the boundary layer displacement thickness defined as $1.74/\sqrt{Re}$; L is the half-width of the combustion chamber (rotary kiln) duct; and r_0 is the radius of the jet nozzle.

While this parameter varies between unity and infinity, experience has shown that for short and intense flames encountered in cement kilns, the recirculation is such that the Craya–Curtet parameter $M \geq 2.0$. Flames with M between 1 and 2 are characterized as long flames with the intensity suitable for processes such as rotary limestone calcination kilns. For flames with $M \leq 1.0$, the entrainment is such that the flame tends to be long and lazy. The Craya–Curtet parameter (Equation (3.33)) is essentially the ratio between the primary air jet momentum and the secondary air momentum and is a measure of the aerodynamic mixing of the two streams. High primary air velocities will return more combustion products to the burner nozzle and will result in attached flames associated with high Craya–Curtet parameters. The parameter can also be changed by altering the secondary air conditions. We will return to the use of the Craya–Curtet parameter for qualitative assessment of mixing, flame shapes, and flame intensity later on in the discussion of freeboard combustion.

3.5 Swirling Jets

The use of swirling jets has long been one of the practical ways of inducing mixing and improving burner effectiveness in rotary kilns. The increased flame stability and intensity associated with swirls are due to improved recirculation vortex (vortex shedding) which, like high velocity primary air jets, returns hotter combustion gases to the flame front where they become entrained in the primary air fluid prior to ignition and this enables the transfer of energy to the incoming reactants. Swirl burners typically seek to stabilize the flame by establishing a central recirculation zone and an external recirculation zone, as depicted in Figure 3.5.

Pulverized fuel combustion swirl jets improve particle residence time in the combustion zone, and consequently improve combustion efficiency. Increasing the swirl number and thereby the jet angle moves the external recirculation eddy (vortex shedding) closer to the burner in most kiln flames and results in flame attachment. Essentially, swirl can provide conditions that can approach a well-stirred reaction zone in

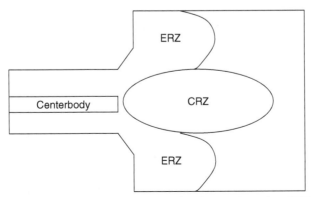

Figure 3.5 Schematic diagram of the effect of swirl on jet aerodynamics. CRZ, central recirculation zone; ERZ, external recirculation zone.

combustion systems. The centrifugal effect of the swirling jet induces a reduced pressure along the axis close to the burner and an increased pressure further downstream resulting in an axial recirculation eddy. The intensity of the swirling jet is characterized by the swirl number, defined as the ratio of the angular momentum to the linear momentum flux:

$$S = \dot{L}_0 / (\dot{G}_0 d_0) \tag{3.34}$$

The nozzle flux of angular momentum $\dot{L}_0$ and the flux of linear momentum $\dot{G}_0$ might be calculated as

$$\dot{L}_0 = 2\pi\rho \int_0^{r_0} (\vec{u}\vec{w})_{\text{nozzle}} r^2 dr \tag{3.35}$$

$$\dot{G}_0 = 2\pi \int_0^{r_0} \left(P + \rho \vec{u}^2\right)_{\text{nozzle}} r dr \tag{3.36}$$

where d_0 and r_0 are the respective nozzle diameter and radius, and P is the static or gauge pressure. P is included to allow for pressure variations due to centrifugal forces. Oftentimes, geometrical swirl number is defined in an experimental setup as

$$S_g = \frac{r_0 \pi r_e}{A_t} \frac{Q_\tau^2}{Q_T^2} \tag{3.37}$$

where r_e is the radius at which the tangential inlets of the swirl vanes are attached with respect to the center axis and A_t is the total area of the tangential inlets. Q_τ and Q_T are the tangential flow rate and total flow rate, respectively. It is found that S is directly proportional to S_g (Chatterjee, 2004). By altering the flow field and hence the strength of the recirculating zones, the flame can be induced to reside predominantly in either the recirculation zones or the shear layer between the zones.

3.6 Precessing Jets

Instead of using swirl generators, another method of generating vortex shedding, which has been found to increase radiant heat transfer, is by precessing the jet stream. A precessing jet flow can be generated either by an axisymmetric nozzle that utilizes a natural fluid flow instability (Figure 3.6) or by a motor-driven mechanical nozzle (Nathan et al., 1996). The precessing jet nozzle provides a new method of mixing fuel and air together in a way similar to variable turbulent swirl generators. It has been established that this method of mixing has significant beneficial effects on combustion and on processes requiring combustion by significantly reducing pollution and

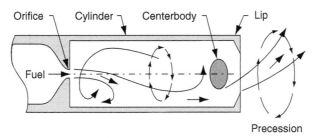

Figure 3.6 Precessing jet nozzle.

simultaneously increasing radiant heat transfer. The invention is described in a patent (Luxton et al., 1988) held by Adelaide Research and Innovation Pty. Ltd, of the University of Adelaide, and marketed to the rotary kiln industry as Gyro-Therm™. In its simplest form, the nozzle is a cylinder with a small concentric inlet orifice at one end, and an axisymmetric lip at the other end. When the dimensions of the pressing jet nozzle fall within certain criteria defined by the Strouhal number, the jet flow through the inlet orifice is subjected to an unstable lateral deflection and attaches asymmetrically to the internal wall of the cylinder.

The exit lip at the nozzle exit then deflects the emerging eccentric jet so that it wobbles or precesses at a large angle to the nozzle axis (Nathan et al., 1998). The dimensionless Strouhal number of precession is defined as

$$\text{St}_p = f_p d_e / u_e \tag{3.38}$$

where f is the frequency of precession, d_e is the exit diameter, and u_e is the mean exit velocity. In the lower Strouhal number flow regime, $\text{St}_p < 0.01$, the cold flow mixing characteristics are somewhat analogous to that of a fully pulsed jet. Here, the timescale of the precession is large in comparison with the timescales associated with the movement of the entrained ambient fluid, so that there are no significant asymmetries in the pressure field surrounding the local jet. Under these conditions, the dominant effect of the precessing motion of the jet is the superimposition of additional shear stress on the stresses that already exists in the stationary jet. It has been established (Nathan et al., 1996) that the additional shear relates to the product of the rotation rate and sine of the jet deflection angle. Cold flow velocity measurements within 10 nozzle diameters of a precessing jet operating in the low-St_p ($\text{St}_p = 0.002$) regime have shown that the radial spread of the instantaneous jet, which is based on the jet half-width and phase-averaged at the precession frequency, is about three times that of a non-precessing jet. Schneider (1996) showed that at the low-St_p regime, a precessing jet flame has been observed to be shorter and less luminous than either conventional jet flames or those at higher St_p. By contrast, for high Strouhal number flow ($\text{St}_p \geq 0.01$), the timescales associated with the precessing motion of the jet is comparable to timescales associated with the motion of entrained fluid within the region of space through which the jet precesses. Under these conditions, a low pressure zone is generated in the region between the jet and the nozzle axis, which also precesses with the jet. In such cases, large streamline curvature can be observed in the phase-averaged flow field, along with a rate of decay in the mean velocity field, which is an order of magnitude higher than that of a nonprecessing jet flow (Schneider, 1996). Here, the largest scale of the turbulent flow is that of the path described by the motion of the precession, so that the phase-averaged radial spread of the instantaneous jet is now some seven times that of a nonprecessing jet. Nathan et al. (1993) and Nathan and Luxton (1993) found an influence of the modified mixing characteristics on both the radiation heat flux and on NO_x emissions for both open and confined flames. They claim to have demonstrated a reduction in NO_x emissions on the order of 50–70% and a simultaneous increase in flame luminosity, an attribute that is important in radiation heat transfer in rotary kilns. It is proposed that due to the large eddy created by the precessing action described herein, natural gas can undergo cracking prior to combustion to generate soot particles that, in turn, proceed to combustion in a manner similar to that of pulverized fuel. In so doing, the soot-laden flame emits radiant energy, a component of the heat transfer mechanism that is enhanced by the rotary kiln curvature. For example, it has been claimed that the first installation of a PJ burner at the Ash Grove Cement Plant in Durkee, Oregon, increased the rotary kiln product output by 11%, increased specific fuel efficiency by 6%, and reduced NO_x emissions by 37% (Videgar, 1997). This prompted the promotion and positioning of the Gyro-Therm™ as a low-NO_x burner; however, subsequent installations at other locations have obtained mixed results in NO_x emissions.

3.7 The Particle-Laden Jet

An air jet laden with particles such as that found in primary air issuing from a pulverized fuel pipe for combustion in cement and lime kilns may be synonymous with a jet of fluid with a density greater than that of air provided the particles are small enough that one can consider the fluid to be homogeneous. Under such conditions, the effect of the solid burden may be accounted for by simply assuming an increase in the gas density and a reduction in the kinematic viscosity. A concomitant result will be an accelerated turbulence and an intensification of mixing and the entrainment phenomena associated with it. Equation (3.32) applies in such situations whereby m_0 might be increased by the factor ρ_0/ρ_a owing to the presence of suspended solid so that the effective change in air entrained per unit volume of jet fluid might increase by a factor of $(\rho_0/\rho_a)^2$. When the particles are not small enough to behave like a homogeneous fluid, a relative motion occurs between the particles and the surrounding air as a result of gravity or as a result of inertial forces resulting in the damping of the turbulence since the drag between the dust and the air will extract energy from the turbulent fluctuations. One important estimate is the distance at which a particle in a particle-laden jet will travel before coming to rest. This distance is defined as the range λ, a product of the initial velocity of the particle and the relaxation time τ_R:

$$\lambda = U_0 \tau_R \tag{3.39}$$

The relaxation time is defined here as the time taken for the relative velocity between particle and gas to fall to 36.8% of its initial value. For a perfect spherical particle, the relaxation time is defined as

$$\tau_R = \frac{m}{6\mu\pi r_p} \qquad (3.40)$$

where m and r_p are the mass and radius of the particle, respectively, and μ is the dynamic viscosity of the surrounding fluid. With these definitions, one can estimate that coal particles with a diameter of 80 μm injected at 60 m/s will have a range of about 150 cm, some 150 nozzle diameters for a 1-cm nozzle pipe, and will have little effect on the jet (Field et al., 1967). However, if the particles were finer, for example, 40 μm in size, then the range would only be 30 cm, which would have a damping effect on the jet due to turbulent energy transfer. The relaxation time is a measure of the shortest timescale of turbulence to which the particle could respond. As mentioned earlier, smaller eddies would have rapid velocity fluctuations and the particles would not have time to accelerate to the velocities within the eddies. However, if the eddies are large, then the particles can follow the streamlines without any appreciable slip and the suspension would tend to behave as a homogeneous fluid. It has been shown that increasing the fluid temperature shortens the relaxation time and thereby reduces the size of the eddy to which particles respond. When it falls within the same range as the timescale of the eddies, some damping of the turbulence can be expected, thereby reducing the eddy viscosity (Field et al., 1967). The concomitant result will be a decrease in the rate of entrainment and the rate of spread of the turbulent jet.

3.8 Dust Entrainment

We have examined the effect of kiln aerodynamics on fluid mixing and combustion. It is equally important to look at the aerodynamic effect on dust carryover from rotary kilns processing mineral materials. Although the principles behind particle motion are related to granular flow, which will be covered in Chapter 4, the interaction of the flow of fluid in the freeboard and the active layer surface of the kiln bed is an aerodynamic phenomenon.

The principle behind dust pickup is known as "saltation" and was first established by Bagnold in his study of sand dunes in deserts (Bagnold, 1941). If the particles are heavy enough and the gas velocity very high, the gas flow over the bed surface will induce a motion known as saltation in which individual grains ejected from the surface follow distinctive trajectories under the influence of gas velocity, resistance, and gravity. Because of their large size, these particles fail to enter into suspension. Instead, once lifted from the surface, they rise a certain distance, travel with the freeboard gas velocity, and descend, either to rebound on striking the surface or embed themselves in the surface and eject other particles (Figure 3.7).

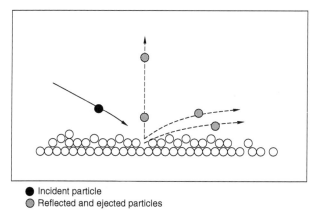

Figure 3.7 Concept of particle saltation. After Owen (1964).

The two principal underlying phenomena for saltation are (1) that the layer behaves, so far as the flow outside it is concerned, as an aerodynamic roughness whose height is proportional to the thickness of the layer, and (2) that the concentration of particles within the layer is governed by the condition that the shear stress borne by the fluid decreases, as the surface is approached, to a value just sufficient to ensure that the surface grains are in a mobile state (Owen, 1964).

The first principle follows the analogy of the aerodynamic behavior of surface roughness and the velocity profile in the fluid outside the saltation layer follows the friction law:

$$\frac{U}{u_\tau} = 2.5 \log\left(\frac{2gy}{u_\tau^2}\right) + D' \quad (y \geq h) \tag{3.41}$$

where U is the freeboard velocity at height y above the free surface, h is the thickness of the saltation layer, and D' is a constant. For wind over sand dunes, Owen (1964) fitted several experimental data presented earlier by Bagnold and others and established that $D' = 9.7$. The second principle postulates that the closer one gets to the surface, the greater the amount of horizontal momentum transported vertically by the particles. Since the shear stress (the momentum transfer rate) must be constant throughout the saltation layer, the proportion carried by the fluid decreases. The skin friction τ_0 required to initiate movement among otherwise stationary grains is defined by the ratio β relating the hydrodynamic force on the surface to the weight of the particle:

$$\beta = \frac{\tau_0}{\rho_p g d_p} \tag{3.42}$$

Bagnold (1941) determined two threshold values of β, a lower threshold value for which grain movement is first detected ($\beta = 0.01$), and an impact threshold at which saltation would be maintained ($\beta = 0.0064$). Hence, the shear stress required near the surface to sustain an equilibrium saltating flow is $\tau_0 = 0.0064\, \rho_p g d_p$. Finally, it may be stated that the saltation can be prescribed by the relationship

$$O(10^{-2}) < \frac{\rho_g u_\tau^2}{\rho_p g d_p} < O(1) \tag{3.43}$$

where the lower limit is required for surface mobility and the upper limit by the condition that grains do not enter into suspension.

The active layer surfaces encountered in rotary kilns are not stationary; rather, there is already a surface mobility due to the continuous surface renewal imposed by the particle cascades or excursions. The situation is even harder to model when kiln internals such as lifters and tumblers are installed. Nonetheless the aerodynamic condition expressed in Equation (3.43) is applicable by making modifications to the surface mobility through the density, perhaps, proportional to the granular dilation, which we will discuss later in Chapter 4. Dust entrainment in rotary kilns requires several regimes of increasing freeboard velocity for a specified cut size distribution of particles. The velocity regimes include the two threshold saltation velocities, the minimum velocity required to pickup a saltating particle and transport it. Sood et al. (1972) used 2–2.5 times saltation velocity as the dust pickup velocity to estimate the entrainment for the design of a rotary coke calciner. They estimated the saltation velocity (in units of feet per second) using what they termed a functional form of a dimensionless analysis relating the field parameters as

$$U_{\text{salt}} = 3.2 u_\tau \left(\frac{m_s}{m_g}\right)^{0.2} \left(\frac{D}{d_p}\right)^{0.6} \left(\frac{\rho_g}{\rho_p}\right)^{0.7} \left(\frac{u_s^2}{gD}\right)^{0.25} \tag{3.44}$$

where $u_s = m_s/A_s$ and the terminal velocity given by

$$u_\tau = \left[\frac{4}{3} g \frac{d_p(\rho_p - \rho_g)}{C_D \rho_g}\right]^{0.5} \tag{3.45}$$

where for the turbulent flow in the freeboard, $10^3 \leq \text{Re} \leq 10^5$, $C_D = 0.44$, that is, Newton's law regime. The dust pickup can be estimated by these correlations in the absence of CFD or without developing a true granular flow for the active layer as $U_{\text{dust}} = (2-2.5) U_{\text{salt}}$. That is, the dust entrainment velocity can be approximated by 2–2.5 times saltation velocity.

Other correlations for dust entrainment in rotary kilns have appeared in the literature (Li, 1974), but these are not easy to use for estimating dust generation. Tackie et al. (1990) attempted to model dust entrainment in a rotary kiln by coupling a simplified form of the Navier–Stokes fluid dynamic equations in the freeboard and the saltating friction factor assuming a saltating layer of 2 cm. The predicted entrainment rate (defined, R_s) for a 3.5-m diameter kiln was presented. The model validation is not known and, to the author's knowledge, has not found any practical application in industrial rotary kilns.

3.9 ID Fan

The total flow through the kiln is induced by the large ID fan located at the outlet of the kiln system and prior to the stack. ID fans are called dirty fans because unlike forced draft fans for primary air, and other uses, ID fans handle the gases produced by combustion, dust that might be entrained and not collected in the bag filter, the excess air, and any infiltration air that occurs up to the fan inlet. As mentioned earlier, infiltration air comes from all sources including leaks directly into the rotary kiln and can be as high as 20% of the total gas mass flow through the kiln. This makes determining ID fan requirements more of an art than an exact science. Knowing the total mass flow, volume flow rate (e.g., actual cubic feet per minute (ACFM)) can be estimated and with some knowledge of the pressure drop (static pressure, given in units of in-H_2O) across the system beginning from the hood to the fan inlet, the fan's performance or the required size can be estimated from readily available fan curves and/or calculations.

References

Bagnold, R.A., 1941. Physics of Blown Sands and Desert Dunes. William Morrow, New York.

Becker, H.A., 1961. Concentration Fluctuation in Ducted Jet Mixing (Sc.D. thesis). MIT, Cambridge, MA.

Chatterjee, P., 2004. A Computational Fluid Dynamics Investigation of Thermoacoustic Instabilities in Premixed Laminar and Turbulent Combustion Systems (Ph.D. thesis). Virginia Polytechnic Institute and State University, Blacksburg, VA.

Craya, A., Curtet, R., 1955. Sur L'evolution d'un jet en espace confine. C. R. Acad. Sci. 241 (1), 621−622.

Field, M.A., Gill, D.W., Morgan, B.B., Hawksley, P.G.W., 1967. Combustion of Pulverized Coal. British Coal Utilization Research Association (BCURA), Leatherhead, UK.

Haas, J., Agostini, A., Martens, C., Carrea, E., Kamp, W.L.v.d., 1998. The Combustion of Pulverized Coal and Alternative Fuels in Cement Kilns, Results on CEMFLAME-3 Experiments. A Report, IFRF Doc. No. F97/y4.

van der Hegge Zijnen, B.G., 1958. Measurement of the distribution of heat and matter in a plane turbulent jet of air. Appl. Sci. Res. A.7, 256−313.

Jenkins, B.G., Moles, F.D., 1981. Modelling of heat transfer from a large enclosed flame in a rotary kiln. Trans. IChemE 59, 17−25.

Li, K.W., 1974. Application of Khodrov's and Li's entrainment equations to rotary coke calciners. AIChE J. 20 (5), 1017−1020.

Luxton, R.E., Nathan, G.J., Luminis Pty. Ltd, 1988. Mixing of Fluids. Patent No. 16235/88, International Patent No. PCT/AU88/00114. Australian Patent Office.

Nathan, G.J., Brumale, S., Protor, D., Luxton, R.E., 1993. NO_x reduction in flames by modification of turbulence with jet precession. In: Syred, N. (Ed.), Combustion and Emissions Control. The Institute of Energy, pp. 213−230.

Nathan, G.J., Hill, S.J., Luxton, R.E., 1998. An axisymmetric fluidic nozzle to generate jet precession. J. Fluid Mech. 370, 347–380.

Nathan, G.J., Luxton, R.E., 1993. A low NO_x burner with a radiant flame. In: Pilavacji, P.A. (Ed.), Energy Efficiency in Process Technology. Elsevier, Inc., New York, pp. 883–892.

Nathan, G.J., Turns, S.R., Bandaru, R.V., 1996. The influence of fuel jet pressessing on the global properties and emissions of unconfined turbulent flames. Combust. Sci. Tech. 112, 211–230.

Owen, P.R., 1964. Saltation of uniform grains in air. J. Fluid Mech. 164 20 (2), 225–242.

Ricou, F.P., Spalding, D.B., 1961. Measurements of entrainment by axisymmetrical turbulent jets. J. Fluid Mech. 11, 21–32.

Schlichting, H., 1979. Boundary-layer Theory. McGraw-Hill, New York.

Schneider, G.M., 1996. Flow Structure and Turbulence Characteristics in a Precessing Jet (Ph.D. thesis). University of Adelaide.

Sood, R.R., Stokes, D.M., Clark, R., 1972. Static Design of Coke Calcining Kilns. Internal Report. Alcan International, Arvida, Canada.

Tackie, E.N., Watkinson, A.P., Brimacombe, J.K., 1990. Mathematical modeling of the elutriation of fine materials from rotary kilns. Can. J. Chem. Eng. 68, 51–60.

Thring, M.W., Newby, M.P., 1953. Combustion length of enclosed turbulent jet flames. In: 4th Int'l Symposium on Combustion, Baltimore, pp. 789–796.

Videgar, R., November 1997. Gyro-Therm Technology Solves Burner Problems. World Cement.

Zhou, L., 1993. Theory and Numerical Modeling of Turbulent Gas-Particle Flows and Combustion. CRC Press, Boca Raton, FL.

Granular Flows in Rotary Kilns

4

We have stated in Chapter 2 that axial flow of a particulate material in a rotary kiln must pass through the active layer. Chemical reaction, for example, the dissociation of limestone in lime kilns, is initiated and takes place there. Hence, any quantification of the transport processes in a rotary kiln must begin with establishing the depth of this layer. This notwithstanding, only a handful of quantitative predictions for the depth of the active layer have appeared in the literature, probably because of the complexity of analyzing granular flows. Granular flow is a form of two-phase flow consisting of particles and interstitial fluid (Hunt, 1997). When sheared, the particulates may either flow in a manner similar to a fluid, or resist the shearing like a solid. The dual nature of these types of flows makes them difficult to be analyzed. There is no deterministic description of the granular flow behavior unless it is approximated by the laws governing conventional non-Newtonian fluid flow as has been done for slurry flows. However, such an approach does not adequately represent the flow behavior because the flow properties such as consistency and the power law exponent associated with non-Newtonian flows are not easily measurable in rotary kilns. The early attempts to model granular flow in kilns include Pershin (1988), who, using results from a series of experimental trials on a small rotary drum, was able to mathematically model the shape of a cascading bed and, as a result, predicted the boundary between the active layer and the plug flow region with a reasonable degree of accuracy. Pershin's model was based on the fundamental principles of the equilibrium theory that hypothesizes that, if motion is steady in a gravity field, then the system should assume the position of minimum potential energy and, as a result, the system's mass must be reduced by moving some of the material beyond the boundary of the system. The model was unique at the time because, for the first time, it provided explicit mathematical expressions that could be used to calculate the centroid of the plug flow region, its area, and consequently the mass of material in the active layer. Although the model fell short of predicting the flow in the active region, where the material is in some form of kinetic motion, an application of mass balance in the transverse plane could offer knowledge of the average mass velocity in the active layer. Unfortunately, Pershin's experiments were conducted at very high speeds of rotation (about 20% of critical speed) and the accuracy of using the model to predict slower flows, for example, the rolling or slumping modes found in industrial kilns, is questionable. Additionally, the problem of quantifying the flow in the active layer for purposes of determining heat transfer cannot really be addressed by this methodology.

4.1 Flow of Granular Materials (Granular Flows)

The rapid deformation of bulk solids—such as sand, mineral ore, coal, grains, ceramic and metal powders, and so on—is termed granular flow. This definition covers the movement

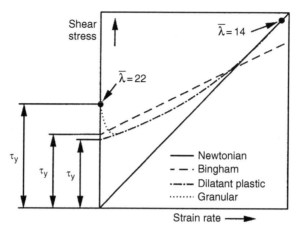

Figure 4.1 Comparison of granular flow behavior with other flow types (Davies, 1986).

of materials in a wide array of mineral and materials processing applications including heat treatment, drying, cooling, ore reduction, and others. The description of the dynamic behavior of these materials involves aspects of traditional fluid mechanics, plasticity theory, soil mechanics, and rheology (Savage, 1989). Despite the fact that granular flow has been applied to the study of gravity flow in hoppers; rockfalls; snow avalanches; mudflows; and so forth in recent years, it has not found application in rotary kilns. However, it has long been recognized (Singh, 1978) that particle diffusion in rotary kilns proceeds by interparticle collision suggesting that granular flow behavior based on either the dilute or dense phase kinetic theory might be a logical modeling step. Ferron and Singh (1991) showed experimentally that the magnitude of the axial diffusion coefficients lies between that of liquids and gases and that the modeling of the axial flow goes beyond Seaman's equilibrium approach. They employed the dilute gas kinetic theory analogy to describe rotary kiln transport phenomena such as mass and heat transfer. Although similar models such as the dense gas theory were being explored at the same time in solving problems associated with chute flows, debris flow, and avalanches (Savage, 1989), the approach did not catch on for rotary kiln characterization.

The primary challenge in granular flow modeling is not in setting up the conservation equations for mass, momentum, and energy, but in establishing the stress/strain relationship for the particulate mass as this relationship depends on the flow regime and vice versa. Davies (1986) has compared the observed behavior of granular materials, subjected to shear stress, with other common types of flow behavior (Figure 4.1). The figure depicts the shear stress as a function of the dilation factor $\bar{\lambda}$ expressed as

$$\bar{\lambda} = \left(\frac{v^*}{v}\right)^{1/3} - 1 \qquad (4.1)$$

Here, v is the volume concentration of solids (or solids fraction) and v^* is the corresponding value at the minimum possible void fraction that the material can maintain. For granular materials in a static condition, the particles fit together into a rigid grid.

This means that some degree of stress can be sustained without inducing a flow. However, as the stress approaches some critical level, the particles begin to ride up on one another and the grid commences dilation. At the critical stress, the dilation $\bar{\lambda}$ reaches a maximum and the material begins to flow. Once this occurs, the shear stress shows an incipient steep decline with increasing strain rate and it is this initial behavior that distinguishes granular flow from Bingham or dilatant plastic flow behaviors. Beyond a certain rate of strain the stress begins to increase again and the granular flow behaves like a dilatant plastic flow, that is, a nonlinear relationship between the shear stress and the rate of strain.

These fundamental aspects of the flow of granular material or bulk solids, similar to the flow in rotary kilns, have been reviewed by Savage (1979). Many theories have been considered in the effort to establish appropriate constitutive relations for such flows. As mentioned earlier, these theories extend from traditional soil mechanics to all types of viscous flows. The most important conclusions that can be drawn from these works to describe granular flow include the following:

1. Granular flow behavior extends beyond the critical state as defined by traditional soil mechanics literature. As a result of rapid deformation associated with the flow, inertia as well as shear-rate effects must be considered.
2. The dominant effect on the two-phase flow arises from particle-to-particle interactions whereas the interstitial fluid plays a minor role.
3. Because of particle-to-particle interaction, the transport processes are assumed to be governed by a field quantity called granular temperature, $\tilde{T}$, which can be defined as the kinetic energy per unit mass contained in a random motion of particles.

It is therefore common to assume that the state variables that describe the rapid deformation response of granular materials would border on the parameters that describe the behavior of fluids and Coulomb-type dissipation of energy (static). In view of the above it is common to find that the theories governing granular flow are formulated around the assumption of a continuum similar in some regard to viscous fluids; however, the equilibrium states of the theories are not states of hydrostatic pressure, as would be in the case of fluids, but are rather states that are specified by the Mohr–Coulomb criterion (Cowin, 1974). The advantage of continuum formulations over alternative particulate (stochastic) formulations is that use of continuum is more capable of generating predictive results. Mathematically one can argue that any state variable, such as the solid fraction, v, can only be either zero or unity depending upon whether a granule is present or not. By using the continuum approach it is common to represent v by a continuous function of position so that its value might represent the average in the neighborhood of that position.

The mechanisms of momentum transfer (and hence stress generation) for granular flows include the following:

1. Static stresses resulting from the rubbing between particles (dry Coulomb-type rubbing), which is independent of strain rate.
2. Translational stresses resulting from the movement of particles to regions having different velocity.
3. Collisional stresses resulting from interparticle collision, which result in transfer of both momentum and kinetic energy.

The relative importance of these three mechanisms will depend on both the volume concentration of solids within the bed, that is, the dilatancy factor $(\bar{\lambda})$, and the rate of strain. The static contribution dominates at high particle concentration and low strain rates. In this situation the particles are in close contact and the shear stresses are of the quasi-static, rate-independent Coulomb type as described in soil mechanics literature (de Jong, 1964; Spencer, 1964; Mandl and Luque, 1970; Roscoe, 1970). Conversely, at low particle concentrations and high strain rates, the mean free path of the particles is large compared with particle diameters and the interchange of particles between adjacent layers moving at different mean transport velocities may dominate stress generation. This situation is analogous to turbulent viscosity in fluid flow. At moderate particle concentration and high strain rate, collision between particles as against translation of particles to layers will dominate stress generation. This happens because there are rarely any void spaces of sufficient size for the interchange of particles over any significant distances. The case pertaining to low and moderate particle concentration was extensively studied by Bagnold (1954), a condition called the grain inertia regime where the dynamics of the actual particle collision become important. The applicable kinetic theories follow either hard sphere models, which assume that the interparticle collisions are instantaneous and therefore the collision trajectory is determined by the rules governing rigid body collisions (i.e., elastic, inelastic, and so on), or soft sphere models, which assume that particle collisions are of finite duration. Both models have been used to describe the collisional interactions that give rise to the transport of momentum and kinetic energy in granular flows. In this grain inertia regime, the stress tensor is said to be strain rate dependent (Campbell and Gong, 1986), and might be expressed as

$$\tau_{ij} \propto \left(d_p \frac{du}{dy}\right)^2 \tag{4.2}$$

Equation (4.2) includes translational and collisional effects but not static effects. However, most engineering applications (e.g., chute flows) and other natural flow situations (e.g., mudflows, snow avalanches, and debris flow) appear to fit into a regime for which the total stress must be represented by a linear combination of a rate-independent static component plus the rate-dependent viscous component just described (Savage, 1989). The flow patterns observed for material flow in rotary kilns appear to fit these descriptions and the constitutive equations for granular flow may apply within the relevant boundary conditions.

4.2 The Equations of Motion for Granular Flows

The equations of motion for granular flows have been derived by adopting the kinetic theory of dense gases. This approach involves a statistical—mechanical treatment of transport phenomena rather than the kinematic treatment more commonly employed to derive these relationships for fluids. The motivation for going to the formal approach (i.e., dense gas theory) is that the stress field consists of static, translational, and collisional components and the net effect of these can be better handled by statistical mechanics because of its capability for keeping track of collisional trajectories. However, when the static and collisional contributions are removed, the equations of motion derived from dense gas theory should (and do) reduce to the same form as the continuity and momentum equations derived using the traditional continuum fluid dynamics approach. In fact, the difference between the derivation of the granular flow equations by the kinetic approach described above and the conventional approach via the Navier–Stokes equations is that, in the latter, the material properties, such as viscosity, are determined by experiment while in the former the fluid properties are mathematically deduced by statistical mechanics of interparticle collision.

Equations of motion and the pertinent constitutive equations for the flow of granular materials have been developed by Lun et al. (1984) using the hard sphere kinetic theory of dense gas approach. In this derivation, a fixed control volume was considered in which a discrete number of smooth but inelastic particles are undergoing deformation. The resulting system of equations was given as follows:

1. Conservation of mass:

$$\frac{\partial \rho}{\partial t} = -\nabla \cdot (\rho \vec{u}) \tag{4.3}$$

where $\vec{u}$ is the velocity of the bulk material, ρ (equal to $\nu \rho_p$) is the bulk density, and ν is the bulk solid fraction.

2. Conservation of momentum:

$$\rho \frac{du}{dt} = \rho \mathbf{b} - \nabla \cdot \mathbf{P} \tag{4.4}$$

where $\mathbf{b}$ (a vector) is the body force, and $\mathbf{P}$ (a vector) is the total stress tensor which, unlike continuous fluid flow, comprises the three stress components mentioned earlier: (1) the static (frictional) stress $\mathbf{P}_f$, (2) the kinetic stress $\mathbf{P}_k$ that arises from the translation (or streaming) of particles, and (3) the stresses resulting from particle collisions $\mathbf{P}_c$. It is perhaps worth pointing out that the translational stress $\mathbf{P}_k$ is analogous to the Reynolds stresses for turbulent flow of fluids.

In addition to mass conservation, Equation (4.3), and momentum conservation, Equation (4.4), a third relationship that is required to describe the flow is some form of energy conservation equation. The total energy per unit mass of the granular material, E, may be broken into three components (Johnson and Jackson, 1987):

1. The kinetic energy, E_k, associated with the local average velocity u

$$E_k = \frac{1}{2}|u|^2 \tag{4.5}$$

2. The pseudo-thermal energy, E_{PT}, associated with deviations of the motion of individual particles from the local average; E_{PT} can be represented by the kinetic energy definition of temperature as

$$E_{PT} = \frac{1}{2}C^2 = \frac{3}{2}\tilde{T} \tag{4.6}$$

where c is the local velocity; $C = c - u$ is called the peculiar velocity. C^2 is the mean square of the velocity fluctuations about the mean. $\tilde{T}$ in Equation (4.6) is the kinetic theory definition of temperature called granular temperature (Johnson and Jackson, 1987) or grain temperature (Lun et al., 1984) which was defined earlier as the kinetic energy per unit mass contained in the random motion of particles (Zhang and Campbell, 1992). This must not be confused with the sensible heat or the true thermal internal energy of the solid material, that is, the enthalpy that might be required for heat transfer rather than momentum transfer. The total kinetic energy flux q will therefore be composed of the sensible heat flux q_h, and the flux of the pseudo-thermal energy, q_{PT}. The former is related to the thermodynamic temperature gradient and the effective thermal conductivity of the assembly of solid particles, while the latter is related to the gradient in the kinetic theory definition of temperature, that is, the granular temperature or grain temperature.

The conservation equation for the true thermal energy (sensible heat) is now distinguished as

$$\rho \frac{DE_h}{Dt} = -\nabla \cdot q_h - \mathbf{P}_f : \nabla u + \gamma \tag{4.7}$$

where D/Dt is the material derivative. $\mathbf{P}_f : \nabla u$ represents the rate of working of the frictional component of the stress tensor, while γ is the rate of dissipation due to the inelasticity of collisions between particles. In Equation (4.7), it is implied that work done by the frictional component of the stress tensor is translated directly into sensible heat and does not contribute to the pseudo-thermal energy (granular temperature) of the particles. It is worth mentioning that Equation (4.7) is the energy equation for the sensible heat or the true thermal energy transport and, since it does not influence the granular flow, is usually treated separately. In addition, the magnitude of the last two terms in Equation (4.7) (i.e., dissipation of frictional energy) is small compared with the thermal (thermodynamic) energy contribution to energy intensive process devices such as rotary kiln and may therefore be neglected during heat transfer calculations. Hence the conservation of kinetic energy in the absence of the terms that do not influence the flow field might be given as (Johnson and Jackson, 1987; Ahn et al., 1991; etc.):

$$\frac{3}{2}\rho \frac{D\tilde{T}}{Dt} = -\nabla \cdot q_{PT} - (\mathbf{P}_k + \mathbf{P}_c) : \nabla u - \gamma \tag{4.8}$$

In order to solve for the foregoing conservation equations to establish the granular flow field they must be closed by plausible constitutive relations for the stress terms, $\mathbf{P}_k$, $\mathbf{P}_c$, and $\mathbf{P}_f$, the kinetic energy flux, q, and rate of dissipation by inelastic collision, γ, along with suitable boundary conditions. Applying these equations to describe the flow of material in the transverse plane of the rotary kiln will require a true quantification of the actual flow properties, for example velocity, in the various modes of rotary kiln observed and described earlier in Chapter 2.

4.3 Particulate Flow Behavior in Rotary Kilns

The rotary kiln is often considered to be a black box into which materials are fed at one end and discharged at the other without knowledge of what happens with the flow behavior in between. It has therefore been a great research curiosity to observe the flow patterns at the various zones through experiments. But, unlike gas flow, which can be visualized on a full scale setup, most particulate flow experiments can only be observed through an end piece. Nevertheless, they have successfully provided adequate information for flow modeling. Recent noninvasive measurement techniques such as use of NMR (Nakagawa et al., 1993), positron emission particle tracking (Parker et al., 1997), and photonic sensors (Hsiau and Hunt, 1993; Boateng and Barr, 1997) have provided sufficient evidence and boundary conditions for flow modeling.

Using photonic sensors, and a 1 m rotary drum, Boateng (1993) characterized the flow behavior of high-density spherical polyethylene particles (elastic material), nonspherical long grain rice (inelastic material), and irregular limestone particles (industrial material). Visual observations were made with regard to the slipping, slumping, rolling, and cataracting as previously observed (Henein et al., 1983a). Henein et al. had experimentally studied bed behavior diagrams and identified various types of bed motion. Boateng and Barr (1997) successfully quantified active layer thickness, bed expansion, and dynamic angle of repose within a wide range of the operational variables. The active layer was quantified in terms of its shape, symmetry, and depth. Dilation within the active layer was quantified by determining the solid fraction there and comparing it with the plug flow region. In analyzing the flow behavior some of the key parameters that characterize rheological behavior of granular solids, specifically the velocity parallel to the bed surface, the granular temperature, the solid fraction, and the active layer depth, were computed. One objective of the exercise was to compare the similarities between the rapidly flowing active layer and other granular flow systems so as to establish and justify the use of the constitutive equations developed for granular flow in kilns. This was accomplished by estimating the mean value of the velocity from instantaneous velocities of particles flowing past an optical fiber probe unit, and the variance of the velocity fluctuation with which the granular temperature, $\tilde{T}$, could be computed. Other granular flow parameters including the solid fraction in the transverse plane, also known as the solids linear concentration in the direction of the flow (Ahn et al., 1991), could be measured. Also calculated was the mean shear rate $\Delta u/\delta$, where Δu was the local velocity difference between the surface and the active layer/plug flow interface and δ is the local active layer depth normal to the surface plane. An overview of the experimental results is discussed herein (Boateng and Barr, 1997).

4.4 Overview of the Observed Flow Behavior in a Rotary Drum

The flow behavior patterns of polyethylene particles at various percent fills (i.e., 3.3, 8.5, 15, and 29%) and operated at drum speeds ranging between 1 and 5 rpm and also for rice grains loaded at 3.3, 8.5, and 10% fills and operated from 3 through 5 rpm have been discussed in published works including Boateng and Barr (1997). The results show that the surface velocity profile is parabolic (Figure 4.2), indicating that particles

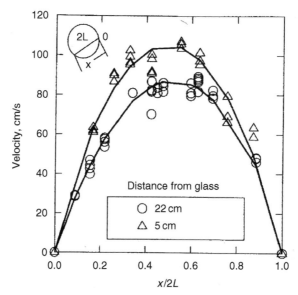

Figure 4.2 Measured exposed surface velocity profile in the active layer (15% fill and 5 rpm).

accelerate rapidly from the apex (top corner with maximum potential energy) up to a location around the mid-chord and then decelerate. The deceleration is a result of the impact of the rotating wall on the material at the base, as would be expected in a confined flow. The mid-chord velocity could be observed to be as high as 110 cm/s, about seven times the circumferential velocity of the drum wall, and an indication of a very rapid flow of particles over the exposed bed surface.

The parabolic nature of the surface velocity is consistent with observations made by Singh (1978) who also used similar polyethylene pellets in a rotary kiln experiment. One observation about the free surface flow is that the symmetry in the velocity profile is quick to disappear with particle inelasticity and operational conditions. Hence velocity profiles can be symmetric or asymmetric due to variations in bed depth (or percent fill) and drum speed. For irregular and inelastic materials, a considerable amount of energy is required for the particles to rearrange themselves and in doing so pile up at the apex in order to minimize energy. This potential energy buildup is subsequently released very rapidly in a manner similar to an avalanche. For inelastic industrial materials such as limestone, bifurcations in the velocity profiles are quickly developed at the free surface. With regard to the active layer depth several works have reported qualitative characterization of it but Boateng (1993) was first to quantify it. For example Henein et al. (1983a) reported the thickness of the active layer for relatively deep beds to be about 10% of the bed depth at mid-chord. However, with the help of optical fiber probes Boateng (1993) showed that 10% is not always the case but, rather, depends on the material. Measurements with polyethylene pellets as the bed material showed the thickness of the active layer could be as high as 30% of the bed depth at mid-chord (Figure 4.3). However, active layer depths for most inelastic materials are generally less. Hence the extent of the active layer thickness can be expected to depend

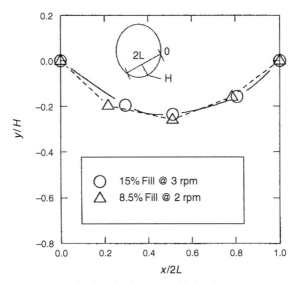

Figure 4.3 The shape and depth of active layer at mid-chord.

on the physical properties of the material that determine the ability of the granular material to shear under stress. Nonintrusive measurements that have recently appeared have substantiated these findings (Parker et al., 1997).

Analysis of the velocity profile as a function of depth shows some agreement with inclined chute flow experiments with polyethylene pellets (Savage, 1979). The behavior can be characterized as typical of flows over rough surfaces depicting drag flows that are mathematically classified as tractrix (from Latin tractum). Except for the regions near the free surface, a constant shear rate ($du/dy =$ const) is deduced implying a uniform simple shear flow (Figure 4.4). Measurement of the granular temperature, the velocity fluctuation, and the linear concentration could similarly be made (Figures 4.5 and 4.6).

Like its counterpart in the kinetic theory of dense gases, that is, thermodynamic temperature, granular temperature can either conduct away from the free surface into the bed or vice versa. The profile shown here indicates that there can be a granular temperature gradient between the bed surface and the bulk bed in agreement with the computer simulation of Zhang and Campbell (1992) for Couette flow of granules. It is observed that the granular temperature is high in the regions where there is a mean velocity gradient, and therefore it is not surprising to see granular conduction into the bed burden. An increase in the drum speed from 1 to 3 rpm results in an order of magnitude increase in the granular temperature. The profile for the linear concentration (Figure 4.6) shows that there is also a gradient between the surface and the bulk bed. Material dilation in the direction normal to the bed surface is not very significant according to visual observations through the glass end piece. This is not surprising for rolling bed behavior; material balance calculations have shown that bed expansion in the normal direction does not exceed 5%.

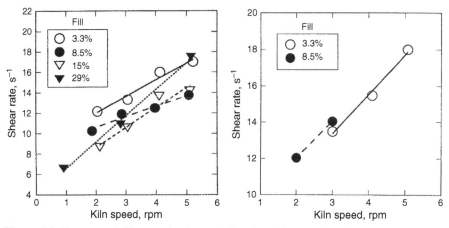

Figure 4.4 Shear rate, du/dy, at active layer. Left: polyethylene pellets; right: limestone.

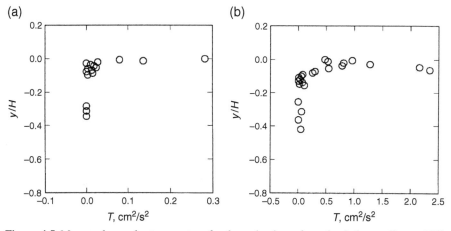

Figure 4.5 Measured granular temperature for the active layer for polyethylene pellets at 29% fill: (a) 1 rpm; (b) 3 rpm.

4.4.1 Modeling the Granular Flow in the Transverse Plane

The flow behavior described can be modeled in several ways including the use of the kinetic theory approach employed for avalanche and debris flow. Ferron and Singh (1991) were first to apply such theory to describe the observations in a rotary kiln using the dilute phase kinetic theory. However, the experimental work described herein (Boateng and Barr, 1997) suggests that dense phase kinetic theory might be most suitable. The model described here employs the dense phase kinetic theory constitutive equations developed for elastic or slightly inelastic granular materials and applied to the boundary conditions of the rotary drum. Following the experimental observations and measurements described (Boateng and Barr, 1997), plausible solutions are found by exploring the boundary layer flow similarities with the rolling bed active layer flow.

Granular Flows in Rotary Kilns 59

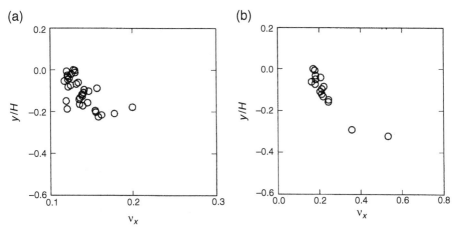

Figure 4.6 Measured linear concentration profile in the active layer for polyethylene pellets: (a) 15% fill and 3 rpm; (b) 29% fill at 5 rpm.

4.5 Particulate Flow Model in Rotary Kilns

Plausible arguments can be made about the similarities of the active layer flow and flows of isotropic fluid over flat plates. Additionally, previous works on granular flow on inclined planes (e.g., chute flow) suggest that the constitutive equations derived for rapidly shearing (slightly inelastic) granular material and based on the analogy of kinetic theory of dense gases (Lun et al., 1984) are adequate in interpreting the experimental results. Although other theories exist, for example, plastic formulations (Mandl and Luque, 1970), they have not been widely tested for rapidly shearing granular flows.

4.5.1 Model Description

The domain for which a solution to the flow problem is sought is depicted in Figure 4.7. As shown in the figure, two distinct regions, (1) the nonshearing (plug flow) region, and (2) the shearing region that forms the active layer near the bed surface, can be discerned by an interfacial boundary which is a few particles away from the zero velocity line. At this boundary, particles are sustained by the dynamic angle of repose. In the plug flow region the particles rotate with the kiln as a rigid body and the strain rate in this region is zero. The flow of particles within the active layer near the upper bed surface is rather more complex since it involves all the aspects of granular flow discussed earlier. In this region the material can acquire any of the several modes described earlier, such as slumping, which is a slow flow occurring when the bed inclination just exceeds the static angle of repose of the material; rolling bed, for which the material is continuously sheared and the flow, as well as diffusion, is the result of interparticle collisions. To model the flow in the cross-section we will focus on the rolling mode behavior because of its importance to industrial kiln operation and later discuss the mode-to-mode transition regimes using the rolling bed results. For the rolling

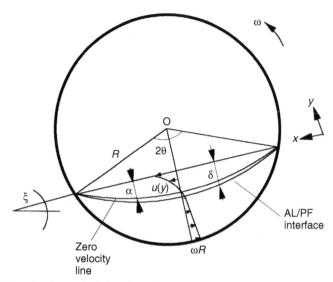

Figure 4.7 Granular flow calculation domain.

mode, the plug flow region might be anticipated to behave as a rigid lattice of particles rotating with the kiln wall without slippage. The velocity within the plug flow region is a linear function of radius. Hence a solution for the flow field is therefore required only for the active layer near the upper surface of the bed. However, the location of the interface between this region and the active layer is not known a priori and the model must allow prediction of its location.

4.5.2 Simplifying Assumptions

The primary assumptions for modeling granularity in the transverse plane of a rotating cylinder are as follows:

1. The bed material consists of cohesionless particles that possess a relatively high coefficient of restitution. This assumption places important emphasis on the role of interparticle collision on momentum transfer and permits the use of the equations of Lun et al. (1984).
2. Particles are spherical, rigid, and slightly inelastic, such as the polyethylene pellets used in the experimental work.
3. The bed motion is in the rolling mode and the active layer is considered to be thin relative to the bed depth (the ratio of the active layer depth at mid-chord to the chord length was less than 0.04 in the experiments). The granular temperature in the active layer is assumed constant in the radial direction at each bed surface position.
4. The motion is essentially two-dimensional in the transverse plane since the transverse velocity is several orders of magnitude greater than the axial velocity. Also particle flux into the active layer at the right quadrant is assumed to equal the particle flux into the plug flow region at the left quadrant (symmetry) and therefore only half of the two-dimensional domain is considered.
5. The particle ensemble behaves as a continuum and the flow properties, for example, solids concentration, are assumed to be continuous functions of position.

4.5.3 Governing Equations for Momentum Conservation

The governing equations for the flow field are similar to those derived for conventional fluids, for example the Navier–Stokes equations for Newtonian isotropic material. However, in the latter case, flow properties such as viscosity are experimentally determined. For the case of the dense gas theory and, by extension, granular flows, collisions between particles play a significant role in the exchange of both energy and momentum. Under these conditions kinetic energy conservation must also be considered in the formulation along with momentum and mass transport. The governing equations (Equations (4.3)–(4.6)) can be restated as (Lun et al., 1984)

$$\frac{\partial \rho}{\partial t} + \nabla \cdot (\rho u) = 0 \tag{4.9}$$

$$\rho \frac{Du}{Dt} = \rho g - \nabla \cdot \mathbf{P} \tag{4.10}$$

$$\frac{3}{2}\rho \frac{D\tilde{T}}{Dt} = -\nabla \cdot q_{\mathrm{PT}} - \mathbf{P} : \nabla u - \gamma \tag{4.11}$$

where u is the bulk velocity, $\rho = v\rho_p$ is the bulk density, and v is the solids volume fraction. $\mathbf{P}$ is the total stress tensor, which consists of both static and kinetic (streaming and collision) components. The term q_{PT} is the flux of pseudo-thermal energy defined by the kinetic energy definition of temperature, $\tilde{T}$, also known as granular temperature (Johnson and Jackson, 1987), while γ is the dissipation of pseudo-thermal energy due to inelastic collision of particles. The kinetic contribution of the stress tensor might be given as (Lun et al., 1984)

$$\mathbf{P} = \left[\rho\tilde{T}(1+4\eta v g_0) - \eta\mu_b \nabla \cdot u\right]I - \left\{\frac{2\mu}{\eta(2-\eta)g_0}\left(1+\frac{8}{5}\eta v g_0\right)\right. \\ \left. \times \left[1+\frac{8}{5}\eta(3\eta-2)v g_0\right] + \frac{6}{5}\mu_b \eta\right\}S \tag{4.12}$$

derived by considering the pair distribution function in collision theory. In Equation (4.12), S is the deviatoric stress which is given by

$$S = \frac{1}{2}\left(u_{i,j} + u_{j,i}\right) - \frac{1}{3}u_{k,j}\delta_{ij} \tag{4.13}$$

where δ_{ij} is the kronecker delta; that is, $\delta_{ij} = 1$ for $i = j$, and $\delta_{ij} = 0$ for $i \neq j$.

The flux of pseudo-thermal energy is expressed as (Johnson and Jackson, 1987)

$$q_{PT} = -\frac{\lambda_i}{g_0}\left\{\left(1+\frac{12}{5}\eta v g_0\right)\left[1+\frac{12}{5}\eta^2(4\eta-3)v g_0\right]+\frac{64}{25\pi}(41-33\eta)(\eta v g_0)^2\right\}\nabla \tilde{T}$$
$$-\frac{\lambda_i}{g_0}\left(1+\frac{12}{5}\eta v g_0\right)\frac{12}{5}\eta(2\eta-1)(\eta-1)\frac{d}{dv}\left(v^2 g_0\right)\frac{\tilde{T}}{v}\nabla v$$

(4.14)

The dissipation of energy due to inelastic collisions may be written as (Johnson and Jackson, 1987)

$$\gamma = \frac{48}{\sqrt{\pi}}\eta(1-\eta)\rho_p\frac{v^2}{d_p}\tilde{T}^{3/2} \tag{4.15}$$

The nomenclature is consistent with existing granular flow literature (Lun et al., 1984; Johnson and Jackson, 1987; Natarajan and Hunt, 1998), but key parameters in these constitutive equations can be defined using the analogy of fluid flow as follows:

μ_b [$= 256\mu v^2 g_0/5\pi$] bulk viscosity for perfectly elastic particles; $\eta\mu_b$ is the bulk viscosity for inelastic particles;

λ [$= 75m\sqrt{\tilde{T}/\pi}/64d_p^2$] is the granular conductivity;

λ_i [$= 8\lambda/\eta(41-33\eta)$] is the granular conductivity for inelastic particles;

η [$= 1/2(1+e_p)$] is the average value between the coefficient of restitution of the particle, e_p, and that of a perfectly elastic particle, $e_p = 1$.

μ [$= 5m\sqrt{\tilde{T}/\pi}/16d_p^2$] is the shear viscosity, where m and d_p are particle mass and diameter, respectively;

μ_i [$= \mu/\eta(2-\eta)$] is the shear viscosity for inelastic particles;

g_0 [$= 1/(1-v/v^*)^{1/3}$] is a radial distribution function at contact during binary collision or the dilation factor. In this term v^* is the maximum shearable solids volume fraction.

The reason for assuming that the active layer is thin relative to the bed depth is to confine the domain for which a solution to the flow problem is sought to the active layer, thus avoiding the computational demands of solving for the already-known velocities in the plug flow region. The obvious task is to develop approximate solutions analogous to those of other thin flows, in other words, of boundary layer flows. To do so it is necessary to normalize the governing equations according to the dimensions of the kiln cross-section and establish whether these equations can be reduced to parabolic equations similar to those for flow over a flat plate as suggested earlier. Recasting Equations (4.9)–(4.12) into primitive variables yields the steady-state continuity, momentum, and kinetic energy equations:

$$\frac{\partial u}{\partial x}+\frac{\partial v}{\partial y} = 0 \tag{4.16}$$

$$\rho\left[u\frac{\partial u}{\partial x}+v\frac{\partial v}{\partial y}\right] = \rho g\sin\xi - \frac{\partial P_{xx}}{\partial x} - \frac{\partial P_{xy}}{\partial y} \tag{4.17}$$

Granular Flows in Rotary Kilns

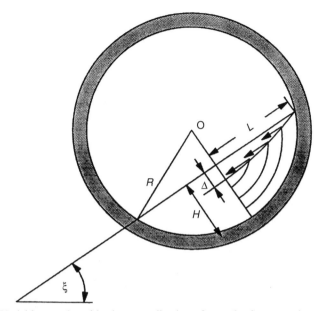

Figure 4.8 Variables employed in the normalization of granular flow equations.

$$\rho\left[u\frac{\partial v}{\partial x}+v\frac{\partial v}{\partial y}\right]=-\rho g\cos\xi-\frac{\partial P_{xy}}{\partial x}-\frac{\partial P_{yy}}{\partial y} \quad (4.18)$$

$$\rho\left[u\frac{\partial \tilde{T}}{\partial x}+v\frac{\partial \tilde{T}}{\partial y}\right]=-\left[\frac{\partial q}{\partial x}+\frac{\partial q}{\partial y}\right]-\left[P_{xx}\frac{\partial u}{\partial x}+P_{xy}\frac{\partial v}{\partial x}+P_{yx}\frac{\partial u}{\partial y}+P_{yy}\frac{\partial v}{\partial y}\right]-\gamma \quad (4.19)$$

We can now invoke the simplifications derived from the thin flow assumption to solve for the velocity distribution in the active layer. This requires some normalization using Δ and H (Figure 4.8).

The coordinates appropriate to the active layer may be defined in cartesian coordinates with field variables taken with respect to x and y where, from assumption (4), $-L \leq x \leq L$ and $0 \leq y \leq -H$. These variables are Δ, the depth of the active layer at mid-chord of the free surface plane, and L, half of the mid-chord length. If the angle subtended by the boundary interface is the dynamic angle of repose, ξ, then $\tan\xi$ is the coefficient of dynamic friction and the stresses can be normalized with the gravity term as (Savage and Hutter, 1989)

$$(x,y) \to ([L]x^*, [\Delta]y^*)$$
$$(u,v) \to \left([(gL)^{1/2}]u^*, [\Delta/L(gL)^{1/2}]v^*\right) \quad (4.20)$$
$$(P_{xx}, P_{yy}, P_{xy}) \to [\rho g \cos\xi\Delta]\left(P^*_{xx}, P^*_{yy}, \tan\xi P^*_{xy}\right)$$

For a detailed derivation, interested readers are referred to the pertinent literature (Boateng, 1998). As $\Delta/L \to 0$ the equations reduce to boundary layer equations, which deleting the * from the nondimensional terms has the form

$$\frac{\partial u}{\partial x} + \frac{\partial v}{\partial y} = 0 \tag{4.21}$$

$$u\frac{\partial u}{\partial x} + v\frac{\partial v}{\partial y} = \sin \xi - \sin \xi \frac{\partial P_{xy}}{\partial y} \tag{4.22}$$

For this thin flow, the y-wise momentum equation becomes the overburden pressure

$$\frac{\partial P_{yy}}{\partial y} = 1, \text{ and } P_{yy} = \int_0^\Delta v\, dy \tag{4.23}$$

It should be noted that, so far as nothing is said about the stresses, which depend on the boundary conditions, Equations (4.21) and (4.22) are similar to those derived for flow over a flat plate (Schlichting, 1979). To solve for these equations the continuity equation can be rearranged as

$$v = -\int_0^y \left(\frac{\partial u}{\partial x}\right) dy \tag{4.24}$$

Substituting Equation (4.24) into the corresponding momentum equation, an approximate solution similar to that obtained by von Karman for flow over a flat plate can be sought (Schlichting, 1979). Having demonstrated the similarity between the active layer flow and boundary layer flow equations, one convenient solution approach is to recast them into momentum integral equations and find simple solutions by integration over the appropriate control volume. Such an exercise provides additional highlights of the physical interpretation of the complex granular flow equations as applied to the rotary kiln cross-section. It will also allow the use of some of the experimentally measured parameters for boundary conditions to the specific flow problem.

4.5.4 Integral Equation for Momentum Conservation

So far the problem has been dealt with by stating the continuum equations for material flow and these equations have been reduced to parabolic equations involving unknown stresses by using the geometry of a rolling bed. It can now be shown that, if the flow behaves as a continuum, then the same equations can be deduced by simply considering material and momentum balance over a control volume in the active layer of the bed (Figure 4.9).

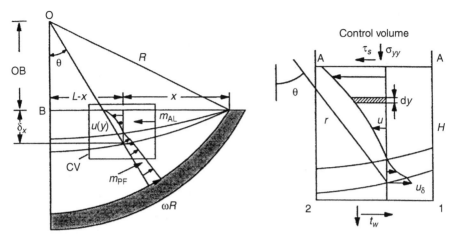

Figure 4.9 Control volume for deriving integro-momentum equations.

By proceeding in a manner similar to that employed in deriving the von Karman equation for a developing boundary layer, the system of partial differential equations (Equations (4.21) and (4.22)) can be reduced to the ordinary differential equation

$$\rho \frac{d}{dx}\left(\int_0^H (u^2 - u_\delta u)\,dy\right) dx + \frac{du_\delta}{dx}\left(\int_0^H \rho u\,dy\right) dx = \sum F_x \qquad (4.25)$$

It can be readily shown that Equation (4.25) can also be obtained by substituting Equation (4.24) into the continuity and momentum equations directly and integrating by parts. It should be emphasized that, provided no statement is made about the stresses or the net forces acting on the control volume, the net momentum equation is the same as that governing fluid flow, otherwise known as the Blasius problem (Schlichting, 1979). In Equation (4.25), u_δ, the velocity at the transition from the active layer to the plug flow, is a function of radius only ($u_\delta = \omega r$) but the radius for this transition requires knowledge of the active layer depth, δ_x, at that distance, x, from the apex. From the geometry in Figure 4.9 it can be shown that

$$r^2 = (L-x)^2 + (\mathrm{OB} + \delta_x)^2 \qquad (4.26)$$

$$\cos\theta = \frac{(\mathrm{OB} + \delta_x)}{r} \qquad (4.27)$$

where OB is the distance from the kiln's centerline to the bed surface. These geometric relationships will introduce nonlinearities in Equation (4.25) and an iteration procedure would be required in order to solve for the active layer depth. Having now derived the momentum conservation (Equation (4.25)) in the active layer the next step is to proceed with the evaluation of the force terms. The forces acting on the control volume

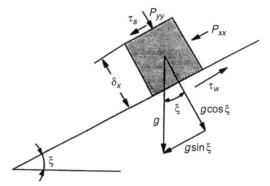

Figure 4.10 Force (pressure) distribution on control volume.

(Figure 4.10) include the body force and the forces that are generated by the stresses described earlier. The net force is equal to

$$-\tau_w - dx + \sigma_N dx + \rho g dx \tag{4.28}$$

where σ_N is the overburden pressure and is given as

$$\sigma_N = P_{yy} + P_{xx} \tag{4.29}$$

The top face of the control volume can be considered the free surface and therefore the shear stress is zero (plane A–A in Figure 4.9). The normal stress at this surface is due to the forces exerted by the freeboard gas. If there is no significant saltation of particles on the free surface, then the normal stress at this plane is also zero and equating the net momentum to the net force yields the integral momentum equation

$$\rho \frac{d}{dx}\left(\int_0^\delta (u^2 - u_\delta u) dy\right) + \frac{du_\delta}{dx}\int_0^\delta \rho u dx = -\tau_w + P_{xx} dy + \rho g \sin \xi \tag{4.30}$$

The second term on the left of Equation (4.30) would, in fluid flow, be equal to the hydrostatic pressure according to Bernoulli's equation. For granular flow on an incline this term is equivalent to the driving force parallel to the inclined plane and hence may be equated to the overburden pressure in the x-direction, that is,

$$\frac{du_\delta}{dx}\int_0^\delta \rho u dy = P_{xx} \tag{4.31}$$

Assembling the results obtained thus far yields the expression

$$\rho \frac{d}{dx}\left(\int_0^\delta (u^2 - u_\delta u) dy\right) = \rho g \sin \xi - \tau_w \tag{4.32}$$

The final task before proceeding with its solution is to derive an appropriate expression for the shear stress acting over the bottom surface of the control volume, τ_w.

The shear stress is a combination of contributions from static and kinetic stresses. The extent of each contribution might be determined by the operational parameters, that is, the rotational speed, degree of fill, and so on. The various modes of bed behavior can therefore be related to the stresses acting on the material within the depth of the active layer. These modes can be mathematically described as follows:

1. **Slumping bed:** When the rotation and/or collision of particles are constrained and if $\partial u/\partial y > 0$ then the shear stress is static and is simply given as

$$\tau_w = \rho g \cos \xi \tan \phi \tag{4.33}$$

where ϕ is the static angle of repose. This situation occurs for slumping bed behavior where there is no interparticle collision contribution to the shearing of the active layer (Henein et al., 1983b). The corresponding momentum equation is

$$\rho \frac{d}{dx}\left(\int_0^\delta (u^2 - u_\delta u) dy \right) = \rho g \sin \xi - \rho g \cos \xi \tan \phi \tag{4.34}$$

$$= g \cos \xi (\tan \xi - \tan \phi)$$

As Equation (4.34) indicates, if $\xi > \phi$, the flow is accelerated, while, if $\xi < \phi$, the flow is damped. When $\xi = \phi$, it can be said that the flow is indeterminate (Kanatani, 1979). Suffice it to say that damped flow is the mode for slipping bed behavior.

2. **Rolling bed:** From the flow experiments discussed earlier, it can be said that the rolling bed is the situation where the kinetic stress is the driving force for material flow, that is, when the bed is in the rolling mode the material in the active layer is in a continuous shearing mode. In this case all the aspects of granular flow come into play and the shear stress is deduced from the constitutive equations we have described earlier. Many forms of the shear stress expressions exist in the literature, however, they are all variants of the equation first proposed by Bagnold (1954), which is given as

$$\tau_w = -c_i \rho_p \left(d_p \frac{du}{dy} \right)^2 \tag{4.35}$$

where c_i is Bagnold's constant (Campbell and Gong, 1986). We can now employ the stress/strain rate relationship from the constitutive equations derived through the analogy of kinetic theory of dense gases (Lun et al., 1984), that is,

$$\tau_w = C' \frac{du}{dy} \tag{4.36}$$

where the apparent viscosity, C', is a function of the dilation, $C' = f(\rho_p; d_p; e_p; v; \tilde{T})$ and is related to the properties of the bed (Appendix 4A)

$$C' = -\rho_p d_p g_2(v) \tilde{T}^{1/2} \tag{4.37}$$

where $g_2(v)$ is a term relating the viscosity to flow properties such as the coefficient of restitution of the particles, $\eta\{=(1/e_p)/2\}$, the solids fraction, v, and is derived in Appendix 4A as

$$g_2(v) = \frac{5\sqrt{\pi}}{96}\left[\frac{1}{\eta(2-\eta)g_0} + \frac{8}{5}(3\eta-1)\frac{v}{(2-\eta)} + \frac{64}{25}\left\{\frac{\eta(3\eta-1)}{(2-\eta)} + \frac{12}{\pi}\right\}v^2 g_0\right] \quad (4.38)$$

Inserting this result for the shear stress in Equation (4.32) the momentum conservation equation takes on its final form:

$$\frac{d}{dx}\rho\int_0^\delta \left(u^2 - u_\delta u\right)dy = \rho g \sin\xi + g_2(v)\rho_p d_p \tilde{T}^{1/2}\frac{du}{dy} \quad (4.39)$$

We notice that this equation involves the shear rate, dy/dy, in the active layer, the particle dilation of the bed, the granular temperature, and gravity, all of which are granular flow characteristics that were previously discussed under experimental overview. The equation points to the fact that boundary layer analogy can be combined with the constitutive equations of Lun et al. (1984) to arrive at a single analytical model which can be used to predict the depth and velocity of the active layer. It should be mentioned that although the kinetic energy equation has been avoided in the derivation of the momentum equations, its solution is required in order to obtain the granular temperature for Equation (4.39). As will soon be shown, in this particular case, a corrector–predictor numerical technique may be used to estimate $\tilde{T}$ to simplify the calculations.

4.5.5 Solution of the Momentum Equation in the Active Layer of the Bed

Equations (4.34) and (4.39) represent the integro-differential equations for the bulk material flow in the bed active layer. However, in order to proceed further a suitable form for the velocity profile is required. In choosing this suitable velocity function, it is necessary to account for the boundary conditions (1) at the free surface, and (2) at the interface between the active and the plug flow region of the bed; and also to satisfy the requirement of continuity at the point where the solution in the active layer is joined to the plug flow solution. However, it is necessary to first consider the material balance for the bed section being considered. The material balance at an arbitrary x-position in the free surface plane establishes $\dot{m}_{AL} = \dot{m}_{PF}$ (Figure 4.11) or stated mathematically,

$$\rho_{AL}\int_0^{\delta_x} u_{AL}(x,y)dy = \rho_{PF}\int_{r_x}^R u_{PF}(r)dr \quad (4.40)$$

Recognizing that the bulk density is simply the particle density times the solid fraction ($\rho = \rho_p v$), and that, within the plug flow region $u = \omega r$, this equation simplifies to (dropping the subscript AL for velocity in the active layer)

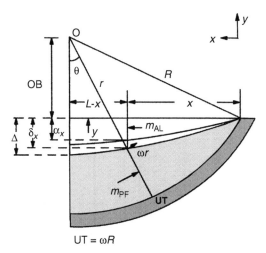

Figure 4.11 Material balance between active layer and plug flow region.

$$v_{AL} \int_0^\delta u\,dy = v_{PF} \int_{r_x}^R \omega r\,dr \qquad (4.41)$$

which, after integration of the right-hand side, gives

$$v_{AL} \int_0^\delta u\,dy = 0.5 v_{PF}\omega \left[R^2 - \left(\frac{H+\delta_x}{\cos\theta}\right)^2\right] \qquad (4.42)$$

In Equation (4.42) $u = u_{AL}(x, y) = f(y)$ is the active layer velocity function which includes the actual active layer depth. At mid-chord, the global material balance for the entire cross-section is satisfied and since the subtended angle there (i.e., θ) goes to zero, Equation (4.42) reduces to

$$v_{AL} \int_0^\Delta u\,dy = 0.5 v_{PF}\omega \left[R^2 - (H+\Delta)^2\right] \qquad (4.43)$$

In solving for the active layer depth and velocity using Equation (4.43) there are two possible constraints that may be used to terminate the iteration, that is, either by ensuring that the mass flow in the active layer is balanced at each x-position from the apex, using Equation (4.42), or by ensuring that global mass in the active layer is balanced at mid-chord using Equation (4.43).

4.5.6 Velocity Profile in the Active Layer

Application of integral methods for solving boundary layer flows involves fitting the velocity profile to a polynomial form

$$U = a_0 + a_1 y + a_2 y^2 \cdots + a_n y^n \tag{4.44}$$

where the degree, n, depends upon the number of conditions imposed on the profile by physical constraints, for example stress or velocities at the boundaries. However, the experimental results indicated that a parabolic profile would be sufficient to describe the shape of the velocity profile in the active layer; hence only three conditions are imposed on the profile, (1) the free surface, (2) the yield line which lies between the active layer and the plug flow region where deformation of material occurs, and (3) the zero velocity line which lies between (1) and (2) as a result of particle flow reversal.

4.5.6.1 Free Surface Boundary Condition

Following the fluid flow analogy, the most obvious choice for a free surface boundary condition is to allow the shear stresses to vanish and thereby force the shear rate to go to zero (i.e., $du/dy = 0$). Although this condition has been used in chute flow calculations (Savage, 1979; Campbell and Brennen, 1983; etc.), it forces the velocity profile to a shape that is inconsistent with the experimental results. This is consistent with instability, a condition imposed by the large particle size causing the continuum assumption to break down. Instead of null shear stress we can impose a known velocity value as this is easily measurable. The free surface velocity depends on the chord length since it is the maximum distance that a particle can travel at a certain fixed speed imposed either by the kiln rotation or by gravity (free fall). Qualitatively, the surface velocity for a bed of material with a given particle size will depend on three parameters: the speed of rotation, the kiln size, and the degree of fill. A correlation equation was derived relating the surface velocity to kiln speed and fill, $u_s = C_0 \omega R$, where the constant of proportionality, C_0, was related to the degree of fill and the rotational speed of the kiln, $C_0 = f(\% \text{ Fill}; \omega)$.

4.5.6.2 Conditions at the Interface between Plug Flow and Active Layer

At the yield line, continuity of flow in both the active layer and the plug flow region requires that $u = u_\delta$. The relationship between α and δ may be established by considering the Coulomb yield criterion, $\sigma_s = \sigma_{yy} \tan \phi$. Since $\tan \phi$ is constant, the ratio between the shear and normal stresses at the interface must also be constant. Because the normal stress is the weight of the overlying burden, the number of particles that the material can sustain between the zero line and the yield line depends on the packing and must, therefore, be related to the degree of fill. Hence, the ratio α/δ represents the yield criterion and must be a constant that is related to the percent fill. Experimental observations indicate that this is indeed true and can be used to establish the two boundary conditions. The argument is analogous to boundary layer flows where the layer depth is usually related

Table 4.1 **Coefficients Resulting from the Velocity Profile**

α/δ	a'_1	a'_2
0.75	3.0−2.33κ	4.0−1.33κ
0.80	4.0−2.25κ	5.0−1.25κ
0.85	5.67−2.18κ	6.67−1.18κ
0.90	9.0−2.11κ	10.0−1.11κ

to a certain percentage of the free stream. Gauthier (1991) gave a value of $\alpha = 0.75\delta$ for a small batch reactor using Ottawa sand as the bed material. However, it has been shown that when larger particle size ranges are involved, the ratio lies between 0.7 and 0.9 depending on the degree of fill (Boateng and Barr, 1997).

Inclusion of these constraints into the parabolic velocity profile provides an analytical expression for the profile (Appendix 4B):

$$\frac{u}{u_\delta} = \kappa + \left(\frac{\kappa\alpha^2 - \kappa\delta^2 + \alpha}{\alpha\delta^2 - \alpha^2\delta}\right)y + \left(\frac{\kappa - \kappa\alpha - \alpha}{\alpha\delta^2 - \alpha^2\delta}\right)y^2 \quad (4.45)$$

where κ is the ratio between the velocity parallel to the bed surface, $u_s(C_0\omega R)$ and $u_\delta(\omega r_x)$, that is, $\kappa = C_0 R/r_x$. The velocity profiles for the range of values of α/δ, may be given as

$$\frac{u}{u_\delta} = \kappa + a'_1\left(\frac{y}{\delta}\right) - a'_2\left(\frac{y}{\delta}\right)^2 \quad (4.46)$$

Pertinent coefficients are presented in Table 4.1.

4.5.7 Density and Granular Temperature Profiles

The analytical model for the momentum conservation equation (Equation (4.39)) applied to the active layer flow also requires density and granular temperature profiles for a solution. The density profile for the bed active layer differs from that in the plug flow region because of material dilation. In the plug flow region the solids volume concentration, v, can be assumed to be constant and equal to the maximum shearable solids concentration, v_*. The numerical value of v_* depends on the material packing, which in turn depends upon the material properties and particle shape. For close packing of spherical particles, the value can be as high as 0.7. However for most practical situations it ranges between 0.59 and 0.62 (Savage, 1989). In the active layer, the solids fraction is not always constant but may vary from that of the plug flow ($\approx v_{PF}$) at the interface to a very small quantity, v_0, at the free surface, as was determined for one of its components (the linear concentration profile) in the experimental results. The actual value for the solids fraction should, therefore, be determined as part of the solution of the granular flow equations because of its interdependency with the granular temperature. For the present application a linear solids concentration profile should be adequate because of the thinness of the active layer. Such an approach has been previously employed in similar calculations pertaining to sedimentation transport problems (Hanes and Bowen, 1985). By assuming a linear solids concentration profile one can use the expression

$$v = v_* - \frac{y}{\delta}(v_* - v_0) \tag{4.47}$$

Here v_* is the "at rest" solids concentration, which should equal the packing in the plug flow region. Ideally, however, v_0 is determined by matching the stress generated by the bed surface velocity and the stress due to the flow of freeboard gas. For sedimentation transport where vigorous saltation of particles occurs due to turbulent ocean flow, the value has been shown to range between 0.05 and 0.16 (Hanes, 1986). In the rotary kiln situation, such vigorous saltation of particles is not observed and the value of v_0 is expected to be high for a rolling bed. Based on the experimental data, the dilation in the active layer is less than 5 percent over that in the plug flow region. Now, with respect to the granular temperature, the experimental results indicate that there is some granular conduction into the bed (Figure 4.5). Although the profile follows the same parabolic behavior as the velocity profile, it is very difficult to establish boundary conditions for the granular temperature and a reasonable isotropic value may be applied. However, with an iteration scheme, the values of granular temperature may be computed as part of the solution of the flow problem.

4.5.8 An Analytical Expression for the Thickness of the Active Layer

Substitution of the velocity profile, Equation (4.46), into the momentum conservation equation, that is, the left-hand side of Equation (4.39), and carrying out the integration yields the result

$$\frac{d}{dx}\rho\int_0^\delta (u^2 - u_\delta u)\,dy = \rho\frac{d}{dx}\left[\left\{(\kappa^2 - \kappa) + \frac{1}{2}(2\kappa a_1' - a_1') + \frac{1}{3}(a_1'^2 - 2\kappa a_2' + a_2')\right. \right.$$
$$\left.\left. - \frac{1}{4}(2a_1'a_2') + \frac{1}{5}a_2''^2\right\}\delta\right]u_\delta^2 \tag{4.48}$$

Substituting this result into Equation (4.39) yields

$$\rho\frac{d}{dx}\left[(b_0 + b_1\kappa + b_2\kappa^2)\delta\right]u_\delta^2 = \rho g \sin\xi + \rho_p d_p g_2(v)\tilde{T}^{1/2}\frac{du}{dy} \tag{4.49}$$

Table 4.2 Coefficients Resulting from the Integro-Momentum Equation

α/δ	b_0	b_1	b_2
0.75	0.170	0.422	0.033
0.80	0.177	1.875	0.0
0.85	0.093	6.768	−0.601
0.90	0.833	−0.487	−0.593

where the coefficients b_i are generated when the terms in a'_i are expanded for the various values of the ratio α/δ. Again by recognizing that $\rho = \rho_p \nu_{AL}$, Equation (4.49) becomes

$$\frac{d}{dx}\left[(b_0 + b_1\kappa + b_2\kappa^2)\delta\right]u_\delta^2 = g\sin\xi + d_p\frac{g_2(\nu)}{\nu_{AL}}\tilde{T}^{1/2}\frac{u_\delta}{\delta}$$

$$= \frac{g\sin\xi}{u_\delta^2} + d_p\frac{g_2(\nu)\tilde{T}^{1/2}}{\nu_{AL}u_\delta\delta} \quad (4.50)$$

In Equation (4.50) (coefficients in Table 4.2), u_δ is a function of active layer depth, δ, that is, dropping the subscript, x,

$$u_\delta = \omega r \cos\theta \quad (4.51)$$

where $r = (H + \delta)/\cos\theta$ which means that the velocity at the active layer depth is given by

$$u_\delta = -\omega(H + \delta) \quad (4.52)$$

Although u_δ is a function of δ, the variables in Equation (4.50) may be separated as if the right-hand side were constant, which is consistent with the boundary layer fluid flow solution (Schlichting, 1979). In a numerical solution scheme, u_δ can be computed with a previous value of δ and then be updated. By carrying out the separation of the variables, Equation (4.50) becomes

$$\int_0^\delta \left[(b_0 + b_1\kappa + b_2\kappa^2)\delta\right]d\delta = \int_0^x \left[\frac{g\sin\xi}{u_\delta^2}\delta + d_p\frac{g_2(\nu)\tilde{T}^{1/2}}{\nu_{AL}u_\delta}\right]dx \quad (4.53)$$

When a boundary condition $\delta = 0|_{x=0}$ is employed at the apex (origin of flow calculation), the final form of the integral equation becomes

$$\frac{1}{2}\left(b_0 + b_1\kappa + b_2\kappa^2\right)\delta^2 - \frac{g\sin\xi}{u_\delta^2}\delta x - d_p\frac{g_2(v)\tilde{T}^{1/2}}{v_{AL}u_\delta}x = 0 \qquad (4.54)$$

This is the quadratic equation required for the prediction of the active layer depth, which in turn is substituted into the velocity profile to obtain the velocity distribution in the two-dimensional domain.

4.5.9 Numerical Solution Scheme for the Momentum Equation

The velocity determination requires the prediction of the active layer depth using Equation (4.54) at any x-position along the bed surface. It also has to be substituted into the velocity profile in order to determine the velocity parallel to the bed surface as a function of bed depth. The local velocity normal to the bed surface is then established by solving the continuity equation given in Equation (4.9). Solving for δ_x also requires the granular temperature, which is found by iteration. The value of the granular temperature obtained after convergence is an average quantity for each x-position in the active layer. The calculation scheme is given in Figure 4.12. The procedure follows the sequence whereby (1) the average granular temperature for the entire depth at any x-position is estimated; with this value $g_2(v)$ is computed; (2) Equation (4.54) is solved for a first approximation of δ_x by neglecting the quadratic term; (3) with the value of δ_x, u_δ is computed and Equation (4.54) solved for an actual value of δ_x; (4) knowing δ_x, the velocity profile is computed, from which follows the calculation of the mass flow for the active layer at the i-position. This mass flow is compared with the value in the plug flow region at the same location using Equation (4.40) and this procedure is repeated using an improved estimate of the granular temperature if the mass flow for the active layer does not balance the mass flow in the plug flow region. Otherwise, the solution is advanced to the next i-position until mid-chord.

The stability of the solution procedure just described depends on the choice of the granular temperature needed to initiate the solution. Experiments have shown that there is an order of magnitude increase in the granular temperature for each increase in kiln speed and therefore instabilities are likely to develop when a solution for successive kiln speeds is required. Nevertheless, this problem is easily rectified by a good initial guess for the granular temperature and the choice of the mass balance convergence criterion. In order to use small convergence criteria, which are required for low granular temperatures (i.e., for low kiln speeds), small mesh sizes are required. In modeling the 1-m diameter experimental rotary apparatus, 24 nodes were used between the apex and the mid-chord for all the kiln speeds solved and the convergence criterion defined as the percent difference between the mass flows in the active layer and the plug flow region was set to 0.1%.

4.6 Model Results and Validation

The prediction of the velocity distribution for a 41-cm diameter pilot kiln is shown in Figure 4.13. Validation of the model is carried out using experimental results of

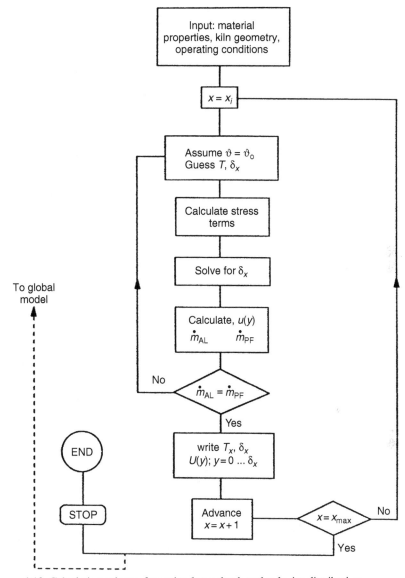

Figure 4.12 Calculation scheme for active layer depth and velocity distribution.

the 1 m rotary drum (Figure 4.14). Here predicted and measured active layer depths are compared for the materials studied. As seen, the model underpredicts at low degree of fill but better agreement is achieved for deeper beds. One likely reason for the underprediction is the fact that the location of the yield line is difficult to measure in shallow beds. Aside from the bed loading, several factors affect the flow,

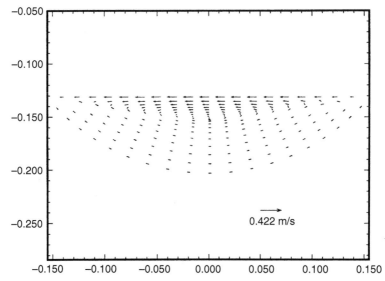

Figure 4.13 Predicted velocity distribution in the cross-section of a 41 cm pilot kiln.

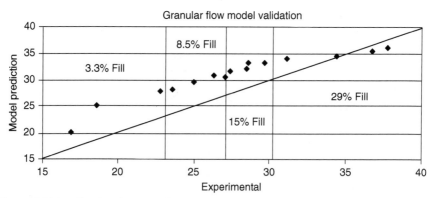

Figure 4.14 Predicted and measured active layer depth for a 1 m rotary drum.

most importantly the rheological properties of the processing materials. Some of these are presented here (Boateng, 1998). The predicted active layer velocity and depth for a 41-cm diameter pilot kiln, a 1-m diameter rotary drum, and a 2.5-m diameter industrial kiln are presented in Figures 4.15 and 4.16 (velocity and depth respectively).

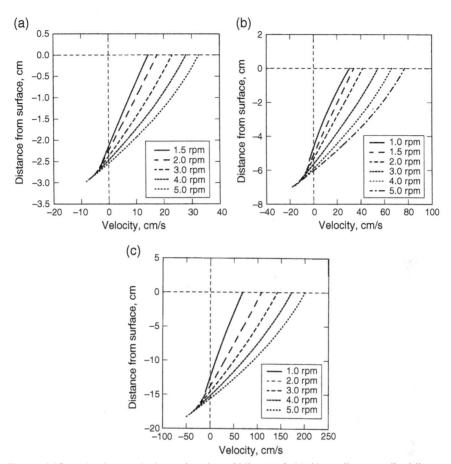

Figure 4.15 Active layer velocity as function of kiln speed. (a) 41 cm diameter pilot kiln, (b) 1 m diameter rotary drum, and (c) 2.5 m diameter industrial kiln (Boateng, 1998).

4.7 Application of the Flow Model

The diffusion coefficient can be calculated from the predicted granular temperature as (Savage, 1983; Hsiau and Hunt, 1993)

$$\tilde{D} = \frac{d_p \sqrt{\pi \tilde{T}}}{8(e_p + 1)\nu g_0(\nu)} \tag{4.55}$$

The predicted results are presented in Figure 4.17. It should be noted that this provides the effect of flow or mixing on advective heat transfer. The convective bed heat transfer can be calculated through knowledge of the flow field. The mass

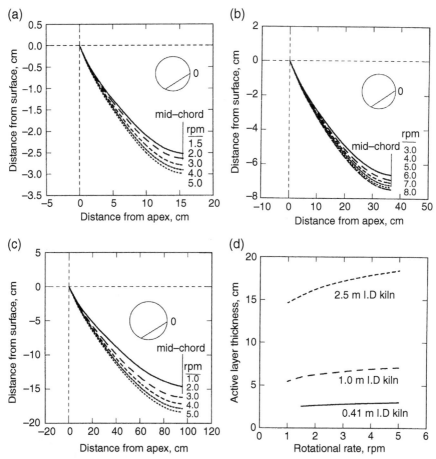

Figure 4.16 Predicted active layer depth for (a) 41 cm diameter pilot kiln, (b) 1 m diameter rotary drum, and (c) 2.5 m diameter kiln; (d) shows asymptotic nature of active layer depth with speed (Boateng, 1998).

transfer-enhanced thermal conductivity can be computed as $\rho c_p \tilde{D}$ where $\tilde{D}$, the kinetic diffusion, is computed from the granular temperature given by the flow model. Hence the enhanced bed effective thermal conductivity can be calculated (Hunt, 1997; Natarajan and Hunt, 1998) and is several orders of magnitude greater than that of a packed bed (Boateng and Barr, 1996). Previously this was not possible to estimate.

Because the active layer velocity is about three or four times greater than that in the plug flow region, mixing effects are confined to this region. However, if fine or denser particles are used as tracers, then the velocity distribution will help to determine the extent to which the tracers will travel before percolating down. These scenarios will be the subject of the next chapter.

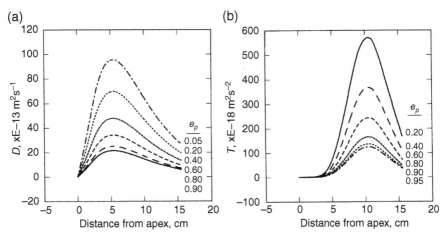

Figure 4.17 Effect of particle coefficient of restitution ($e_p = 0.05$–0.9) on active layer diffusion (12% fill, 2 rpm) (Boateng, 1998); (a) effect on granular diffusion and (b) effect on granular temperature.

References

Ahn, H., Brennen, C.E., Sabersky, R.H., 1991. Measurements of velocity, velocity fluctuation, density, and stresses in chute flows of granular materials. J. Appl. Mech. 58, 792–803.

Bagnold, R.A., 1954. Experiments on a gravity-free dispersion of large solids spheres in a Newtonian fluid under shear. Proc. R. Soc. Lond. A225, 49–63.

Boateng, A.A., 1993. Rotary Kiln Transport Phenomena: Study of the Bed Motion and Heat Transfer (Ph.D. dissertation). The University of British Columbia, Vancouver.

Boateng, A.A., 1998. Boundary layer modeling of granular flow in the transverse plane of a partially filled rotating cylinder. Int. J. Multiphase Flow 24 (3), 499–521.

Boateng, A.A., Barr, P.V., 1996. A thermal model for the rotary kiln including heat transfer within the bed. Int. J. Heat Mass Transfer 39 (10), 2131–2147.

Boateng, A.A., Barr, P.V., 1997. Granular flow behaviour in the transverse plane of a partially filled rotating cylinder. J. Fluid Mech. 330, 233–249.

Campbell, C.S., Brennen, C.E., 1983. Computer simulation of shear flows of granular material. In: Jenkins, J.T., Satake, M. (Eds.), Mechanics of granular materials: new models and constitutive relations. Elsevier pubishers.

Campbell, C.S., Gong, A., 1986. The stress tensor in a two-dimensional granular shear flow. J. Fluid Mech. 164, 107–125.

Cowin, S.C., 1974. A theory for the flow of granular materials. Powder Technol. 9, 62–69.

Davies, T.R.H., 1986. Large debris flow: a macro-viscous phenomenon. Acta Mech. 63, 161–178.

Ferron, J.R., Singh, D.K., 1991. Rotary kiln transport processes. AIChE J. 37 (5), 747–757.

Gauthier, C., 1991. Etude du mouvment granulaire dans cylindre en rotation (M.Sc. thesis). Universite du Quebec.

Hanes, D.M., 1986. Grain flows and bed-load sediment transport: review and extension. Acta Mech. 63, 131–142.

Hanes, D.M., Bowen, A.J., 1985. A granular-fluid model for steady intense bed-load transport. J. Geophys. Res. 90 (C5), 9149–9152.

Henein, H., Brimacombe, J.K., Watkinson, A.P., 1983a. Experimental study of transfer bed motion in rotary kilns. Met. Trans. B 14B (6), 191–205.

Henein, H., Brimacombe, J.K., Watkinson, A.P., 1983b. The modeling of transverse solids motion in rotary kilns. Met. Trans. B 14B (6), 207–220.

Hsiau, S.S., Hunt, M.L., 1993. Shear-induced particle diffusion and longitudinal velocity fluctuations in a granular-flow mixing layer. J. Fluid Mech. 251, 299–313.

Hunt, M.L., 1997. Discrete element simulations for granular material flows: effective thermal conductivity and self-diffusivity. Int. J. Heat Mass Transfer 40 (3), 3059–3068.

Johnson, P.C., Jackson, R., 1987. Frictional-collisional constitutive relations for granular materials, with application to plane shearing. J. Fluid Mech. 176, 67–93.

de Jong, G.D.J., 1964. Lower bound collapse theorem and lack of normality of strain rate to yield surface for soils. In: Rheology and Soil Mechanics Symposium, Grenoble, pp. 69–78.

Kanatani, K., 1979. A continuum theory for the flow of granular materials. Theor. Appl. Mech. 27, 571–578.

Lun, C.K.K., Savage, S.B., Jeffrey, D.J., Chepurniy, N., 1984. Kinetic theories for granular flow: inelastic particles in couette flow and slightly inelastic particles in a general flowfield. J. Fluid Mech. 140, 223–256.

Mandl, G., Luque, R.F., 1970. Fully developed plastic shear flow of granular materials. J. Geotechnique 20 (3), 277–307.

Nakagawa, M., Altobelli, S.A., Caprihan, A., Fukushima, E., Jeong, E.-K., 1993. Non-invasive measurements of granular flows by magnetic resonance imaging. Exp. Fluids 16, 54–60.

Natarajan, V.V.R., Hunt, M.L., 1998. Kinetic theory analysis of heat transfer in granular flows. Int. J. Heat Mass Transfer 41 (13), 1929–1944.

Parker, D.J., Dijkstra, A.E., Martin, T.W., Seville, J.P.K., 1997. Positron emission particle tracking studies of spherical particle motion in rotating drums. Chem. Eng. Sci. 52 (13), 2011–2022.

Pershin, V.F., 1988. Energy method for describing granular motion in a smooth rotating cylinder. J. Teoreticheskie Osnovy Khimicheskoi Tekhnologii 22 (2), 255–260.

Roscoe, K.H., 1970. The influence of strains in soil mechanics. Geotechnique 20, 129–170.

Savage, S.B., 1979. Gravity flow of cohesionless granular materials in chutes and channels. J. Fluid Mech. 92, 53–96.

Savage, S.B., 1983. Granular flow down rough inclines. Review and extension. In: Jenkins, J.T., Satake, M. (Eds.), Mechanics of Granular Materials: New Models and Constitutive Relations. Elsevier, Inc., New York, pp. 261–282.

Savage, S.B., 1989. Granular flow materials. In: Germain, P., Piau, M., Caillerie, D. (Eds.), Theoretical and Applied Mechanics. Elsevier, Inc., New York, pp. 241–266.

Savage, S.B., Hutter, K., 1989. The motion of a finite mass of granular material down a rough incline. J. Fluid Mech. 199, 177–215.

Schlichting, H., 1979. Boundary-Layer Theory. McGraw-Hill, New York.

Singh, D.K., 1978. A Fundamental Study of the Mixing of Solid Particles (Ph.D. dissertation). University of Rochester, New York.

Spencer, A.J.M., 1964. A theory of the kinematics of ideal soils under plane strain conditions. J. Mech. Phys. Solids 12, 337–351.

Zhang, Y., Campbell, C.S., 1992. The interface between fluid-like and solidlike behaviour in two-dimensional granular flows. J. Fluid Mech. 237, 541–568.

Appendix 4A: Apparent Viscosity

The term C' from the constitutive relations of Lun et al. (1984) represents the apparent viscosity if an analogy is drawn from fluid mechanics. The expression C' is derived in terms of the coefficient of restitution of the particles, the solid fraction, particle size, and the granular temperature as follows:

$$P_{xy} = \frac{\partial}{\partial y}\left(C'\frac{\partial u}{\partial y}\right) \qquad (4A.1)$$

where $C' = C/2$ and,

$$C = \frac{2\mu}{\eta(2-\eta)g_0}\left(1 + \frac{8}{5}\eta v g_0\right)\left[1 + \frac{8}{5}\eta(3\eta - 2)v g_0\right] + \frac{6}{5}\mu_b \eta \qquad (4A.2)$$

and

$$\mu_b = 256\mu v^2 g_0/5\pi \qquad (4A.3)$$

$$\mu = 5m\left(\frac{\tilde{T}}{\pi}\right)^{1/2} \bigg/ 16 d_p^2 \qquad (4A.4)$$

$$m = \rho_p V = \rho_p \frac{\pi}{6}d_p^3 \qquad (4A.5)$$

Substitution of these equations into C' yields

$$C' = \frac{5\sqrt{\pi}}{96}\rho_p d_p \sqrt{\tilde{T}}\left[\frac{1}{\eta(2-\eta)g_0}\left(1 + \frac{8}{5}\eta v g_0\right)\left\{1 + \frac{8}{5}\eta(3\eta - 2)v g_0\right\} \right.$$
$$\left. + \frac{768}{25\pi}v^2 g_0\right] \qquad (4A.6)$$

where the term in the square bracket may be expressed as

$$C' = -g_2(v, \varepsilon_p)\rho_p \sqrt{\tilde{T}} \qquad (4A.7)$$

with $g_2(v, \varepsilon_p)$ expressed as

$$g_2(v, \varepsilon_p) = \frac{5\sqrt{\pi}}{96}\left[\frac{1}{\eta(2-\eta)g_0} + \frac{8}{5}\frac{(3\eta-1)v}{2-\eta} + \frac{64}{25}\left\{\frac{\eta(3\eta-2)}{2-\eta} + \frac{12}{\pi}\right\}v^2 g_0\right] \qquad (4A.8)$$

Similarly,

$$P_{xx} = P_{xy} - \rho_p g_1(v, \varepsilon_p)\tilde{T} \tag{4A.9}$$

$$q_y = -\rho_p d_p \left(g_3(v, \varepsilon_p)\sqrt{\tilde{T}}\,\frac{d\tilde{T}}{dy} + g_4(v, \varepsilon_p)\sqrt{\tilde{T}}\,\frac{dv}{dy} \right) \tag{4A.10}$$

$$\gamma = \frac{\rho_p}{d_p} g_5(v, \varepsilon_p)\tilde{T}^{3/2} \tag{4A.11}$$

where g_1 through g_5 follow (Johnson and Jackson, 1987):

$$g_0 = \left(1 - \frac{v}{v^*}\right)^{-1/3} \tag{4A.12}$$

$$g_1(v, \varepsilon_p) = v + 4\eta v^2 g_0 \tag{4A.13}$$

$$g_2(v, \varepsilon_p) = \frac{5\sqrt{\pi}}{96}\left[\frac{1}{\eta(2-\eta)g_0} + \frac{8}{5}\frac{(3\eta-1)v}{2-\eta} + \frac{64}{25}\left\{\frac{\eta(3\eta-2)}{2-\eta} + \frac{12}{\pi}\right\}v^2 g_0\right] \tag{4A.14}$$

$$g_3(v, \varepsilon_p) = \frac{25\pi}{16\eta(41-33\eta)}\left[\frac{1}{g_0} + \frac{12}{5\eta}\{1 + \eta(4\eta-3)\}\eta + \frac{16}{25\eta^2}\right.$$
$$\left. \times \left\{9\eta(4\eta-3) + \frac{4}{\pi}(41-33\eta)\right\}v^2 g_0\right] \tag{4A.15}$$

$$g_4(v, \varepsilon_p) = \frac{15\sqrt{\pi}}{4}\frac{(2\eta-1)(\eta-1)}{(41-33\eta)}\left(\frac{1}{g_0} + \frac{12\eta}{5}\right)\frac{d}{dv}(v^2 g_0) \tag{4A.16}$$

$$g_5(v, \varepsilon_p) = \frac{48}{\sqrt{\pi}}\eta(1-\eta)v^2 g_0 \tag{4A.17}$$

Appendix 4B: Velocity Profile for Flow in the Active Layer

The velocity profile with the appropriate boundary conditions is

$$u = a_0 + a_1 y + a_2 y^2 \tag{4B.1}$$

at $y = \alpha$, $u = 0$

at $y = \delta$, $u = u_\delta$ (4B.2)

at $y = 0$, $u = u_s$

Substituting these boundary conditions gives three equations, that is,

$$u_s = a_0$$
$$0 = a_0 + a_1\alpha + a_2\alpha^2 \tag{4B.3}$$
$$-u_\delta = a_0 + a_1\delta + a_2\delta^2$$

These can be reduced to two equations

$$a_1\alpha\delta + a_2\alpha^2\delta = -u_s\delta \tag{4B.4}$$

$$a_1\alpha\delta + a_2\alpha\delta^2 = -(u_s + u_\delta)\alpha \tag{4B.5}$$

from which

$$a_1 = \frac{-u_s\delta^2 + (u_s + u_\delta)\alpha^2}{\alpha\delta^2 - \alpha^2\delta} \tag{4B.6}$$

$$a_2 = \frac{u_s\delta^2 - (u_s + u_\delta)\alpha^2}{\alpha\delta^2 - \alpha^2\delta} \tag{4B.7}$$

Substituting the coefficients gives

$$u = u_s + \frac{u_s(\alpha^2 - \delta^2) + u_\delta\alpha^2}{\alpha\delta^2 - \alpha^2\delta}y + \frac{u_s(\delta - \alpha) - u_\delta\alpha}{\alpha\delta^2 - \alpha^2\delta}y^2 \tag{4B.8}$$

Recognizing that $u_s = C_0\omega R$ and $u_\delta = \omega r_x$ yields a ratio

$$u_s = \frac{C_0 R}{r_x}u_\delta = \kappa u_\delta \tag{4B.9}$$

$$\frac{u}{u_\delta} = \kappa + \frac{\kappa\alpha^2 - \kappa\delta^2 + \alpha^2}{\alpha\delta^2 - \alpha^2\delta}y + \frac{\kappa\delta - \kappa\alpha - \alpha}{\alpha\delta^2 - \alpha^2\delta}y^2 \tag{4B.10}$$

for $\alpha = 0.75\delta$

$$\frac{u}{u_\delta} = \kappa + (3 - 2.33\kappa)\left(\frac{y}{\delta}\right) - (4 - 1.33\kappa)\left(\frac{y}{\delta}\right)^2 \tag{4B.11}$$

or

$$\frac{u}{u_\delta} = \kappa + a_1'\left(\frac{y}{\delta}\right) - a_2'\left(\frac{y}{\delta}\right)^2 \tag{4B.12}$$

Mixing and Segregation

Thorough mixing of particles in the transverse plane of a rotary kiln is fundamental to the uniform heating or cooling of the charge and, ultimately, to the generation of a homogeneous product. However, differences in particle size and density result in a demixing process whereby smaller or denser particles segregate to form an inner core or kidney of segregated material which may never reach the bed surface to be exposed to freeboard temperatures. In this chapter, an analytical model and its numerical solution is developed which relates particle segregation rates to primary operating parameters such as rotary kiln diameter, bed depth, and rotational speed. The model considers a binary mixture of small and large particles in the continuously shearing active region of the kiln bed. Continuum equations are employed to describe the mixing and segregation rates in the transverse plane of the bed that result from both particle percolation and diffusional mixing. The granular flow model developed in Chapter 4 provides the flow field needed for the convective terms for material concentration and the diffusion coefficients. The percolation velocities are generated using existing models that relate percolation to the probability of void formation in the shear plane.

The rheological properties of the bed material can be expected to change during the passage of a charge through a rotary kiln. Severe changes can result in alterations to material properties, such as particle size, shape, and surface character, and these ultimately may result in distinct changes to bed behavior. One such behavioral phenomenon is segregation which, since it acts as a mechanism of demixing, may influence heat transfer within the bed. Segregation may also influence the rate at which particles are elutriated from the exposed bed surface when, for example, large amounts of gas are being released from the bed. Additionally, the effect of segregation on heat transfer is of considerable practical importance since it may significantly influence the degree of product homogeneity.

The main causes of segregation are differences in particle size, density, shape, roughness, and resilience (Williams and Khan, 1973). Although any of these may produce segregation under certain circumstances, most rotary kiln segregation issues arise from differences in particle size (Pollard and Henein, 1989), and the work described herein is focused on this phenomenon. The mechanisms by which size segregation occurs might be classified as trajectory segregation or percolation (Williams and Khan, 1973; Bridgwater et al., 1985):

1. Trajectory segregation: This is due to the fact that, for certain modes of kiln operation, particles being discharged from the plug flow region into the active layer may be projected horizontally from the apex onto the exposed bed surface. This situation may apply in the slumping, rolling, and cataracting modes, whereby different sized particles are emptied onto the surface during material turnaround. It has been suggested that the distance over

which these particles travel is proportional to the square of the particle diameter (Bridgwater, 1976), which means that finer particles will tend to be concentrated at the mid-chord section.

2. Percolation: When a bed of particles is disturbed so that rearrangement takes place (rapid shearing), the probability that a particle will find a void into which to fall depends on the size of the particles (Savage, 1983). Thus smaller particles will tend to filter downward through a bed of flowing granular material while large particles will simultaneously tend to be displaced upward.

Trajectory segregation has been identified (Bridgwater et al., 1985) as the main cause of axial segregation or "banding" whereby particles of different sizes are selectively collected into bands occurring over the kiln length. This axial segregation is not considered in the present work and therefore not critically reviewed; rather, attention is focused on segregation in the transverse plane, specifically, percolation. Although percolation theory, also known as inverse sieving in granular flow models (Savage, 1989), is reasonably well developed, it has seldom been employed to model segregation patterns encountered in rotary kilns. Instead, most of the rotary kiln literature characterizes the rate of segregation by a first or second order type kinetic expression such as Nityanand et al. (1986) and Pollard and Henein (1989).

$$S_n = A \times (Fr)^b \tag{5.1}$$

where S_n is defined as the normalized rate of segregation, Fr is the rotational Froude number, and A and b are experimentally determined kinetic parameters.

Size segregation in failure zones, for example, the active (shearing) region of a rolling bed in a rotary kiln or gravity flow on an incline (also known as free surface segregation), has been described by the mechanism of percolation (Bridgwater et al., 1985; Savage, 1989). Size segregation in such systems is considered as a random continuous network whereby voids are randomly formed and distributed (Figure 5.1).

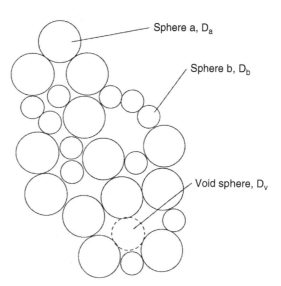

Figure 5.1 Random continuous network of segregating particles (Savage, 1989).

For gravity flow on an incline Savage and Lun (1988) approached the problem by considering three events:

1. The probability of voids forming that are of sufficient size for the smaller particles to percolate into. The probability of such voids forming has been given as

$$p(E) = \frac{1}{\overline{E} - E_m} \exp\left\{\frac{E - E_m}{\overline{E} - E_m}\right\} \tag{5.2}$$

   where E is the void diameter ratio which is defined as the ratio of the void diameter to the average void diameter, that is, $E = (D_v/\overline{D})$. E_m is the minimum possible void diameter ratio and $\overline{E}$ is the mean void diameter ratio. $E_m = 0.154$ is the value for which voids in a packing will result in spontaneous percolation (Bridgwater et al., 1978).
2. The capture of particles by voids in the underlying layer. For this event the number of particles captured by a void per unit time is dependent on the velocity differences between two neighboring shear layers.
3. The establishment of a mass flux of small particles. When this occurs, the average percolation velocities in the plane normal to the bed surface can be determined by material balance.

The proposed sequence of events is useful in exploring the possibility of employing the physically derived continuum equations required to solve for the concentration gradients with a minimum probabilistic input. The stochastic approach tends to conceal the detailed behavior and the mechanisms that are the source of industrial problems (Bridgwater, 1976). This may also be said about the characterization of segregation by degree of mixing or by kinetic expressions such as the one given in Equation (5.1). Such expressions will conceal all the interparticle mechanisms in a single parameter that serves little or no industrial purpose.

Although classification as applied to solids is synonymous with different sizes of the same material, it should be pointed out that other classifications, such as particle density or mass, and even particle shape, can also cause mixing or segregation. It has recently been established (Alonso et al., 1991) that size and density differences can, indeed, be combined to reduce segregation through mutual compensation.

5.1 Modeling of Particle Mixing and Segregation in Rotary Kilns

It is evident from operator experience that thorough mixing of particles in the transverse plane of a rotary kiln is fundamental to achieving uniform heating or cooling of the charge and, ultimately, to the production of a homogeneous product. This is particularly so for processes involving granular materials. In the modeling of granular flows one generally assumes that particles are evenly sized and that mixing effectively means that (statistically) exposure to freeboard will be the same for all particles. Unfortunately, when significant variation in particle sizing occurs, there will be the tendency of small particles in the active layer to sieve downward through the matrix of larger particles. Therefore, the bed motion presented in Chapter 4 tends

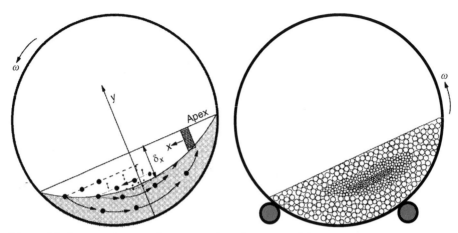

Figure 5.2 Segregated core at the cross section of a rotary calciner.

to concentrate finer material within the core (Figure 5.2). Material within the core, because it has very little chance of reaching the exposed bed surface for direct heat transfer from the freeboard, tends to be at a lower temperature than the surrounding material. Thus segregation can counteract advective transport of energy and thus can promote temperature gradients within the bed. However, the net effect might not necessarily be negative because for a process such as limestone calcination where smaller particles react faster than larger ones (at the same temperature), segregation of fines to the cooler core may be essential to obtaining uniform calcination of all particles. This suggests that particle size distribution in the feed material might be optimized, if our predictive capabilities for determining material mixing and segregation were well developed.

With an adequate granular flow model, a logical extension is to construct a mathematical model that will describe the phenomenon of segregation and predict the extent and dimensions of the segregated core which has become an integral part of processing real materials in rotary kilns. It is not fair to say that the modeling of segregation has not been addressed because it is a well-known phenomenon in applications such as rotary mixers and has been characterized in many ways. Fan and Wang (1975) reviewed over 30 forms of mixing indices that have appeared in the literature to characterize particulate mixing and segregation in drum mixers. However, most of these characterization parameters are probabilistic or statistical in nature and, although often helpful, tend to conceal the details of the phenomenon and yield little information on, for example, the effect of material properties on flow and hence on segregation. Scaling factors are also difficult to evaluate without a good grasp of the physical phenomena that drive segregation.

We will consider a segregation model composed of a binary mixture of small and large particles in the continuously shearing active layer of the kiln bed. We will employ continuum equations to describe the mixing and segregation rates in the transverse plane of the bed that result from both particle percolation and diffusional mixing. The diffusion coefficients and the convective terms for material concentration in the continuum equations can be obtained from the granular flow model in Chapter 4. Percolation velocities are calculated along with void formation in the shear plane (Savage, 1988). Finally, the developed segregation model is applied in order to predict the size and extent of the segregated core as well as to show the effect of segregation on material mixing and on the effective thermal conductivity of the bed.

5.2 Bed Segregation Model

The mechanism of segregation as observed in rotary kilns indicates that percolation is the primary cause of driving smaller size particles to the core of the bed. To model the phenomenon, several sequential processes must be used, including the following observed conclusions about the phenomenon:

1. From a condition of uniform mixing of particles within the bed, radial segregation proceeds very rapidly and is fully implemented within 2–10 kiln revolutions (Rogers and Clements, 1971; Pollard and Henein, 1989). The mechanism of segregation can therefore be considered as a steady-state problem.
2. The segregation process is continuous and there is a constant discharge of fines from the plug flow region into the active layer. This discharge of fines occurs in the upper part of the bed toward the apex and is followed by percolation normal to the bed surface as material is sheared in the active layer.
3. The resultant segregated core, popularly known in the industry as the "kidney" (or "tongue") does not only consist entirely of fine particles but also contains a small amount of coarse particles (i.e., there are concentration gradients even in the core).
4. The bulk velocity distribution in the active layer does not change with addition of fines and the bed behavior (e.g., rolling and slumping) remains unchanged with fines (Henein, 1980).
5. The percolation velocity of fine particles depends on the size of the voids formed in an underlying layer; these voids are formed in a random manner (Savage, 1988).
6. For particles below some critical size, spontaneous percolation may also occur in the plug flow region thereby resulting in a possible collection of fines near the bed/wall interface (Bridgwater and Ingram, 1971).
7. Downward movement of segregating particles in the active layer is compensated by an equal volumetric upward movement of bulk particles in the active layer in a phenomenon known as "squeeze expulsion mechanism," Savage, 1988).

Based on this information a credible mathematical model can be constructed. We consider a control volume in the active layer that accounts for the mass conservation of the sinking and/or floating particles. We restrict the analysis to the active layer because, in view of (6) above, the probability of fines moving through the plug flow region is very low compared to the dilated shearing flow in the active layer. The plug flow region can therefore be assumed to be impermeable. It serves only as the circulation path by which particles are fed back to the active layer. This assumption therefore precludes spontaneous percolation from the model. The situation to model is shown schematically in Figures 5.3(a) and (b). For convenience, the coordinate system used here is consistent with that in the flow model where a Cartesian system is allocated to the active layer such that $0 \leq x \leq 2L$ where $2L$ is the chord length, and the origin is at the apex of the bed.

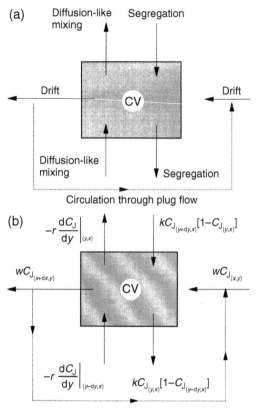

Figure 5.3 (a) Control volume for material conservation in the active layer. (b) Mathematical description for material balance in control volume; the convective flux.

We consider a simple binary system of two particle sizes, each of the same density in the control volume. Since the calculation domain is restricted to the active layer, the fine particles are assumed to be larger than the critical size that causes spontaneous percolation. The situation occurs when the diameter ratio of the small to large particles exceeds the critical value, which, for a closely packed bed, has been given as (Savage, 1988)

$$\bar{\sigma} = \frac{d_{ps}}{d_{pL}} \leq 0.1547 \tag{5.3}$$

where d_{ps} and d_{pL} are, respectively, the sizes of the small and large particles for the binary system. By choosing the size ratio to be greater than the critical value it is implicitly assumed that percolation will occur only when the voids formed are larger than the smaller size particle, d_{ps}, of the binary mixture. For the active layer of the rotary kiln where continuous shearing is occurring, it will be further assumed that void formation is a random occurrence and follows a probability law. For generality we assign the term "jetsam" to sinking particles, either due to size or density differences, and "flotsam" to floating particles which the fines displace (Gibilaro and Rowe, 1974). The model is derived on a volume balance basis and the concentration terms refer to volumetric fraction of jetsam in a given volume of solids. The relationship between the volume fraction jetsam concentration and the number of particles in the control volume can, therefore, be expressed as

$$C_J = v \frac{\bar{\eta}\bar{\sigma}^3}{1 + \bar{\eta}\bar{\sigma}^3} \qquad (5.4)$$

where v is the solids concentration, $\bar{\eta}$ is the particle number ratio, n_J/n_F, with n_J and n_F being the respective jetsam and flotsam number particles. This relation is developed in Appendix 5A for jetsam loading of the kiln.

Following on with the continuum assumption, the control volume required for the material balance is shown in Figure 5.3(a). The equilibrium concentration of jetsam within the control volume depends on the interaction of three components: (1) convection (drift) caused by the bulk velocity, (2) diffusion-like mixing, and (3) segregation associated with movement through voids. The rate at which jetsam is spread over the cross section is mathematically depicted in Figure 5.3(b). Of the three mechanisms shown, segregation is the only one that distinguishes jetsam from flotsam and it depends on the percolation of jetsam into the underlying layer and subsequent displacement of flotsam from the underlying layer as compensation. This compensation is what is termed the "squeeze expulsion" mechanism (Savage, 1988). Because the upward flow of material that compensates percolation of jetsam may itself contain jetsam, the rate of jetsam concentration due to the segregation mechanism is represented by a nonlinear concentration gradient.

5.3 The Governing Equations for Segregation

The governing equations for mixing and segregation are derived by considering an equilibrium balance of material for the control volume (Figure 5.3). First, particles drift into the control volume by convection as a result of the bulk velocity in the active layer. The rate of jetsam dispersion into and out of the control volume may be represented, respectively, as $AuC_{J(x,y)}$ and $AuC_{J(x+dx,y)}$, where A is the area normal to the bulk flow, and u is the bulk velocity. The rate of diffusional mixing is proportional to the concentration gradient and the effect of this component in the x-wise direction of the active layer may be neglected relative to the large advection term. The rate of diffusion-like mixing at each i-position in the active layer is therefore given as $-\bar{r}(\partial C_J/\partial y)$, where $\bar{r}$ is the proportionality constant equal to the product of the diffusion coefficient and the participating area in the control volume, that is, $\tilde{D}_y A [s^{-1}]$. The rate of segregation for jetsam particles is given by a nonlinear quantity, $\bar{k}C_J(1 - C_J)$ where $\bar{k}$ is the product of the area, and the percolation velocity is given as $Av_p[m^3 \, s^{-1}]$.

By employing the Taylor series expansion, the rates of jetsam outflow from the control volume may be expressed as follows:

1. Bulk flow:

$$Au(y)C_J = Au(y)\left[C_{J|_x} + \frac{\partial}{\partial x}(C_J)dx + \frac{\partial^2}{\partial x^2}(C_J)dx^2 + \cdots\right] \quad (5.5)$$

2. Diffusion:

$$-\bar{r}\frac{\partial C_J}{\partial y}\bigg|_{y-dy,x} = -\bar{r}\left[\frac{\partial C_J}{\partial y}\bigg|_{y,x} - \frac{\partial}{\partial y}\left(\frac{\partial C_J}{\partial y}\right)dx + \cdots\right] \quad (5.6)$$

3. Segregation:

$$\bar{k}C_{J|_{y+dy,x}}\left[1 - C_{J|_{y,x}}\right] = \bar{k}\left[1 - C_{J|_{y,x}}\right]\left[C_{J|_{y,x}} + \frac{\partial}{\partial y}(C_J)dy + \cdots\right] \quad (5.7a)$$

$$\bar{k}C_{J|_{y,x}}\left[1 - C_{J|_{y-dy,x}}\right] = \bar{k}C_{J|_{y,x}}\left[1 - \left\{C_{J|_{y,x}} - \frac{\partial}{\partial y}(C_J)dy + \cdots\right\}\right] \quad (5.7b)$$

By expanding the terms given in Equations (5.5–5.7) and substituting the rate of jetsam inflow of particles for the control volume, the net change of jetsam concentrations becomes:

$$\tilde{D}_y \frac{\partial^2 C_J}{\partial y^2} dx\, dy\, dz + v_p(1 - 2C_J)\frac{\partial C_J}{\partial y}dx\, dy\, dz = u(y)\frac{\partial C_J}{\partial x}dx\, dy\, dz \quad (5.8)$$

The differential equation describing the movement of jetsam concentration in the active layer may be written as:

$$\tilde{D}_y \frac{\partial^2 C_J}{\partial y^2} + v_p(1 - 2C_J)\frac{\partial C_J}{\partial y} - u(y)\frac{\partial C_J}{\partial x} = 0 \quad (5.9)$$

In arriving at Equation (5.9), the boundary layer condition whereby $u_x \gg u_y$ has been imposed; the y-component of the species convection term has been ignored and thus the vertical movement of jetsam occurs only by percolation or diffusion.

The diffusion flux in the active layer occurs as a result of particle collision in the continuously shearing active layer. The diffusion coefficient and the bulk velocity are determined by the flow model detailed in Chapter 4. $\tilde{D}_y$ is the kinetic diffusivity which had been computed from the granular temperature as (Savage, 1983; Hsiau and Hunt, 1993)

$$\tilde{D}_y = \frac{d_p\sqrt{\pi \tilde{T}}}{8(e_p + 1)v g_0} \quad (5.10)$$

As well, $u(y)$ is the velocity profile for the active layer which is obtained from the flow model described in Chapter 4. At this point the percolation velocity is the only remaining unknown component required to reach a solution to the segregation problem. In order to determine this velocity, the model developed by Savage (1988) for segregation in inclined chute flow might be applicable to the situation under consideration. It considers the probability for formation of a void in an underlying layer with a size sufficiently large enough to capture the smaller particles within the overlying layer. The net percolation velocity for the smaller particles in the neighborhood might be determined as (Savage, 1988):

$$v_p = d_{ps} \left(\frac{du}{dy}\right) \frac{1}{(1+\bar{\eta}\bar{\sigma}^3)} (v_{ps} - v_{pL}) \qquad (5.11)$$

Here the percolation velocities for smaller (jetsam), v_{ps}, and larger (flotsam) particles, v_{pL}, are given by the following equations:

$$v_{ps} = d_{pL} \left(\frac{du}{dy}\right) G(\bar{\eta},\bar{\sigma}) \left[\bar{E} - E_m + 1 + \frac{(1+\bar{\eta})\bar{\sigma}}{(1+\bar{\eta}\bar{\sigma})}\right]$$
$$\exp\left\{\frac{(1+\bar{\eta})\bar{\sigma}/(1+\bar{\eta}) - E_m}{\bar{E} - E_m}\right\} \qquad (5.12)$$

$$v_{pL} = d_{pL} \left(\frac{du}{dy}\right) G(\bar{\eta},\bar{\sigma}) \left[\bar{E} - E_m + 1 + \frac{(1+\bar{\eta})}{(1+\bar{\eta}\bar{\sigma})}\right]$$
$$\exp\left\{\frac{(1+\bar{\eta})/(1+\bar{\eta}) - E_m}{\bar{E} - E_m}\right\} \qquad (5.13)$$

The function $G(\bar{\eta},\bar{\sigma})$, in Equations (5.12) and (5.13), relates the packing of particles around a void to particle size ratio σ and particle number ratio, $\bar{\eta}$, and is given by the expression

$$G(\bar{\eta},\bar{\sigma}) = \frac{4k_{LT}^2 (M/N)(1+\bar{\eta}\bar{\sigma})}{\pi(1+\bar{\eta})\left\{\frac{(1+\bar{\eta})/(1+\bar{\eta}^2\bar{\sigma}^2)}{(1+\bar{\eta}\bar{\sigma})^2} + \frac{\bar{E}^2}{k_{AV}(M/N)}\right\}} \qquad (5.14)$$

where $\bar{E}$ is the mean void diameter ratio and E_m is the minimum possible void diameter ratio when spontaneous percolation occurs as defined earlier. Here, M is the total number of voids in the neighborhood, N is the total number of particles in the same region, and k_{AV} is the ratio of the mean voids sphere projected area to the mean projected total area (Savage, 1988). The parameters M/N, E_m, and k_{AV} are constants that depend on particle packing, and for which appropriate values can be chosen for the particle assembly. For example, for the closest packing of spherical particles these values are: $M/N = 2$, $E_m = 0.1547$, and $k_{AV} = 0.466$ while for a simple cubic array they are, respectively, 1.0, 0.414, and 0.63 (Savage, 1988). The parameter k_{LT} in Equation (5.14) depends on the geometry of the grid chosen for the control volume and is defined as $\delta y = k_{LT} \overline{D}$ (Savage, 1988), which is the mean particle diameter in the neighborhood. The number of particles per unit area, that is, the number density, is computed as a function of the voids area ratio, e_A, as

$$N_p = \frac{(1+\overline{\eta})}{A_s(1+e_A)(1+\overline{\eta}\,\overline{\sigma}^2)} \tag{5.15}$$

In the application of such a model to the rotary kiln, it must be pointed out that, as a result of jetsam segregation, the values for M/N, E_m, and k_{AV} are susceptible to changes because of rearrangement of the particle ensemble. Nevertheless, it is possible to alter these constants dynamically with respect to both time and space (e.g., for each kiln revolution or material turnover in the cross section). With the altered values of the constants the solid fraction for the segregated core may be computed with the following relationship (Savage, 1988)

$$v = \frac{2(1+\overline{\eta})(1+\overline{\eta}\,\overline{\sigma}^3)}{3k_{LT}(1+e_A)(1+\overline{\eta}\,\overline{\sigma}^2)(1+\overline{\eta}\,\overline{\sigma})} \tag{5.16}$$

It should be recalled that, although a constant value of the solids concentration had been employed in the granular flow model, Equation (5.16) provides a means of determining changes in void fraction due to segregation.

5.4 Boundary Conditions

The calculation domain for jetsam segregation and the percolation process were shown in Figure 5.3. Owing to kiln rotation, an initially well-mixed binary mixture will follow a specific path in the plug flow region until it crosses the yield line into the active layer. For the active layer, material enters from the plug flow region with a given jetsam concentration and then travels down the inclined plane in a streaming flow. During this journey, jetsam particles sink when the voids in the underlying layer are large enough for the particles to percolate. If this does not occur they will pass the yield line again and recirculate. The plug flow region serves only as an "escalator" and within this region particles do not mix or percolate unless small enough to undergo spontaneous percolation; a condition that is precluded from the model. The percolation process in the active layer is repeated for each material turnover, and as the jetsam content in the core increases, fines will no longer be visible at the exposed bed surface. Henein (1980) had observed that the only time fine particles are seen at the top of the bed is when the vessel is loaded with 40–50% fines. The boundary conditions for Equation (5.9) will, therefore, depend on the operational conditions of the kiln. For a dilute mixture of jetsam particles, for example, the boundary conditions will be as follows:

$$\text{at} \quad x = 0; \quad C_J = C_{J0} \tag{5.17a}$$

at $y = 0$; $C_J = 0$ (5.17b)

at $y = \delta_x$; $C_J(1 - C_J) = 0$ (5.17c)

where C_{J0} is the influx of jetsam particles at the apex (bed/wall boundary). Condition (5.17b) indicates that, at the free surface, there are no jetsam particles as all the fines in such a dilute mixture will percolate to the core region, whereas condition (5.17c) is the result of the nonlinear concentration term which will render pure jetsam ($C_J = 1$) at any boundary where particles are finally settled (Gibilaro and Rowe, 1974). It is assumed that this latter boundary condition can be applied at the interface between the active layer and the plug flow region, thus rendering the yield line impermeable to flotsam/jetsam percolation. Nevertheless, the percolation process described above allows particles at the interface to be replaced by those escalated by the plug flow and, as a result, the most appropriate boundary condition for the interface will be

$$\frac{\partial C_J(x, \delta_x)}{\partial y} = 0 \qquad (5.17d)$$

5.5 Solution of the Segregation Equation

The basic expression describing segregation, Equation (5.9), with the appropriate boundary conditions, can be solved when the bulk velocity, the percolation velocity, and the diffusion coefficients are all determined a priori. The solution of the differential equation can be achieved by considering the problem in terms of several specific cases.

5.5.1 Strongly Segregating System (Case I)

For a strongly segregated binary mixture of different sized particles the diffusion of jetsam particles in the vertical plane can be ignored. This situation will pertain to a very dilute mixture where $\bar{\eta} \to 0$ and from Equation (5.4), although the gradient does not go to zero, the inference is that $C_J \to 0$. The differential equation for segregation thus becomes

$$v_p \frac{\partial C_J}{\partial y} - u_x(y) \frac{\partial C_J}{\partial x} = 0 \qquad (5.18)$$

The required boundary conditions are given in Equation (5.17). It should be noted that Equation (5.18) is the same as that employed to describe segregation in chute flows. It can be solved analytically by the method of characteristics (Bridgwater, 1976; Savage, 1988). The characteristic solution would normally involve choosing a characteristic value, s, say, such that

$$\frac{dy}{ds} = v_p \qquad (5.19a)$$

$$\frac{dx}{ds} = -u(y) \tag{5.19b}$$

Substituting Equation (5.19b) into (5.19a) gives

$$-u(y)dy = v_p dx \tag{5.19c}$$

Recalling that the velocity profile for bulk flow in the active layer was given in the flow model by the parabolic equation, that is,

$$\frac{u(y)}{u_\delta} = \kappa + a_1'\left(\frac{y}{\delta}\right) - a_2'\left(\frac{y}{\delta}\right)^2$$

This result can be substituted into Equation (5.19c), which, after integration, gives

$$\kappa y + a_1'\frac{y^2}{\delta} - a_2'\frac{y^3}{\delta^2} = -v_p x + B_1 \tag{5.19d}$$

If the integration constant B_1 is set to zero, the result is given as

$$\kappa y + \frac{a_1'}{\delta}y^2 - \frac{a_2'}{\delta^2}y^3 + v_p x = 0 \tag{5.19e}$$

The solution of Equation (5.19e) at various x-positions will yield the characteristic lines for equal jetsam concentration in the active layer.

5.5.2 Radial Mixing (Case II)

When the system contains mono-sized particles (i.e., particles are identified only by color differences) of uniform density, the percolation term in the differential equation can be ignored and the problem reduces to that of diffusional mixing with drift. In this case, $C_J = C$ (color) and the resulting differential equation may be given by

$$\tilde{D}_y \frac{\partial^2 C}{\partial y^2} - u_x(y)\frac{\partial C}{\partial x} = 0 \tag{5.20}$$

This is the linear diffusion problem of Graetz (Aparci, 1966). Analytical solutions to Equation (5.20) exist for several boundary conditions. By employing the boundary condition discussed for the rotary kiln,

$$C(0, y) = C_{J0}$$

$$\frac{\partial C(x, \delta_x)}{\partial y} = 0 \tag{5.21}$$

$$C(x, 0) = 0$$

The solution for the diffusional mixing may be given as (Aparci, 1966)

$$\frac{C(x,y)}{C_{J0}} = \frac{2}{\delta_x}\sum_{n=0}^{\infty}\frac{(-1)^n}{\lambda_n}\exp\{-\lambda_n^2 x/2s\}\cos\lambda_n y \qquad (5.22)$$

where $s = u/2D$ and

$$\lambda_n = \frac{(2n+1)\pi}{2\delta_x}, \quad n = 0, 1, 2, \ldots$$

5.5.3 Mixing and Segregation (Case III)

This is the complete solution to the mixing and segregation problem and it describes the movement of jetsam particles by the mechanism of mixing as well as segregation. The differential equation, as was given earlier, is

$$\tilde{D}_y\frac{\partial^2 C_J}{\partial y^2} + v_p(1-2C_J)\frac{\partial C_J}{\partial y} - u(y)\frac{\partial C_J}{\partial x} = 0 \qquad (5.23)$$

Although Equation (5.23) is nonlinear, solutions can be found by functional transformation (Ames, 1977). Analytical methods leading to the solution of the equation are presented in Appendix 5B. The concentration of jetsam particles in the active layer is given by this solution as

$$C_J(x,y) = \frac{1}{2}\left[1 - 2\frac{\tilde{D}_y}{v_p}\frac{\partial}{\partial y}(\ln\tilde{Q})\right] \qquad (5.24)$$

where $\tilde{Q}$ represents the solution for the special case of diffusional mixing (Case II).

5.6 Numerical Solution of the Governing Equations

In solving the governing equations by analytical methods, advantage may be taken of the symmetry of the problem as was employed in the flow model earlier in Chapter 4. Although the analytical methods suggested provide one avenue of approach to the problem, factors such as geometry preclude their ultimate exploitation for various reasons. For example, a recirculation term is required to furnish jetsam particles from the plug flow region into the active layer as was shown in Figure 5.3(a). Therefore, a finite difference numerical scheme may be easily employed. We demonstrate the development of a numerical solution of the derivative terms in the governing equations by discretizing the differential equations as follows (Anderson et al., 1984):

1. For Equation (5.18):

$$C_{i,j} = \frac{1}{[v_p/\Delta y_j + u_{i-1,j}/\Delta x]}\left\{\frac{v_p}{\Delta y_j}C_{i,j-1} + \frac{u_{i-1,j}}{\Delta x}C_{i-1,j}\right\} \qquad (5.25)$$

2. Equation (5.20):

$$-\left\{\frac{2\tilde{D}_i}{[\Delta y_{j-1}+\Delta y_{j+1}]}\frac{1}{\Delta y_{j-1}}+\frac{2\tilde{D}_i}{[\Delta y_{j-1}+\Delta y_{j+1}]}\frac{1}{\Delta y_{j+1}}+\frac{u_{i,j}}{\Delta x}\right\}C_{i,j}$$
$$+\left\{\frac{2\tilde{D}_i}{[\Delta y_{j-1}+\Delta y_{j+1}]}\frac{1}{\Delta y_{j-1}}\right\}C_{i,j-1}+\left\{\frac{2\tilde{D}_i}{[\Delta y_{j-1}+\Delta y_{j+1}]}\frac{1}{\Delta y_{j+1}}\right\}C_{i,j+1} \quad (5.26)$$
$$=-\frac{u_{i,j}}{\Delta x}C_{i+1,j}$$

3. Equation (5.23) is solved numerically either by linearizing the nonlinear term and discretizing the resulting equation or by discretizing Equation (5.24) as an extension of the mixing problem. In the former case, the resulting equation is

$$C_{i,j}=\frac{1}{[2A_1+A_2+A_3]}\left(A_1 C_{i,j+1}+A_1 C_{i,j-1} A_3 C_{i,j-1}+A_2 C_{i-1,j}-\mathrm{d}C^2/\mathrm{d}y\right) \quad (5.27)$$

where,

$A_1 = \tilde{D}_i/\Delta y_j$
$A_2 = u_{i,j}/\Delta x$
$A_3 = v_{i,j}/\Delta y_j$

The nonlinear term, $\partial C^2/\partial y$, may be discretized as

$$\frac{\partial C^2}{\partial y} = \frac{[C_{i,j}+C_{i,j-1}]^2-[C_{i,j+1}+C_{i,j}]^2}{4\Delta y_j}$$
$$+\frac{\gamma|C_{i,j}+C_{i,j-1}|(C_{i,j}-C_{i,j-1})-\gamma|C_{i,j+1}+C_{i,j}|(C_{i,j+1}-C_{i,j})}{4\Delta y_j}$$
(5.28)

As in the case of discretizing fluid flow equations, Equation (5.27) requires the appropriate "upwinding" and as a result Equation (5.28) represents the upstream donor cell difference whereby $\gamma=1$ gives a full upstream effect. For $\gamma=0$ the equation becomes numerically unstable (Anderson et al., 1984).

It might be noted in the preceding development that Equation (5.25) is an explicit algebraic formulation because of the parabolic nature of the differential equation. Thus, once the mixture concentration at the apex is given, the jetsam concentration along the chord length can be computed by marching down the incline. Equation (5.26) is the algebraic form of a one-dimensional diffusion/convection equation (Graetz problem) and may be solved numerically using the tri-diagonal method algorithm (TDMA; Anderson et al., 1984). Equation (5.27) is an implicit algebraic equation for the calculation of two-dimensional jetsam concentration in the cross section;

it may be solved by an iterative procedure, for example, the Gauss Siedel method, whereby the nonlinear term, which is expressed by Equation (5.28), is computed using previous values of $C_{i,j}$. In all the scenarios, a solution technique is employed whereby a set of calculations is carried out by marching down the inclined plane starting from the apex to the base. The solution of this set of calculations represents the concentration of jetsam particles for a single pass or material turnover in the cross section. Because there is no diffusion in the plug flow region, particles are allowed to drift (or recirculate) from the lower section of the plug flow/active layer interface to the upper section interface. The second set of calculations for the next pass is initiated with the convected concentration as the initial condition (boundary condition). The calculation is repeated until the overall jetsam concentration in the cross section equals the jetsam loading. Because the bed material circulates for about three or four times per each kiln revolution, this approach allows for the estimation of the number of revolutions required to accomplish complete mixing or complete segregation. The solution method, therefore, represents a pseudo-transient solution in a two-dimensional plane.

5.7 Validation of the Segregation Model

As was said earlier in the chapter, the objective of the segregation model was to determine the extent and dimensions of the segregated core and, as a result, estimate the jetsam concentration gradient. Validation of the model has been carried out against the experimental data of Henein and coworkers (Henein, 1980; Henein et al., 1985). In that work, a 40 cm I.D. drum loaded with a prescribed jetsam concentration was rotated for some desired number of times and then stopped. The bed was then sectioned using discs inserted normal to the drum axis. In each section the fines concentration was measured, beginning from the apex to the base by sieving and weighing, or by simply counting, thereby mapping out a one-dimensional representation of jetsam concentration as a function of chord length. In order to use the data we converted the two-dimensional model result into the one-dimensional representation in the experiment by averaging the jetsam concentration for all radial nodes at each x-location as

$$C_{J_{x,av}} = \sum_{j=1}^{j\max} C_{J_{i,j}} A_{i,j} / \sum_{j=1}^{j\max} A_{i,j} \qquad (5.29)$$

Predicted and measured radial segregation patterns determined for the case of a strongly segregated system (Case I) based on relatively low values of the jetsam loading is shown in Figure 5.4. The ratio of the fine particle diameter to the coarse particle diameter was about 0.125, which is below the threshold mark at which spontaneous percolation could occur. However, in an experiment, fines will always be sifted through the matrix of the plug flow region although this condition is precluded from the model.

Figure 5.4 Predicted and measured profiles for jetsam concentration for a 40 cm drum: limestone, 3.11 rpm, 16% fill.

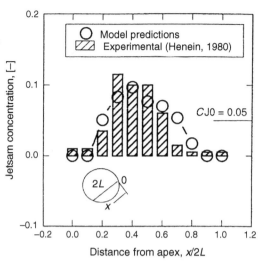

5.8 Application of Segregation Model

The model has been applied for the calculation of particle concentration profiles at the mid-chord plane of a 0.41 m pilot kiln section (Figure 5.5) for which granular flow predictions were made in Chapter 4. Because Case II is for complete mixing, no further discussion on this scenario is carried out. Some results developed for the segregated core (Cases I and III) are shown in Figures 5.6 and 5.7. Notice the difference between a strongly segregated system (Case I) and combined mixing and segregation (Case III). The result shows that if diffusion is present then it will tend to spread

Figure 5.5 Predicted jetsam concentration in the active layer at the mid-chord position for the three cases described in the text: 0.41 m drum, 2 rpm, 12% fill, polyethylene pellets.

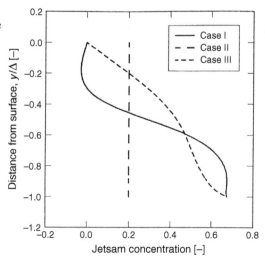

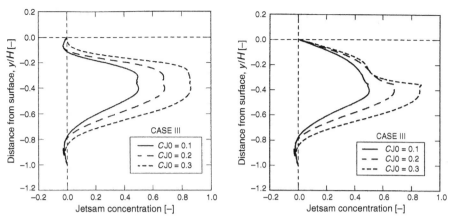

Figure 5.6 Predicted jetsam concentration in both active layer and plug flow region at mid-chord position (bed material: polyethylene; 2 rpm, 12% fill; $d_{pF}/d_{pJ} = 2$).

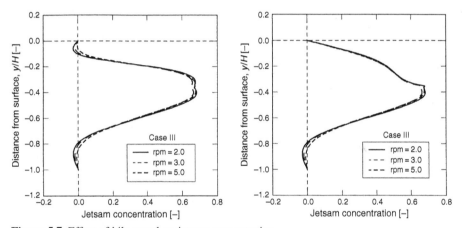

Figure 5.7 Effect of kiln speed on jetsam concentration.

jetsam concentration by moving fines toward the top; and when percolation ceases (Case II) the bed will be well mixed. For higher jetsam loading, the strongly segregated solution (Case I) is no longer applicable; rather, Case III must give a more reasonable result. Also, there is symmetry between the concentration gradient in the active layer and that in the plug flow region for Case I (Figure 5.6) due to the escalator role played by the plug flow region. This symmetry is distorted when Case III is employed due to the effect of the diffusion term in the governing equations, which tends to spread jetsam in the radial direction of the active layer (active layer mixing). The effect of kiln speed on segregation (Figure 5.7) indicates there is very little effect of kiln speed on the concentration profiles. This is because segregation results very rapidly and is complete by a few revolutions since, for each kiln revolution, there are three to four material exchanges between the plug flow and the active layer region.

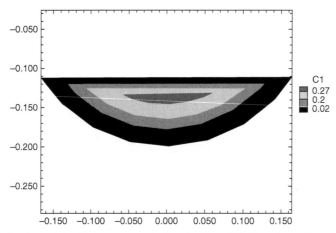

Figure 5.8 Contour plots of jetsam concentration for Case I: 0.41 m drum; 2 rpm; 12% fill; jetsam loading at 20%; polyethylene.

The segregated tongue in a 2.5 m diameter industrial size kiln operating at 2 rpm is shown in the contour plot (Figure 5.8). The model shows that smaller kilns tend to concentrate the fines to the center of the kiln more than larger kilns. In other words, the distribution of jetsam in a larger kiln tends to skew more to the apex, depicting a more defined segregated tongue. The reason for the difference in jetsam distribution in the two kilns may be attributed to the fact that, for the same degree of fill, the chord length in the industrial kiln is longer than that of the pilot kiln and, as a result, most of the percolation process occurs between the apex and the mid-chord.

References

Alonso, M., Satoh, M., Miyanami, K., 1991. Optimum combination of size ratio, density ratio and concentration to minimize free surface segregation. Powder Technol. 68, 145–152.

Ames, W.F., 1977. Numerical methods for partial differential equations, second ed. Academic Press, NY.

Anderson, D.A., Tannehill, J.C., Pletcher, R.H., 1984. Computational Fluid Mechanics and Heat Transfer. Hemisphere, New York.

Aparci, V.S., 1966. Conduction heat transfer. Addison-Wesley Publishing Company (ISBN-0201003597), Reading, Mass.

Bridgwater, J., 1976. Fundamental powder mixing mechanisms. Powder Technol. 15, 215–236.

Bridgwater, J., Ingram, N.D., 1971. Rate of spontaneous inter-particle percolation. Trans. Instn. Chem. Engrs. 49, 163–169.

Bridgwater, J., Cooke, M.H., Scott, A.M., 1978. Inter-particle percolation: equipment development and mean percolation velocities. Trans. IChemE 56, 157–167.

Bridgwater, J., Foo, W.S., Stephens, D.J., 1985. Particle mixing and segregation in failure zones—theory and experiment. Powder Technol. 41, 147–158.

Fan, L.T., Wang, R.H., 1975. On mixing indices. Powder Technol. 11, 27–32.

Gibilaro, L.G., Rowe, P.N., 1974. A model for segregating gas fluidized bed. Chem. Eng. Sci. 29, 1403–1412.

Henein, H., 1980. Bed Behavior in Rotary Cylinders with Applications to Rotary Kilns (Ph.D. dissertation). University of British Columbia, Vancouver.

Henein, H., Brimacombe, J.K., Watkinson, A.P., 1985. An experimental study of segregation in rotary kilns. Met. Trans. B 16B, 763–774.

Hsiau, S.S., Hunt, M.L., 1993. Shear-induced particle diffusion and longitudinal velocity fluctuations in a granular-flow mixing layer. J. Fluid Mech. 251, 299–313.

Nityanand, N., Manley, B., Henein, H., 1986. An analysis of radial segregation for different sized spherical solids in rotary cylinders. Met. Trans. B 17B, 247–257.

Pollard, B.L., Henein, H., 1989. Kinetics of radial segregation of different sized irregular particles in rotary cylinders. Can. Met. Quart. 28 (1), 29–40.

Rogers, A.R., Clements, J.A., 1971. The examination of granular materials in a tumbling mixer. Powder Technol. 5, 167–178.

Savage, S.B., 1983. Granular flow down rough inclines—review and extension. In: Jenkins, J.T., Satake, M. (Eds.), Mechanics of Granular Materials: New Models and Constitutive Relations. Elsevier, Inc., New York, pp. 261–282.

Savage, S.B., 1988. Symbolic computation of the flow of granular avalanches. In: ASME Annual Meeting, Chicago.

Savage, S.B., Lun, C.K.K., 1988. Particle size segregation in inclined chute flow of dry cohesionless granular solids. J. Fluid Mech. 189, 311–335.

Savage, S.B., 1989. Granular flow materials. In: Germain, P., Piau, M., Caillerie, D. (Eds.), Theoretical and Applied Mechanics. Elsevier, Inc., New York, pp. 241–266.

Williams, J.C., Khan, M.I., 1973. The mixing and segregation of particulate solids of different particle size. Chem. Eng. 269, 19–25.

Appendix 5A: Relationship between Jetsam Loading and Number Concentration

Let the respective jetsam and flotsam loading be W_J and W_F.

$$W_J = \frac{1}{6} n_J \pi d_{pJ}^3 \rho_J \tag{5A.1}$$

$$W_F = \frac{1}{6} n_F \pi d_{pF}^3 \rho_F \tag{5A.2}$$

The jetsam concentration at kiln loading becomes

$$C_{J0} = \frac{W_J}{W_J + W_F} \tag{5A.3}$$

$$C_{J0} = \frac{\frac{1}{6} n_J \pi d_{pJ}^3 \rho_J}{\frac{1}{6} n_J \pi d_{pJ}^3 \rho_J + \frac{1}{6} n_F \pi d_{pF}^3 \rho_F} \tag{5A.4}$$

For equal density particles where $\rho_J = \rho_F$,

$$C_{J0} = \frac{n_J d_{pJ}^3}{n_J d_{pJ}^3 + n_F d_{pF}^3} \tag{5A.5}$$

By invoking a definition for number ratio as $\bar{\eta} = n_G/n_F$ and $\bar{\sigma} = d_{pJ}/d_{pF}$, we have

$$C_{J0} = \frac{\bar{\eta}\bar{\sigma}^3}{\bar{\eta}\bar{\sigma}^3 + 1} \tag{5A.6}$$

Appendix 5B: Analytical Solution for Case III Using Hopf Transformation

The differential equation for Case III can be solved using Hopf transformation (Ames, 1977) as follows:

$$D\frac{\partial^2 C}{\partial y^2} + v(1 - 2C)\frac{\partial C}{\partial y} - u\frac{\partial C}{\partial x} = 0 \tag{5B.1}$$

Let, $R(x,y) = 2C(x,y) - 1$ so that

$$\frac{\partial R(x,y)}{\partial y} = 2\frac{\partial C(x,y)}{\partial y}$$

$$\frac{\partial^2 R(x,y)}{\partial y^2} = 2\frac{\partial^2 C(x,y)}{\partial y^2} \tag{5B.2}$$

$$\frac{\partial R(x,y)}{\partial x} = 2\frac{\partial C(x,y)}{\partial x}$$

Substituting Equation (5B.2) into (5B.1) gives,

$$D\frac{\partial^2 R(x,y)}{\partial y^2} = vR(x,y)\frac{\partial R(x,y)}{\partial y} + u\frac{\partial R(x,y)}{\partial x} \tag{5B.3}$$

We can set $R = \partial S/\partial y$ such that

$$\frac{\partial R}{\partial y} = \frac{\partial^2 S}{\partial y^2}$$

$$\frac{\partial^2 R}{\partial y^2} = \frac{\partial^3 S}{\partial y^3} \tag{5B.4}$$

$$\frac{\partial R}{\partial x} = \frac{\partial^2 S}{\partial y \partial x}$$

Substitution of Equation (5B.4) into (5B.3) gives the Hopf transformation as

$$D\frac{\partial^3 S}{\partial y^3} = v\frac{\partial S}{\partial y}\cdot\frac{\partial^2 S}{\partial y^2} + u\frac{\partial^2 S}{\partial y\partial x} \tag{5B.5}$$

for which the first two terms can be rearranged as

$$\frac{\partial}{\partial y}\left[D\frac{\partial^2 S}{\partial y^2} - \frac{v}{2}\left(\frac{\partial S}{\partial y}\right)^2\right] = 0 \tag{5B.6}$$

We can integrate this equation by arbitrarily setting the function in x to zero to yield a solution of the differential equation as

$$D\frac{\partial^2 S}{\partial y^2} - \frac{v}{2}\left(\frac{\partial S}{\partial y}\right)^2 = 0 \tag{5B.7}$$

and when the function with respect to x is introduced, the original differential Equation (5B.1) becomes

$$D\frac{\partial^2 S}{\partial y^2} = \frac{v}{2}\left(\frac{\partial S}{\partial y}\right)^2 + u\frac{\partial S}{\partial x} \tag{5B.8}$$

It is desired to make Equation (5B.8) a linear equation of the form

$$D\frac{\partial^2 \tilde{Q}}{\partial y^2} = u\frac{\partial \tilde{Q}}{\partial x} \tag{5B.9}$$

where

$$\begin{aligned}
T(x,y) &= F(S(x,y)) \\
\frac{\partial T}{\partial y} &= F'(S(x,y))\cdot\frac{\partial S}{\partial y} \\
\frac{\partial T}{\partial x} &= F'(S(x,y))\cdot\frac{\partial S}{\partial x} \\
\frac{\partial^2 T}{\partial y^2} &= F''(S(x,y))\frac{\partial S}{\partial y}\cdot\frac{\partial S}{\partial y} + F'(S(x,y))\cdot\frac{\partial^2 S}{\partial y^2}
\end{aligned} \tag{5B.10}$$

Substituting Equation (5B.10) into (5B.9) yields

$$D\left[F''(S)\left(\frac{\partial S}{\partial y}\right)^2 + F'(S)\frac{\partial^2 S}{\partial y^2}\right] = uF'(S)\frac{\partial S}{\partial x} \quad (5B.11)$$

Divide by $F'(S)$ for $F'(S) \neq 0$ so that

$$D\left[\frac{F''(S)}{F'(S)}\left(\frac{\partial S}{\partial y}\right)^2 + \frac{\partial^2 S}{\partial y^2}\right] = u\frac{\partial S}{\partial x} \quad (5B.12)$$

By identifying Equation (5B.12) with (5B.8) we can write

$$D\frac{F''(S)}{F'(S)} = -\frac{v}{2} \quad \text{or} \quad F''(S) = -\frac{v}{2D}F'(S)$$

which can be integrated to give

$$F(S) = A\exp(-vS/2D) + B \quad (5B.13)$$

Set $A = 1$ and $B = 0$ and invert it to give

$$S = -\frac{2D}{v}\ln\{F(S)\} = -\frac{2D}{v}\ln \tilde{Q}$$

but $R = \partial S/\partial y$, therefore

$$R(x,y) = -(2D/v)\frac{\partial}{\partial y}\left(\ln \tilde{Q}\right) \quad (5B.14)$$

Substituting Equation (5B.14) into (5B.2) yields the solution for the jetsam concentration for mixing and segregation in the active layer as

$$C(x,y) = \frac{1}{2}[R(x,y) + 1]$$

that is,

$$C(x,y) = \frac{1}{2}\left[1 - \frac{2D}{v}\frac{\partial}{\partial y}\left(\ln \tilde{Q}\right)\right] \quad (5B.15)$$

where $T(x, y)$ is the solution for the special case of diffusional mixing given for Case II.

Combustion and Flame

6

The main function of the rotary kiln in the minerals and materials industry is to convert raw materials (ore) into useful product materials. The kiln also has become the workhorse in waste destruction and remediation in the environmental industry. Most or all of these processes involve some chemical or physical reactions that will occur at economically rapid rates, if at all, only at high temperatures. We must therefore get energy into the process as well as later extract it. In materials processing, it is often impractical to supply energy mechanically, that is, by work, which means that heat transfer is usually the way to drive these operations. We will treat kiln heat transfer in the chapters that follow but for now we will examine the energy source and supply of the required heat. Although a handful of small-size indirectly heated kilns might use electrical energy to process material, for most large processing operations the dominant source of energy is fossil fuel combustion. Therefore the flame that develops from fuel firing is the heart of the direct-fired rotary kiln.

6.1 Combustion

Combustion is the conversion of fossil fuel into chemical compounds (or products) by combining it with an oxidizer, usually oxygen in air. The combustion process is an exothermic chemical reaction, that is, a reaction that releases heat energy as it occurs.

$$\text{Fuel} + \text{Oxidizer} \rightarrow \text{Products of combustion} + \text{Energy} \quad (6.1)$$

Here the fuel and the oxidizer are the reactants, that is, the substances that were present before the reaction took place. The relation indicates that the reactants produce combustion products and energy but as it is known in the "fire triangle," there must also be an ignition source for it to proceed. The amount of heat released during combustion depends upon the type of fuel. Fuels are evaluated based on the amount of energy or heat they release per unit mass, per volume, or per mole during combustion of the fuel. This quantity is known as the fuel's heat of reaction or its heating value and is usually expressed in specific terms such as J/kg (Btu/lb), kcal/L (Btu/gal), kcal/m^3 (Btu/ft^3), and so on. Physically, the heat released may be characterized by the intensity of the flame and also by its luminosity. The effectiveness of dissipating this heat may be judged by the flame shape. Since the kiln operator has no control over the heat content of the fuel (usually dependent on its source), the success of all kiln operations lies in the effectiveness of controlling the flame to ensure that the temperature profile in the freeboard matches the intended material process.

In most combustion systems and for that matter in rotary combustion chambers, the oxidizer is usually air but it could be either pure oxygen or an oxygen mixture. For practical reasons we will limit our attention to combustion of fuel with oxygen or air as the main source of the oxidant. Chemical fuels exist in various forms including gaseous, liquid, and solid forms. Common fuels that are burnt in most kilns are either solids (e.g., coal or coke), liquids (e.g., residual oils), or gases (e.g., natural gas, coke oven gas). Some kiln operations are permitted to burn liquid hazardous wastes. In the United States these may require either an incinerator's permit or the Boiler and Industrial Furnace (BIF) permit that allows use of some waste fuels to supplement fossil fuels to produce useful industrial product. Waste fuels include all types of solids such as used tires, contaminated grease rags, and liquid solvents such as alcohols, esters, and so on, that may collectively be called liquid burnable materials (LBM). Although some of these waste fuels have energy contents that are comparable to coal or diesel oil, solvents are usually lighter liquid fuels that can result in low luminosity flame upon combustion. Some of the negative attributes of LBM are high moisture content, low carbon content, and high concentration of noncombustible impurities.

In the case of gaseous fuels it is common practice to analyze the mixture for the component gases and to report the analysis in terms of volume (or mole) percent. This is essential because most gaseous fuels are mixtures of only a few chemical compounds. On the other hand, naturally occurring or commercially available organic liquids or solids (including coal) can contain thousands of compounds, many of which have very complex molecular structure. These are usually analyzed for the weight percentages of the elements of carbon, hydrogen, nitrogen, oxygen, and sulfur that are present in the fuel, that is, C:H:N:O:S. In kiln process design and operation it is common practice to establish mass and mole fractions of the intended fuel so as to design the appropriate burners for the kiln process and to keep track of emissions during kiln operations.

6.2 Mole and Mass Fractions

The amount of a substance in a fuel sample may be indicated by its mass or by the number of moles of that substance. A mole is defined as the mass of a substance equal to its molecular mass or molecular weight. Gram-mole or pound mole of C:H:N:O:S is 12:2:28:32:32 kg or lb_m. The mass fraction of a component i, mf_i is defined as the ratio of the mass of the component, m_i, to the mass of the mixture, m, that is,

$$mf_i = \frac{m_i}{m} \tag{6.2}$$

where the sum of the mass fractions of all components must equal 1, as

$$mf_1 + mf_2 + mf_3 + \cdots = 1 \tag{6.3}$$

An analogous definition for the mole fraction of a component, i, x_i, is the ratio of the number of moles of i, n_i, to the total number of moles in the mixture, n, that is,

$$x_i = \frac{n_i}{n} \tag{6.4}$$

where $n = n_1 + n_2 + \cdots$ and $n_1 + n_2 + n_3 + \cdots = 1$

Thus, the mass of a component i of a mixture is the product of the number of moles of i and the molecular weight, M_i, that is, the total mass is therefore the sum

$$m = n_1 M_1 + n_2 M_2 + \cdots$$

Dividing and multiplying the right-hand side by the total number of moles, n, and invoking Equation (6.4) defines the average molecular weight, that is,

$$M = \frac{m}{n} = x_1 M_1 + x_2 M_2 + \cdots \tag{6.5}$$

The mass fraction of component i is defined as follows:

$$mf_i = \frac{n_i M_i}{n_1 M_1 + n_2 M_2 + \cdots} \quad \text{or} \quad mf_i = \frac{x_i M_i}{x_1 M_1 + x_2 M_2 + \cdots} \tag{6.6}$$

For a mixture at a given temperature and pressure, the ideal gas law shows that $pV_i = n_i \Re T$ for any component, and $pV = n \Re T$ for the mixture as a whole. The ratio of these two equations gives

$$x_i = \frac{V_i}{V} = \frac{n_i}{n} \tag{6.7}$$

Similarly, for a given volume of mixture of gases at a given temperature $p_i V = n_i \Re T$ for each component and $pV = n \Re T$ for the mixture. The ratio of these two equations shows that the partial pressure of any component i is the product of the mole fraction and the pressure of the mixture, that is,

$$p_i = \frac{p n_i}{n} = p x_i \tag{6.8}$$

Air, usually used as the oxidant in combustion, contains approximately 21% oxygen, 78% nitrogen, and 1% other inert constituents by volume. Assuming 21% O_2 and 79% N_2 by volume for ease of combustion calculations, we have 21 mol of O_2 and 79 mol of N_2 present in combustion air. Hence $79/21 = 3.76$ mol of nitrogen accompany every mole of oxygen that reacts with the fuel in combustion. At very high temperatures some of the nitrogen will react to form NO_x but at moderate temperatures where the nitrogen is treated as inert, the 3.76 mol of the accompanying nitrogen will end up in the combustion products. Based on the foregoing definitions and equations, the molecular weight of air can be estimated based on the molar fractions on O_2 and N_2 as follows:

$$M_{air} = \sum n_i M_i = 0.79 M_{N_2} + 0.21 M_{O_2} = 0.79(28) + 0.21(32) = 28.84 \tag{6.9}$$

The mass fractions of oxygen and nitrogen are

$$mf_{O_2} = \frac{n_{O_2} M_{O_2}}{M_{air}} = \frac{(0.21)(32)}{28.84} = 0.233;$$

$$mf_{N_2} = \frac{n_{N_2} M_{N_2}}{M_{air}} = \frac{(0.79)(28)}{28.84} = 0.767 \tag{6.10}$$

6.3 Combustion Chemistry

Following Equation (6.1), combustion will be a chemical reaction between fossil fuel atoms (predominantly carbon and hydrogen, i.e., hydrocarbon) and oxygen in air to form carbon dioxide and water vapor. Methane, CH_4, is the major constituent of most natural gases and undergoes combustion reaction

$$CH_4 + 2O_2 = CO_2 + H_2O \Delta H (192 \text{ kcal/mol}) \tag{6.11}$$

For coal with C:H:N:O:S as constituents, the individual combustible fuel fraction molecules will undergo combustion described by the following reactions:

$$C + 2O_2 = CO_2 + \Delta H(393.7 \text{ kJ/mol}) \tag{6.12}$$

$$2H_2 + O_2 = 2H_2O + \Delta H(285 \text{ kJ/mol}) \tag{6.13}$$

$$C + O_2 = 2CO + \Delta H(110.6 \text{ kJ/mol}) \tag{6.14}$$

$$S + O_2 = SO_2 + \Delta H(297 \text{ kJ/mol}) \tag{6.15}$$

Suffice it to say that carbon dioxide is the product formed by complete combustion. Incomplete combustion will yield CO, a toxic compound, which can further be oxidized to CO_2. The toxicity of CO comes from the fact that when inhaled it forms oxy-hemoglobin in the bloodstream which prevents hemoglobin from absorbing oxygen.

The coefficients in the chemical reactions of combustion may be interpreted as the number of moles of the substances required for the reactions to occur. For example, in the methane combustion reaction, 1 mol of methane reacts with 2 mol of oxygen to form 1 mol of carbon dioxide and 2 mol of water. Although the number of atoms of each element must be conserved during a reaction, the total number of moles or molecules need not. Because the number of atoms of each element cannot change, it follows that the mass of each element and the total mass must be conserved. Hence, using the atomic weights or the masses of each element, the sums of the masses of the reactants and the products must be balanced.

$$CH_4 + 2O_2 = CO_2 + 2H_2O$$
$$[12 + 4(1)] + 2(32) = [12 + 2(16)] + 2[2(1) + 16] = 80 \tag{6.16}$$

Additionally, there are 2 mol of water and 1 mol of CO_2 in the 3 mol of combustion products. Therefore the mole fraction of water and carbon dioxide in the combustion products are

$$X_{H_2O} = 0.667; \quad X_{CO_2} = 0.333 \tag{6.17}$$

But there are 2(18) mass units of water and 44 mass units of CO_2 in the 80 mass units in the products, so the mass fractions are 36/80 and 44/80, respectively, that is,

$$mf_{H_2O} = 0.45; \quad mf_{CO_2} = 0.55 \tag{6.18}$$

Analogous fractions can be expressed for the other substances by balancing their combustion equations. Because we know that in air every mole of oxygen is accompanied by 3.76 mol of nitrogen, the methane reaction in air can be written as

$$CH_4 + 2O_2 + 2(3.76)N_2 = CO_2 + H_2O + 2(3.76)N_2 \tag{6.19}$$

Hence with air as an oxidizer instead of oxygen, there are 2 mol of water per 10.52 mol of product, compared with 2 mol of water per 3 mol of products.

Because the combustion products exit the atmosphere through the stack, they are known as the stack gases or flue gases. Often the flue gas composition is stated on a wet basis (wfg) or on a dry basis (dfg) because under certain conditions the water vapor in the flue gas may condense and escape as liquid water. When the water vapor condenses, the mass of the dry combustion product is used in estimating the mass or mole fraction. The term higher heating value (HHV) of the fuel, frequently used in the United States, refers to a heating value measurement in which the product water condenses. As a consequence the latent heat of vaporization of the water is released and becomes part of the heating value. The lower heating value (LHV), frequently used in Europe, corresponds to the heating value in which water remains as vapor and does not yield its latent heat of vaporization, that is,

$$HHV = LHV + \left(\frac{m_{H_2O}}{m_{fuel}}\right) h_{fg} \quad [kJ/kg] \tag{6.20}$$

where h_{fg} is the latent heat of vaporization of water.

6.4 Practical Stoichiometry

When designing a burner for rotary kiln combustion, or for any combustion chamber for that matter, it is important to establish the amount of air required for the combustion reaction. This requirement includes sizing the burner nozzle to an air velocity that will induce all the jet properties appropriate to the operation, and ensure optimum energy savings with minimum environmental pollutants. The aim of practical stoichiometry is to determine exactly how much air must be used to completely oxidize the fuel to the combustion products. A stoichiometrically correct mixture of fuel and air is defined as one that would yield exactly the products and have no excess oxygen if combustion were complete. Knowing the stoichiometric amount, the burner designer can decide how much excess air would be needed to ensure the appropriate jet character and flame temperature, among others. In general, one can write a balanced stoichiometric relationship for any CHNOS fuel-air system as follows:

$$C_u H_v O_w N_x S_y + \left(u + \frac{v}{4} - \frac{w}{2} + y\right)(O_3 + 3.76 N_2) \rightarrow u CO_2 + \frac{v}{2} H_2 O \\ + y SO_2 + \left[3.76\left(u + \frac{v}{2} - \frac{w}{2} + y\right) + \frac{x}{2}\right] N_2 \quad (6.21)$$

The negative sign associated with w indicates that less oxygen from air is needed for complete oxidation because atoms of oxygen already exist in the fuel itself. In most situations it is convenient to normalize the actual mixture composition to the stoichiometric mixture composition for that fuel-oxidizer system. The normalization yields a dimensionless number whose magnitude tells us how far the mixture composition has deviated from stoichiometric conditions. The most convenient dimensionless number used in practical combustion is the equivalence ratio, Φ, defined as

$$\Phi = \frac{m_{\text{fuel}}/m_{\text{air}}}{(m_{\text{fuel}}/m_{\text{air}})_{\text{stoich}}} \quad (6.22)$$

or, on a molar basis,

$$\Phi = \frac{n_{\text{fuel}}/n_{\text{air}}}{(n_{\text{fuel}}/n_{\text{air}})_{\text{stoich}}} \quad (6.23)$$

With this definition, mixtures with $\Phi < 1$ are said to be fuel-lean while those with $\Phi > 1$ are said to be fuel-rich. Two other dimensionless ratios that are commonly used to specify the composition of a combustible mixture relative to the stoichiometric composition are the percent theoretical air and percent excess air. These are defined respectively as $100/\Phi$ and $100(1/\Phi - 1)$. Thus, a mixture that has $\Phi = 0.8$ can also be said to contain 125% theoretical air or 25% excess air.

6.5 Adiabatic Flame Temperature

The adiabatic flame temperature is one that occurs when the combustion chamber is well insulated with no heat losses (adiabatic conditions). The peak adiabatic flame temperature occurs at around $\Phi = 1$ in an ideally insulated combustion chamber. Figure 6.1 is a typical graph of flame temperature for a natural gas–air mixture. As percent combustion air increases, that is, as we deviate from the stoichiometric condition, some of the heat generated is used to heat up the excess air. As a result, the flame temperature will drop. By the same token, it is important to note that increasing the fuel at stoichiometric conditions will reduce the flame temperature as is indicated by the left-hand side of the temperature peak in Figure 6.1. Therefore, under controlled conditions, flame temperature can be a useful measure of air-fuel ratio, that is, how far we deviate from stoichiometric conditions and whether the combustion is fuel-lean or fuel-rich. Note that although we have illustrated flame temperature with stoichiometric air (100%), the same general rules apply to any combustion condition. This is helpful because it allows temperature control by changing the air-fuel mixture. It also gives us a sense of whether unburned fuel is being released into the atmosphere. In order to maintain the flame temperature at any set of conditions, one must increase or decrease the fuel and air proportionately. Increasing fuel alone or air alone will result in a change in flame temperature. Such a temperature control mechanism is no different from that encountered in the carburetor of spark-ignition automobile engines.

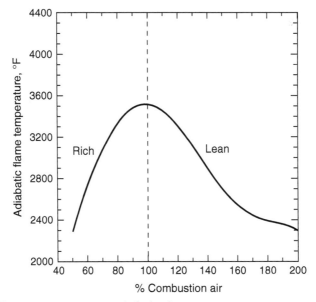

Figure 6.1 Flame temperature versus air-fuel ratio.

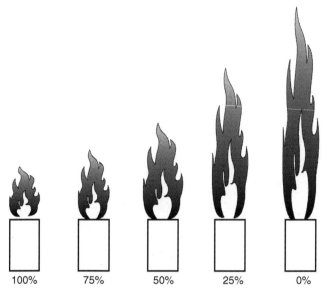

Figure 6.2 Aeration effect on flame shape.

Controlling the air-fuel ratios not only controls the flame temperature but also controls the flame shape or the character of the burning jet. The flame shape can be illustrated by considering the aeration of a simple laminar Bunsen burner jet flame (Figure 6.2). The example shows four systems for introducing air to the fuel in the nozzle. These include nozzle mix, premix, atmospheric, and use of raw fuel without any aeration. Figure 6.2 hypothetically illustrates the expected flame geometry for gaseous fuel as primary aeration diminishes from 100% to 0%; 100% primary air would correspond to a sharp pale blue flame. This is because all of the air for combustion is mixed with the fuel and is ready to ignite as soon as it reaches the nozzle. A flame burning with 75% primary air would be characterized by a double blue cone and would be a little longer than the previous one. This is because 25% of the fuel molecules will need time to find oxygen from the surrounding air before burning. At 25% primary air, only a slight amount of blue color remains in the flame; the flame becomes much longer and predominantly yellow in color. When fuel is burned without any premixing, the flame is normally lazy or ragged; it is very long and all yellow. Assuming that all the fuel molecules are burned in each case, then the same amount of energy would be generated.

6.6 Types of Fuels Used in Rotary Kilns

Kilns can be fired with natural gas, fuel oil, pulverized solid fuels, or a combination of all of these. As mentioned earlier, the fuel can also include waste fuels. Because of the long gas residence time in the freeboard, kilns have become the combustion chamber of choice for hazardous waste incineration, in some cases achieving 99.99% (also referred to as "4-nine") destruction efficiencies. Because most of the rotary kiln operations are high-temperature processes and require intense flames, radiation heat transfer is predominant in the mechanisms of energy transfer to the material being processed. Thermal radiation is also maximized because of the cylindrical enclosure of the freeboard space. Hence fuels having high luminosity as a result of soot formation during combustion tend to improve thermal efficiency and consequently fuel use. Because of the high cost and the transparent nature of natural gas flames, it is normally the last choice of fuel selection for rotary kilns except in places with strict environmental restrictions on CO_2 emissions. Most rotary kilns use pulverized fuel, typically coal or petroleum coke as the primary fuel for combustion. In this section, we will examine the source and types of these fuels and also the factors that maximize their combustion in rotary kilns.

6.7 Coal Types, Ranking, and Analysis

Coal consists of a complex range of materials that differ from deposit to deposit. The differences encountered in coal types result from many factors including the vegetation from which the coal originated, the depths of burial of the vegetation from which the coal was formed, the temperatures and pressures at those depths, as well as the length of time the coal has been forming in the deposit. The varying amount of mineral matter in a coal deposit may also have a significant effect on its properties and classification (Table 6.1). The American Society for Testing Materials (ASTM) has established a ranking system that classifies coals as anthracite (I), bituminous (II), sub-bituminous (III), and lignite (IV). Ranking is determined by the degree of transformation of the original plant material to carbon. High-rank coals are high in carbon and therefore heat value, but low in hydrogen and oxygen. Low-rank coals are low in carbon but high in hydrogen and oxygen content. Coal ranking and analysis of combustion rely on two types of analysis of coal composition: the proximate analysis and the ultimate analysis. The proximate analysis involves thermo-gravimetric analysis (TGA) whereby the sample is continuously heated in the absence of oxygen (devolatilization or pyrolysis) in a continuously weighing system. The weight remaining when the temperature of the sample reaches water evaporation temperature determines the moisture content (M). Further weight loss, up to the temperature when no further weight loss occurs, determines the total volatile matter (VM) and the char

Table 6.1 Proximate Analysis of Some US Coal by Rank (Watson, 1992)

Source	Rank	M (%)	VM (%)	FC (%)	A (%)	S (%)	BTU/lb
Schuylkill, PA	I	4.5	1.7	84.1	9.7	0.77	12,745
McDowell, WV	II	1.0	16.6	77.3	5.1	0.74	14,715
Sheridan, WY	III	25.0	30.5	40.8	3.7	0.30	9345
Mercer, ND	IV	37.0	26.6	32.2	4.2	0.40	7255

remaining. The char is composed of carbon and ash, so when the remaining sample is burned in air or oxygen until only noncombustible minerals remain (A), the weight loss gives the fixed carbon (FC). The proximate analysis is usually reported as percentages (or fractions) of the four quantities, that is, moisture, ash, volatile matter, and fixed carbon. The ultimate analysis is a chemical analysis that provides the elemental mass fractions of C, H, N, O, and S usually on a dry, ash-free basis.

$$m = m_{comp} + m_{ash} + m_{moist}$$
$$\frac{m_{comp}}{m} = 1 - A - M \tag{6.24}$$

Thus an equation for the wet and ashy volatile matter fraction in the proximate analysis may be determined from dry, ash-free proximate analysis using the expression

$$\begin{aligned} VM_{as-fired} &= \left(\frac{m_{comp}}{m}\right)(VM)_{dry,ash-free} \\ &= (1 - A - M)(VM)_{dry,ash-free} \end{aligned} \tag{6.25}$$

6.8 Petroleum Coke Combustion

Petroleum coke (also known as pet coke) is a carbonaceous solid derived from the cracking processes of oil refineries and has been a source of relatively cheap pulverized fuel for the kiln industry. It is called green coke until it is thermally treated into crystalline or calcined pet coke used in the manufacture of electrodes for steel and aluminum extraction. Green coke comes from several sources, all from the petroleum refinery industry. Table 6.2 gives some green coke analyzed by Polak (1991) showing their sources and their elemental analyses.

Table 6.2 Some Sources of Petroleum Coke and Their Analyses (Polak, 1991)

Supplier	Site	VM (%)	H (%)	C (%)	S (%)	A (%)	O (%)	N (%)
Gulf oil	Canada	16.2	4.18	93.4	0.73	0.08	—	—
Collier	IL	15.0	4.12	90.4	3.04	0.31	1.54	1.25
Esso	Argentina	10.6	3.66	90.7	0.76	0.49	1.21	1.67
Standard	OH	8.8	3.73	89.6	2.86	0.12	1.83	1.32
Humble oil	LA	6.6	3.69	91.8	1.50	0.16	—	—

Heating values range between 14,000 and 16,000 Btu/lb.

As seen in the table, some pet cokes are good for their low ash content and high carbon content, however their high sulfur content can present environmental problems by emitting SO_2, a major source of acid rain, into the atmosphere unless measures are taken to scrub the exhaust gas, which can be very expensive. Hence they are used in the kiln industry by blending it with cheaper, low energy content, and high volatile coal to balance emissions and take advantage of their high heating value. Other sources of pulverized fuel include wood and scrap tires, the latter having been used extensively in modern cement kilns.

6.9 Scrap Tire Combustion

There are over 280 million scrap tires produced annually in the United States. Of these over 100 million are used as fuel and most of these are burnt to supplement fuel use in cement and other rotary kiln operations. The cement process is particularly convenient for tire combustion because the reinforced steel wire in the tire tread can be a source of iron for the cement chemistry. Kilns burning tires must comply with the EPA's boiler and industrial furnace act and hence are heavily regulated as a pollution source. Table 6.3 gives a typical average composition of tires supplied by the Rubber Manufacturers Association of America.

Combustion of scrap tires and, for that matter, any pulverized fuels including coal, coke, or biomass proceeds in two phases. First, the organic solid polymer undergoes pyrolysis upon reaching a temperature of about 250–300 °C to release the volatile matter and solid residue (char). Second, these volatiles and char undergo combustion (Figure 6.3).

Table 6.3 **Tire Composition (Rubber Manufacturers Association of America)**

Material	Chemical Formula	Percent Composition	
		Car	Truck
Natural rubber	C_5H_8	12.43	24.57
Styrene	C_8H_8	5.63	2.99
1,3-Butadiene	C_4H_6	18.34	9.75
Fabric	–	15.47	14.85
Carbon black	C	24.86	25.48
Steel (belt and bead)	–	18.13	17.40
Fillers	–	5.16	4.95
Water (in tires)	H_2O	0.00	0.00

Average heating value ≈ 15,000 Btu/lb.

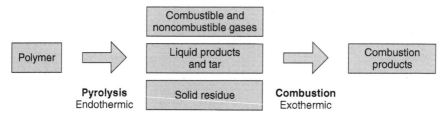

Figure 6.3 The mechanisms of polymer decomposition and combustion.

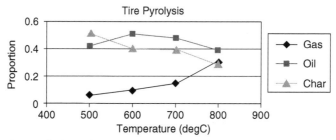

Figure 6.4 Tire pyrolysis products (UCLLNL-DOE—waste tires).

Pyrolysis of tires will yield noncondensable gas (mainly CO and H_2), pyrolytic oil, and solids (char and metal wire). The proportion of these depends on the pyrolysis temperature, as shown in Figure 6.4.

The char contains carbon black, which is very stable and difficult to burn. A simple analysis based on burning only the volatile component of the pyrolysis (syngas and oil) in a lightweight aggregate kiln (LWA) that typically burns one short ton an hour of coal at 26.75 MJ/kg (11,500 Btu/lb) indicates a substantial economic advantage. There is potential savings in replacing 100%, 80%, and 50% (Figure 6.5) of the coal's energy release with tires at two tire pyrolysis temperatures, 950 °F (500 °C) for the low temperature and 1472 °F (800 °C) for the high temperature, at a coal price of $37 per ton and a tire price at $20 per ton, delivered. Burning the entire tire including carbon black or finding an alternative use for the carbon black makes tire utilization a lucrative energy-saving proposition.

6.10 Pulverized Fuel (Coal/Coke) Firing in Kilns

Milling of coal or coke produces a powder called pulverized fuel which contains particles of a wide range of sizes. As we saw in Chapter 3, the distance at which a particle in a particle-laden jet will travel in a combustion chamber plays a role in the damping of the jet's turbulent energy. Therefore theoretical analysis of combustion must take the particle size distribution of the fuel into account. Pulverized fuel fineness is therefore an important parameter in the modeling of coal combustion. An analytical expression of particle size distribution that has found a wide application for expressing the fineness of pulverized fuel is the Rosin–Rammler relation. The relationship is given by (Field et al., 1967)

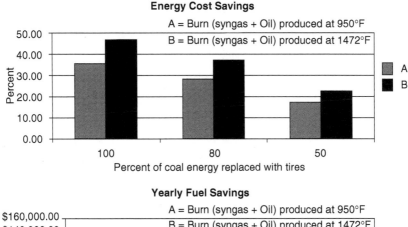
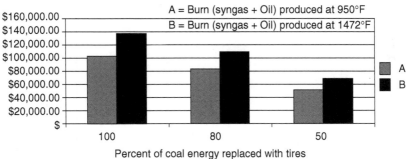

Figure 6.5 Calculated potential energy and yearly cost savings for scrap tire use in LWA kiln.

$$R = 100 \exp\left[-(x/x')^n\right] \tag{6.26}$$

where R is the weight percent of particle size greater than x, and x' and n are adjustable constants. The constant x' is a measure of the fineness of a powder and is the size, x, for which the percent oversize R is 100/e, that is, 36.8%. Although x' is not an average size, it may be related to the weight mean size. The constant n is a measure of size dispersion with a low number indicating a wide size distribution. Equation (6.26) can also be written as

$$\log(\log 100/R) = n \log x + \text{const} \tag{6.27}$$

so that a plot of log(log 100/R) against log x should give a straight line with slope n. With two known mesh sizes of a particle size analysis, an analytical size distribution of pulverized fuel material can be generated iteratively from Equation (6.26) by first guessing x' calculating components of Equation (6.27), and revising x' until R values match. A distribution where 3% weight fraction is retained by 100 US mesh, that is, 150 μm particle size and 77% retained by the 200 mesh (75 μm) gives $x' = 108$ μm, $n = 3.82$, and $R = -0.91$ with the distribution shown in Figure 6.6. The parameters can also be estimated using a 2-parameter estimation in a nonlinear regression analysis.

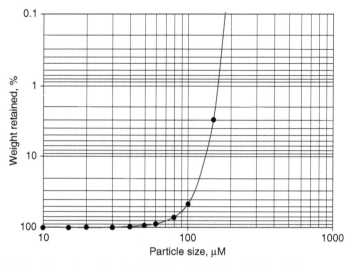

Figure 6.6 Rosin–Rammler pulverized fuel size distribution with 3 and 77 percent passing 150 and 75 μm respectively.

6.11 Pulverized Fuel Delivery and Firing Systems

Pulverized fuel can be delivered and fired in kilns either directly (direct system) or indirectly (indirect system). The simplest and most commonly used system for pulverized fuel firing is the direct-fired system (Figure 6.7).

The wet coal is fed into a pulverizer along with hot gases drawn from the firing hood. The hot gas temperature is controlled by bleed air. The coal or coke is simultaneously dried and ground in the hot-air swept pulverizer. The pulverizer can be a hammer mill, ring-roll mill, or a ball mill type. The ground particles and air are swept by the primary air fan and delivered to the burner as a dilute suspension of pulverized fuel and air. A combination of solid fuel materials can be fired directly or semi-directly

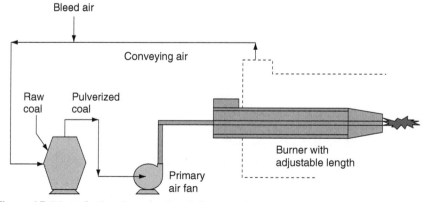

Figure 6.7 Direct-fired coal combustion delivery system.

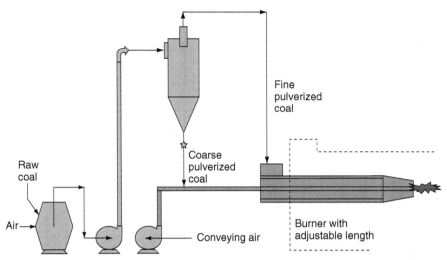

Figure 6.8 Semi-direct-fired coal combustion delivery system.

using the system shown in Figure 6.8. Here we envision the firing of coke and coal in a proportion chosen either by design, for economic, or for environmental consideration. Using two pulverizers provides an opportunity to fire either of the fuels independently in the event of a supply shortfall of either fuel. Also, if a moisture-laden coal provides a cheap source of energy, or if one is considering wood or biomass firing as part of an environmental sustainability program, then the second pulverizer can operate as a semi-direct system whereby the alternative fuel can be delivered into a cyclone collector where it can be separated.

It is important to ascertain the appropriate conveying velocities in the pipes that will prevent settling. Low transport velocities or excessively high solids loading may result in unstable operation. If the gas velocity in a horizontal pipe is reduced progressively, particles will eventually settle along the bottom of the pipe. The minimum velocity at which settling occurs is called the saltation velocity. This is affected by all the resistances to the flow including pipe roughness, pipe diameter, and particle size, but more importantly, by the gas/solid loading. Increased solid loading will violate the gas laws and pressure drop across the pipe may decrease rather than increase with increasing airflow. Minimum transport velocities recommended based on coal milling experience are 15 m/s with a specified minimum air temperature at the mill of 60 °C. The minimum figure for air velocity applies to the lowest coal throughput, which is about 30–40% of the maximum coal throughput depending upon the turndown ratio. As the coal loading is increased, the airflow should also increase but not proportionately. Therefore, as the coal feed is increased, the transport velocity increases but the air-to-coal ratio decreases. At maximum coal flow, the air-to-coal ratio reaches its minimum, around 1.6:1 wt/wt, and the flow velocity reaches its maximum. Hence by a combination of suitably sized transport pipe work and suitably controlled air-to-coal ratios, a system should be operated so that the velocity at minimum throughput is at least 15 m/s and around 25 m/s at maximum throughput (Scott, 1995).

6.12 Estimation of Combustion Air Requirement

We have already discussed under practical stoichiometry how the air requirements can be estimated based on the fuel composition (ultimate analysis). The primary and secondary air requirements for combustion of pulverized coal or coke are best estimated by mass and heat balance at the mill. In Appendix 6A we show a calculation taken from Musto (1978) for the primary and secondary air required for coal pulverizer with 4.5 metric ton per hour (10,000 lb/h) coal feed rate at initial moisture of 15 percent which is required to be ground and dried to 2% with a 200 HP mill. In order to estimate the actual primary and secondary air, one has to make some estimation of the evaporation rate, the amount of gas entering the coal mill, and the bleed air required so that the quantity of air that should be vented from the hood off-take can be properly estimated. It shows that for a take-off gas temperature of 315 °C (600 °F) and vent gas temperature of 76 °C (170 °F) and allowing ambient air infiltration of 10% at 15 °C (60 °F), the primary air will be about 22% of stoichiometric air and 21% of total air. The remaining air (about 79%) will be the secondary air. With this information we can size a burner using a burner pipe diameter based on a Craya–Curtet parameter of choice bearing in mind the conditions that ensure the desired jet recirculation patterns described in Chapter 3.

6.13 Reaction Kinetics of Carbon Particles

In the combustion of pulverized fuel, firing begins with the combustion of volatiles followed by the combustion of the solid particles remaining after pyrolysis, that is, the oxidation of the carbon in the char (carbon and ash). Perhaps the char combustion step is the most important one since it involves heterogeneous reaction of solid and gas and, hence, requires thorough mixing. The burnout of the carbon typically follows a shrinking core model that is depicted in Figure 6.9. In this model a spherical char particle is assumed to be surrounded by a boundary layer of stagnant gas (oxidant and

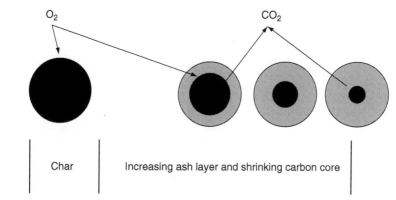

Figure 6.9 Char burnout model.

combustion products) several particle diameters thick, through which oxygen diffuses from the free stream and reacts at the surface to form combustion products, which also diffuse outward into the free stream. The generalized equations for transport can be deduced by assuming that in the boundary layer, oxygen is conserved and that the rate at which oxygen is transported through an imaginary surface at any radius is equal to the rate at which oxygen is transported to the particle surface. The flux of oxygen at the surface of a particle with radius r can be expressed as (Field et al., 1967).

$$rG(r)\frac{R'T_m}{\tilde{D}M} = p_g - p_s \tag{6.28}$$

where $\tilde{D}$ is the diffusion coefficient, R' is the gas constant, M is the molecular weight, T_m is the mean temperature, and p_g and p_s represent the oxygen partial pressures in the free stream (subscript g, for gas) and at the surface (subscript s). Equation (6.28) gives the rate of oxygen transport to the particle surface in terms of the concentration of oxygen in the free stream and at the surface.

Since the carbon particle is porous and also char particle in a stagnant environment will leave a residual ash uninterrupted, the diffusion coefficient involves the resistance to flow of oxygen and combustion products through the ash layer and through the porous carbon itself. The rate of reaction of the carbon with oxygen at the reaction front can be expressed in terms of the oxygen partial pressure at the surface and the surface reaction rate constant k_s as

$$q = k_s p_s \tag{6.29}$$

where q is the rate of removal of carbon per unit external geometric surface area, [g/cm^2 s]. The rate constant k_s may be expressed by the Arrhenius-type equation as

$$k_s = A \exp(-E/RT_s) \tag{6.30}$$

where T_s is the surface temperature in degrees Kelvin, and the gas constant $R = 1.986$ cal/mol °K. Typical values of the frequency factor A, and the activation energy E, for anthracite coal are $A = 8710$ g/cm^2 s and $E = 35{,}700$ cal/mol.

6.14 Fuel Oil Firing

Atomization is to liquid fuels as pulverization is to solid fuels. Atomization is required to break down liquid fuel oil into tiny droplets prior to combustion. Droplet combustion follows the same model as char combustion shown in Figure 6.9. The size of the droplets and the way they are dispersed has a critical effect on burner performance. Like pulverized solid fuel, the ignition temperature of fuel oils is much lower than gas, so the mixing is the rate-controlling step in the combustion process. Although some light fuel oils, for example kerosene, may readily vaporize and mix with the oxidant prior to ignition, heavy fuel oils—which are frequently used in rotary kilns—need to be spray-atomized into the secondary air to ensure adequate mixing. The droplet combustion process is similar to coal combustion, starting with initial heating and flaming pyrolysis, which releases volatiles and a porous sphere of coke particles known as a *cenosphere*. The volatiles burn at the surface of the droplet as oxygen diffuses to the fuel molecules through the combustion products. Like char, the combustion of the coke follows the shrinking core (see Figure 6.9). Hence, liquid fuels are characterized by their droplet size and burning rate. The burning rate may be used to predict the heat and mass release rates of different-sized droplets as a function of residence time in the kiln environment. Additionally, the time required to completely consume a given size droplet may be easily computed from knowledge of the characteristic burning rate. By performing an energy and mass balance over a single-component, spherical droplet vaporizing in a uniform, quiescent atmosphere (i.e., low Reynolds-number droplet vaporization), the instantaneous rate of droplet evaporation can be expressed by

$$\dot{m}_v = -\frac{d}{dt}\left(\frac{\pi}{6}\rho_l d_d^3\right) = -\frac{\pi}{4}\rho_l d_d \frac{d}{dt}\left(d_d^2\right) \quad (6.31)$$

and

$$\frac{d}{dt}\left(d_d^2\right) = -\frac{8(\lambda/c_p)}{\rho_l}\cdot\ln(1 + B_{h,v}) \quad (6.32)$$

where λ is the thermal conductivity of the gas, and $B_{h,v}$ is a characteristic heat transfer number (Law, 1982) defined as

$$B_{h,v} = \frac{c_p(T_a - T_l)}{q_v} \quad (6.33)$$

Here q_v is the latent heat of vaporization. The heat transfer number, $B_{h,v}$, represents the ratio of the vaporization potential of the ambient environment to the vaporization energy requirement of the liquid fuel. Integrating Equation (6.32) yields

$$d_d^2 = d_0^2 - K_v \cdot t \quad (6.34)$$

with the evaporation rate constant, K_v, defined by Equation (6.32) as

$$K_v = -\frac{8(\lambda/c_p)}{\rho_l}\cdot\ln(1 + B_{h,v}) \quad (6.35)$$

Equation (6.34) is the so-called "d^2 law" which shows that the square of the droplet diameter decreases linearly with time. The equation also reveals that the residence time for complete consumption of a droplet may be expressed as

$$\tau = d_0^2 / K_v \quad (6.36)$$

This demonstrates a quadratic dependence of burnout on the diameter of the initial liquid droplet. For burning droplets, the droplet flame front acts as a nearby source of heat for droplet vaporization, otherwise the mechanism is the same as pure vaporization. For a single-component, uniform droplet burning under quasi-steady gas-phase conditions

$$d_d^2 = d_0^2 - K_c \cdot t \tag{6.37}$$

with the combustion rate constant, K_c, defined as

$$K_c = -\frac{8(\lambda/c_p)}{\rho_l} \cdot \ln(1 + B_{h,c}) \tag{6.38}$$

and the combustion heat transfer number given by

$$B_{h,c} = \frac{c_p(T_\infty - T_l) + \left(Y_{O_2,\infty}/\sigma_{O_2}\right)}{q_v} \tag{6.39}$$

where $Y_{O_2,\infty}$ and σ_{O_2} are the oxidizer mass fraction in the far field and the stoichiometric mass fraction of the oxidizer to the fuel, respectively. Therefore, under idealized conditions, the d^2 law holds for droplet combustion. However, in reality, what is not accounted for in the d^2 law are the effects of multicomponent fuels, transient droplet heating, fuel vapor depletion or accumulation between the droplet and the flame, and various gas-phase transport properties (as functions of concentration and temperature). Experiments have shown that droplet combustion often begins with the droplet heating (transient) period during which the droplet diameter is relatively constant because of thermal expansion offsetting initial weak vaporization. After the transient period, whose duration is dependent on size, ambient environment, and fuel properties, single-component fuels and fuels with narrow volatility ranges typically exhibit d^2 law behavior throughout the majority of the droplet burning (Shaddix and Hardesty, 1999). Calculated droplet evaporation and burning rates for diesel No. 2 fuel oil, water, and some pyrolysis oils derived from wood, grasses, and their mixtures are presented in Table 6.4 (Shaddix and Hardesty, 1999). For fuels containing water, such as the pyrolysis liquids derived from biomass, the droplet burning rates are reduced as the water content increases due to the higher heat of vaporization and higher stoichiometric air requirements.

Most atomizers produce a range of droplet sizes with the smallest on the order of a few micrometers in diameter. Large droplets can range anywhere between 100 and 1000 μm (Mullinger and Chigier, 1974). In a typical industrial rotary kiln flame a droplet particle of 100-μm diameter will take about half a second to burn out according to the burning rate calculations presented here. With a residence time of only a few seconds in the freeboard of a rotary kiln, atomization, which produces droplets larger than 200 μm, might not ensure a complete burnout of the fuel. Therefore, in order to ensure carbon burnout and proper flame shape and length in kiln burners, either droplet

Table 6.4 **Calculated Droplet Vaporization and Combustion for Liquid Burnable Fuels (Shaddix and Hardesty, 1999)**

Liquid Fuel	ρ_l (g/mL)	q_v (J/g)	K_v (mm²/s)	σ_{ox}	q_c (J/g)	K_c (mm²/s)
Diesel No. 2	0.86	267	0.56	12.6	41.0	0.99
Water	1.00	2257	0.099	–	–	–
Biomass-Derived Pyrolysis Liquids Produced by National Renewable Energy Lab (NREL)						
NREL 154 (oak)	1.20	613	0.25	5.6	17.6	0.52
NREL 175 (poplar)	1.20	711	0.23	6.3	16.0	0.45
NREL 157 (switchgrass)	1.20	887	0.19	8.2	19.3	0.40
NREL 175 + water	1.18	842	0.20	5.7	14.6	0.42
NREL 175 + methanol	1.16	738	0.23	6.3	16.3	0.45
NREL 175 + ethanol	1.16	720	0.23	6.4	16.8	0.46

sizes are made smaller, or larger spray droplets must undergo microexplosions. Microexplosions occur in multicomponent droplets that have a broad range of volatiles as associated with liquid burnable fuels such as solvents or waste-derived liquid fuels. The driving force for microexplosions is the much faster process of thermal transport (diffusion) within the liquids compared to mass transport. The ratio of the thermal diffusivity, α_l, and the mass diffusivity, D_l, is the dimensionless Lewis number ($Le = \alpha_l/D_l$). Since the droplet surface temperature is at the boiling point of the instantaneous molecular mixture at the surface, it increases rapidly as more volatile components become depleted. The temperature rise at the surface sets up a temperature gradient between the interior and the surface which results in heat transfer to the interior. A microexplosion occurs when the characteristic temperature (superheat limit) is reached by the liquid mixture at any location within the droplet. This limit is about 90 percent of the thermodynamically defined critical temperature at 1 atm pressure (Shaddix and Hardesty, 1999).

6.15 Combustion Modeling

As was shown in Figure 6.9, combustion begins with flaming pyrolysis, which is an endothermic reaction requiring initial heat to proceed when the fuel particle, whether liquid droplets or pulverized, reaches a certain threshold temperature. This is followed by the homogeneous combustion of the released gaseous compounds and the exothermic heterogeneous reaction of the solid carbon particles with the oxygen in the air. Combustion modeling requires knowledge of the kinetics of all the reaction steps involved. However, since the fuel particles are suspended in a surrounding jet fluid in direct-fired systems, the effect of flow on combustion and the combustion chamber boundaries are important considerations. We also discussed that mixing is the rate-controlling step in the combustion of pulverized fuels as well as liquid droplets. For a more accurate estimation of the combustion process one must model the turbulent gas-particle flows and combustion simultaneously. In Chapter 3 we presented a set of equations in a reactive flow system involving mass, momentum, and energy conservation (cf., Equations (3.17–3.28)):

Continuity Equation:

$$\frac{\partial \rho}{\partial t} + \frac{\partial}{\partial x_j}(\rho u_j) = 0$$

Momentum Equation:

$$\frac{\partial}{\partial t}(\rho u_i) + \frac{\partial}{\partial x_j}(\rho u_j u_i) = -\frac{\partial P}{\partial x_i} + \frac{\partial}{\partial x_j}\left[\mu\left(\frac{\partial u_j}{\partial x_i} + \frac{\partial u_i}{\partial x_j}\right)\right] + \rho g_i$$

Concentration Transport Equation:

$$\frac{\partial}{\partial t}(\rho Y_s) + \frac{\partial}{\partial x_j}(\rho u_j Y_s) = \frac{\partial}{\partial x_j}\left(D\rho\frac{\partial Y_s}{\partial x_j}\right) - w_s$$

Enthalpy Transport Equation:

$$\frac{\partial}{\partial t}(\rho c_p T) + \frac{\partial}{\partial x_j}(\rho u_j c_p T) = \frac{\partial}{\partial x_j}\left(\lambda\frac{\partial T}{\partial x_j}\right) + w_s Q_s$$

Arrhenius Equation:

$$w_s = B\rho^2 Y_F Y_{ox} \exp(-E/RT)$$

Equation of State:

$$P = \rho RT \sum Y_s/M_s$$

Ideally, one would numerically model these equations using computational fluid dynamic (CFD) modeling, which we will discuss later. However, before such tools became available there were qualitative assessments based on the simplification of these equations that helped the burner designer or kiln operator evaluate combustion performance based on a set of rules and experiences. In Chapter 3 we explored, based on the dimensionless numbers derived from the reactive flow conservation equations, the importance of mixing in the combustion process and stated that if the reactants are mixed then the fuel is burnt because mixing is the slowest step. The importance of turbulent mixing in the form of recirculation eddies is that it returns combustion products to the flame front and increases the residence time in the combustion chamber. Although the correct estimation and use of the turbulent kinetic energy (TKE), which is a linear function of retention time, through CFD predictions is preferred, a combination of flow visualization techniques and mixing parameters such as the Craya—Curtet parameter calculations are equally powerful tools that are still used in the industry today. We will explore the two methods to determine the flame character before proceeding to a full CFD analysis.

6.16 Flow Visualization Modeling (Acid—Alkali Modeling)

It is said that the technique of modeling of mixing processes using a mixture of dilute acid and alkaline was first employed by Hawthrone in 1939 at MIT in his thesis work on the mixing of gas and air in flames (Moles et al., 1973). He developed a method of accurately modeling physical jet mixing by discharging a jet of dilute sodium hydroxide (NaOH) solution colored with phenolphthalein indicator into a surrounding fluid of dilute hydrochloric acid (HCl), the reaction of which visually simulates the combustion process and thereby the flame boundary. Ruhland (1967) used this technique later to investigate fuel/air mixing and combustion processes in cement kilns. The technique was later employed by Jenkins and Moles (1981) who, with Mullinger, later founded Fuel and Combustion Technology to optimize burner operations in rotary kilns. To this date such techniques are still useful and are often employed by kiln burner providers as a viable alternative to CFD modeling. To model the combustion chamber, a physical model of the hood and the combustion zone of the kiln are constructed to an appropriate scale in clear acrylic plastic. In order to simulate combustion, the fuel is represented by the alkali with the phenolphthalein coloring. Once the alkali is neutralized by the acid, which represents the fuel oxidant (air), the mixture solution becomes clear, thereby providing a visual representation of the physical location where combustion is complete, that is, where all the fuel is presumed consumed. The technique involves isothermal flows. However, in real combustion systems buoyancy effects result due to temperature changes. One way of introducing buoyancy in a cold model such as this is to distort the flows in the physical model so as to make it representative of the full site plant. Hence an essential feature of the technique is to maintain a dynamic similarity. The concentration of the alkali (simulated fuel), and the stoichiometric ratio of the acid (simulated air) and alkali reactions are chosen to represent the correct air-fuel ratio requirement for the particular fuel used in practice. For example, when natural gas is the primary fuel rather than fuel oil, the higher air-fuel ratio requirement for natural gas is represented by higher alkali concentration and the acid flow rate is adjusted to simulate the levels of excess air. Figure 6.10 is an acid—alkali representation of a pulverized fuel flame with a low (Figure 6.10(a)) and high (Figure 6.10(b)) burner momentum conditions.

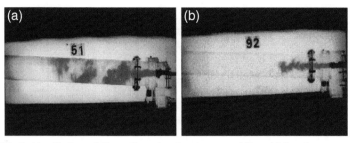

Figure 6.10 Acid–alkali modeling of combustion in rotary kilns; (a) low burner momentum, (b) high burner momentum.

Experience has shown that for a confined flame, sufficient primary air jet momentum is required to create mixing via external recirculation zones. As the acid-alkali model shows, high momentum and, for that matter, intense and perhaps complete mixing, is tantamount to efficient combustion. We discussed in Chapter 4 that, for confined jets, the onset of flame recirculation can be described by the Craya–Curtet parameter M, which is essentially the momentum ratio between the primary and secondary air streams. Although M can be anywhere between zero and infinity ($0 < M < \infty$), practical experience has shown that $M = 1.5$ is the threshold momentum ratio that generates the recirculation associated with good mixing. $M = 1.5$ will ensure that the burner jet will have momentum in excess of that required to entrain all the secondary air in the near field and to cause vortex shedding that returns combustion products to the flame front, resulting in an attached flame (Figure 6.10(b)). Hence Figure 6.10(a) representing a low momentum burner has a Craya–Curtet number $M < 1.5$. Experience has also shown that burners with excessively high jet momentum will suffer poor combustion. Some of the adverse effects include inadequate flame length, unstable flame, pulsating flows, and so on. The optimum Craya–Curtet parameter value for coal-fired kilns is somewhere between 1.5 and 2.5. For cement kilns, experience has shown that the short and intensive flames needed to complete the bed reactions are achieved with a Craya–Curtet parameter value between 2.0 and 2.5. Lime kilns requiring less intense and longer flames operate with M values between 1.5 and 2.0. Employing these techniques provides the kiln engineer or the operator an avenue of evaluating the flame character without resulting in full-blown computational fluid dynamics modeling. However, CFD provides a means of achieving all of the above and more including heat fluxes, combustion product, and other species concentrations that are associated with emissions.

6.17 Mathematical Modeling Including CFD

Computational fluid dynamics is a virtual prototyping that guides in building accurate flow models by solving transport equations. The features of combustion flows can be analyzed in detail with CFD. In particular, mixing, temperature, flow velocity, flame stability, and concentration of combustion species can accurately be computed for different geometry. CFD modeling of gas flow inside the rotary kiln chamber provides three-dimensional analyses of the general flow pattern by mapping the pressure field and velocity vectors that show recirculation zones, mixing, and resident time within the chamber and further provides temperature and species concentration of the combustion products. It therefore makes it possible to evaluate useful operational parameters of interest for design optimization prior to prototyping, or troubleshooting an existing design for operational performance.

The physical and chemical phenomena of the reacting flow may be simulated by numerically solving a set of generalized conservation equations for flow (Navier Stokes equations), and an associated set of equations involving enthalpy, combustion, and so on. A fundamental method for the numerical simulation of the governing equations is the finite-difference or finite-element approximations. The formal steps involved in the application of the methods follow four steps (Vichnevetsky, 1981): (1) the domain of the problem is covered by a simple mesh, (2) values of the numerical solution are labeled at the intersections or nodes of the mesh, (3) a finite-difference or finite-element approximation to the differential equation is formulated in each node resulting in a system of algebraic finite-difference or finite-element equations, and (4) the system of equations approximating the problem is solved to produce a numerical solution. This process generally involves solving numerically large systems of linear algebraic equations and the corresponding computer algorithms are those of numerical linear algebra. Until recently when computer power improved, it was not possible to apply CFD modeling to industrial rotary kilns because of the large aspect ratio and the large number of mesh points involved to accurately represent the problem. A typical 3 m diameter by 30 m long rotary kiln ($L/D = 10$) may require a couple million mesh points for the calculation domain and require several days or even weeks to execute the program depending on the power of the computer used. However, several CFD providers have improved their modeling capabilities and CFD has become a powerful tool in modeling the complex rotary kiln phenomena including combustion and flames. CFD solves the conservation equations involving mass, momentum, combustion, and enthalpy equations within the boundaries of the kiln. The two-phase flow nature of pulverized fuel combustion requires tracking of the particles as localized sources and brings an added complexity to CFD modeling of such operations. To model the combustion in kilns, mathematical expressions are required for the turbulent reactive flow, coal pyrolysis or devolatilization, homogeneous volatile combustion, heterogeneous char reaction, particle dispersion, radiation, and pollutant emission. There are several CFD providers that offer software packages for the simulation of rotary kiln processes. They all treat the mathematical expressions and the numerical schemes required for their solution slightly differently. We will present some of the generalized equations that are common to combustion modeling.

As pointed out earlier in Chapter 3 and will be covered in more detail later on in this chapter, the flow in a rotary kiln is typically gas—solid turbulent flow with chemical reactions, mainly combustion. The building blocks behind the user-defined functions (UDF) in commercial CFD codes applied to rotary kiln combustion modeling consist of "renormalization group" (RNG) k–ε turbulent model for gas phase and, in the case of pulverized combustion particles, the statistical (stochastic) trajectory model for homogeneous volatile and heterogeneous solid-phase char combustion. The underlying equations are discussed in the next section.

6.18 Gas-Phase Conservation Equations Used in CFD Modeling

The set of conservation equations that are solved in most CFD analyses are as presented earlier but expanded to include the stress generation as in Equations (6.40–6.45) (Wang et al., 2006).

Continuity:

$$\frac{\partial \rho}{\partial t} + \frac{\partial}{\partial x_i}(\rho u_i) = S_p \qquad (6.40)$$

where the variable in the continuity equation represents a source term typical to fuel injection or combustion of particles in a control volume. The components of velocity in a three-dimensional coordinate system are represented by the momentum equation
Momentum:

$$\frac{\partial}{\partial t}(\rho u_i) + \frac{\partial}{\partial x_i}(\rho u_i u_j) = -\frac{\partial p}{\partial x_j} + \frac{\partial \tau_{ij}}{\partial x_j} + \rho g_i + F_i + S_p \qquad (6.41)$$

which includes pressure, turbulent shear stresses, gravitational force, that is, buoyancy effects, and the source terms arising from gas–solid interactions. The τ_{ij} term in Equation (6.41) represents Reynolds stress as

$$-\overline{\rho u_i u_j} = \mu_t \left(\frac{\partial u_i}{\partial x_j} + \frac{\partial u_j}{\partial x_i} \right) - \frac{2}{3} \delta_{ij} \left(\rho k + \mu_t \frac{\partial u_i}{\partial x_i} \right) \qquad (6.42)$$

Turbulence modeling is implemented as a closure model for the Reynolds stress with the most commonly used k-ε turbulence model being
k-equation

$$\frac{\partial}{\partial t}(\rho k) + \frac{\partial}{\partial x_i}(\rho k u_i) = \frac{\partial}{\partial x_j}\left[\left(\mu + \frac{\mu_t}{\sigma_k}\right)\frac{\partial k}{\partial x_j}\right] + G - \rho\varepsilon \qquad (6.43)$$

ε-equation

$$\frac{\partial}{\partial t}(\rho\varepsilon) + \frac{\partial}{\partial x_i}(\rho\varepsilon u_i) = \frac{\partial}{\partial x_j}\left[\left(\mu + \frac{\mu_t}{\sigma_\varepsilon}\right)\frac{\partial \varepsilon}{\partial x_j}\right] + C_{1\varepsilon}\frac{\varepsilon}{k} - C_{2\varepsilon}\frac{\varepsilon^2}{k} + S_\varepsilon \qquad (6.44)$$

where the generation of turbulence denoted by G in Equation (6.43) comprises two terms, (1) the generation of turbulence kinetic energy due to the mean velocity gradients, and (2) that due to the generation of turbulence kinetic energy due to buoyancy. S_ε is the turbulence source term in Equation (6.44). The turbulent Navier–Stokes equations are presented with the standard nomenclature and the enthusiastic reader is referred to the appropriate CFD literature.

In order to include temperature distribution, the Navier–Stokes equations are accompanied by an energy equation that solves for enthalpy ($h = c_\text{p}T$). The balance equation for enthalpy is

$$\frac{\partial}{\partial t}(\rho h) + \frac{\partial}{\partial x_i}(\rho u_i h) = \frac{\partial}{\partial x_j}\left(\Gamma_\text{h}\frac{\partial h}{\partial x_j}\right) + S_\text{h} \tag{6.45}$$

The source term, S_h, includes combustion, that is, the heat source and the heat transfer within the system that affect temperature. In rotary kilns, the dominant heat transfer mode is radiation and there are several models to evaluate its value, some of which will be examined in detail later.

CFD providers treat gas-phase combustion by using a mixture fraction model (Wang et al., 2006). The model is based on the solution of the transport equations for the fuel and oxidant mixture fractions as scalars and their variances. The combustion chemistry of the mixture fractions is modeled by using the equilibrium model through the minimization of the Gibbs free energy, which assumes that the chemistry is rapid enough to assure chemical equilibrium at the molecular level. Therefore, individual component concentrations for the species of interest are derived from the predicted mixture fraction distribution.

6.19 Particle-Phase Conservation Equations Used in CFD Modeling

Most CFD providers track particles in the reactive flow field by solving the pertinent equations for the trajectory of a statistically significant sample of individual particles that represents a number of the real particles with the same properties. For example, following the Rosin–Rammler size distribution (Figure 6.6), coal particles are tracked using a statistical trajectory model followed by the modeling of the kinetics of devolatilization and subsequent volatile and char combustion as discussed previously in this chapter (Figure 6.9). Models similar to the d^2 law presented earlier are used for droplet combustion of atomized fuel oil.

In CFD modeling, numerical solutions are sought by discretizing Equations (6.40–6.45) and integrating over each control volume represented by the mesh following:

$$\int_v \frac{\partial \rho \phi}{\partial t} dV + \oint_A \rho \cdot \phi \cdot u \cdot dA = \oint_A \Gamma \cdot \nabla \phi \cdot dA + \int_v S dV \tag{6.46}$$

Figure 6.11 CFD mesh generation for rotary kiln combustion modeling.

where ϕ is the property value, for example, velocity, temperature, species concentration, and so on. Usually, specific to a problem to be analyzed, boundary and initial conditions are provided and these equations are solved iteratively until the convergent criteria are met. A typical mesh generation for the combustion zone of an industrial rotary kiln firing pulverized fuel is shown in Figure 6.11. There is no need to extend the calculation domain to the entire kiln if the focus is to evaluate burner performance.

6.20 Emissions Modeling

6.20.1 Modeling of Nitric Oxide (NO_x)

Nitric oxide (NO_x) is one of the main pollutants in combustion systems, and rotary kilns are no exception particularly for pulverized fuel combustion. NO_x formation depends on three factors, namely (1) the amount of nitrogen present in the fuel, (2) the combustion temperature, and (3) the stoichiometric conditions for the combustion reaction. Hence NO_x production is classified into fuel NO_x, thermal NO_x, and prompt NO_x. Some of the mechanisms for the formation of these species during pulverized coal combustion in rotary cement kilns have been described in commercial CFD packages (e.g., FLUENT, CINAR).

6.20.1.1 Fuel NO_x

Most models assume that the fuel-bound nitrogen that is released by the devolatilization of coal is in the form of HCN, or some instantaneous transforms of HCN, which in turn form the base species of NO formation. It is believed that the HCN not only contributes to fuel NO_x formation but also to some destruction of NO_x and that the net formation might depend on the chemical as well as the thermal state of the mixture. The global chemical reactions involved for coal flames might therefore be expressed as

$$HCN + O_2 = NO + \text{Products} \tag{6.47}$$

$$HCN + NO = \text{Products} \tag{6.48}$$

Kinetic studies have shown that the principal pathways of NO formation and reduction by hydrocarbon radicals are

$$NO + CH_2 \leftrightarrow HCN + OH \qquad (6.49)$$

$$NO + CH \leftrightarrow HCN + O \qquad (6.50)$$

$$NO + C \leftrightarrow CN + O \qquad (6.51)$$

where the reactions proceed mainly in the forward path. Detailed kinetic studies have also established the following set of reactions (Visser, 1991)

$$CH_4 + H = CH_3 + H_2 \qquad (6.52)$$

$$CH_3 + OH = CH_2 + H_2O \qquad (6.53)$$

$$CH_2 + H = CH + H_2 \qquad (6.54)$$

$$CH + H = C + H_2 \qquad (6.55)$$

where, for hydrocarbon diffusion, the flames can be considered, within reasonable accuracy, to be in partial equilibrium.

The concentration of H radicals in the post flame regions of hydrocarbon diffusion flames have been observed to be on the same order as the H_2 concentration, and the OH radical can be estimated by the observed partial equilibrium of the reaction

$$OH + H_2 = H + H_2O \qquad (6.56)$$

in these flames.

These fast chemistry multi-mixture fraction combustion models and similar models have been incorporated into several CFD packages including CINAR, FLUENT, and others.

6.20.1.2 Thermal NO

One source of thermal NO is the oxidation of the molecular nitrogen in the combustion air. The activation energy, the threshold energy required for the formation of thermal NO is high, hence its formation is highly temperature-dependent and important only at high temperatures. Thermal NO is usually modeled by the so-called "extended Zeldovich" mechanism (Westenberg, 1971), which follows these reaction mechanisms:

$$O + N_2 \leftrightarrow NO + N \qquad (6.57)$$

$$N + O_2 \leftrightarrow NO + O \qquad (6.58)$$

$$N + OH \leftrightarrow NO + H \qquad (6.59)$$

Reactions (6.57) and (6.58) are important and reaction (6.59) is significant only under highly fuel-rich conditions. The reaction step (6.59) is usually neglected for coal combustion because coal flames are normally very lean in fuel.

6.20.1.3 Prompt NO

Prompt NO is formed by the attack of hydrocarbon fragments on molecular nitrogen in the flame zone. The contribution of prompt NO to the total NO is normally very small in rotary kilns and is usually ignored.

6.20.2 Modeling of Carbon Monoxide

The predictions of CO are based on a thermodynamic equilibrium computer program in which the chemical equilibrium compositions for assigned thermodynamic states are calculated. Most often temperature and pressure, calculated from the main CFD code, are used to specify the thermodynamic state. The method for evaluating the equilibrium compositions is based on the minimization of Gibbs energy as mentioned earlier (Wang et al., 2006).

6.21 CFD Evaluation of a Rotary Kiln Pulverized Fuel Burner

The combustion of pulverized coal in a 3.66 m (12 ft) rotary kiln was modeled using the commercial CFD code offered by FLUENT, which employs a user-defined function (UDF) that one can program and that can be dynamically loaded with the software's solver. The problem at hand involves a burner hood arrangement with a primary air jet issuing from a burner pipe and secondary air from a product cooler at the bottom and discharged into the combustion chamber. The objective function is to compare the performances of two burner nozzles with two different flow areas, one with a swirl vane (Burner A), and the other with a 7.62 cm insert installed for flame stabilization (Burner B; Figure 6.12). The kiln geometry and operating conditions including pulverized fuel flow rate and tip velocities are presented in Table 6.5. The flow was such that the Craya–Curtet parameters were respectively 1.34 and 1.82 for Burner A and Burner B.

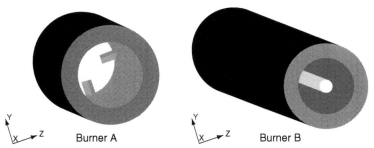

Figure 6.12 Burner nozzle geometry for pulverized fuel combustion in a rotary kiln. Left: swirl vanes; Right: center insert for flame stabilization.

Table 6.5 **Input Variables for CFD Modeling of a Rotary Kiln Combustion System**

	Burner A	Burner B
Kiln diameter (m)	3.66	3.66
Stoichiometric air/fuel ratio(wt/wt)	8.41	8.41
Fuel flow rate (kg/s)	2.38	2.47
Primary air (% of stoichiometric)	22.75	23.02
Secondary air temperature (°C)	760.0	760.0
Primary air temperature (°C)	63.89	73.89
Excess air (%)	4.50	4.50
Fuel specific gravity	1.30	1.30
Nozzle diameter (mm)	336.55	298.18
Burner insert diameter (mm)	None	76.2
Burner pipe flow area (m^2)	0.0889	0.0844
Primary airflow rate (kg/s)	4.55	4.78
Primary air nozzle velocity (m/s)	62.21	83.75
Secondary air velocity (m/s)	5.44	5.62
Momentum ratio	0.10	0.11
Craya–Curtet parameter	1.34	1.82

The burner performance, as predicted by the CFD results, is presented in Figures 6.13–6.16. The model is validated when predicting the interaction of the jet flow from the nozzle and secondary air from the cooler. Figure 6.13 shows slightly asymmetrical flow with the nozzle of Burner A without an insert and an almost symmetrical flow Burner B, which has the flame stabilizer. It has been shown that for self-similar turbulent eddies the vortex rotation period must be a linear function of the diffusion timescale. The effect of flame stabilizers includes increased tip velocity due to lesser flow area and manifests as a higher momentum ratio between the primary and secondary air, thereby inducing stronger recirculation zones as depicted by the vortex shedding. The turbulent kinetic energy (TKE) is a linear function of time. For confined flows, where recirculation eddies return solid particles to the flame front, stronger TKE means longer retention for particles at the flame front for combustion. For a good mixing, one would expect increased diffusion exhibited by internal and external recirculation zones with return of solid particles to the flame region. Hence, Burner B, with the flame stabilizer, provides a better mixing as expected due to a high Craya–Curtet parameter. The Craya–Curtet number for Burner B was 1.82

Combustion and Flame 137

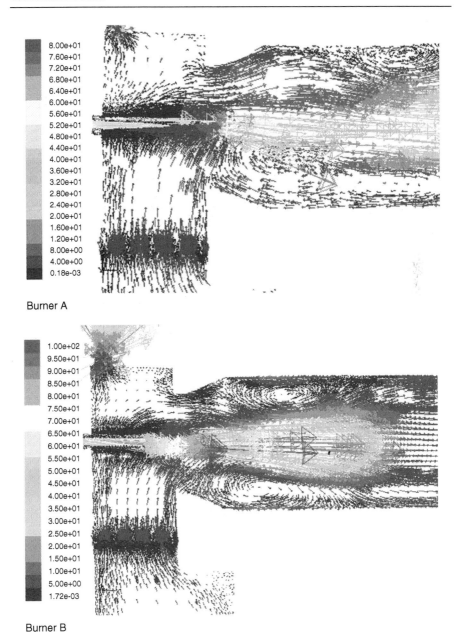

Burner A

Burner B

Figure 6.13 CFD results for the velocity distribution of the flow field in the combustion zone of a rotary kiln with a pulverized fuel burner showing interaction between primary air jet and the entrained secondary air.

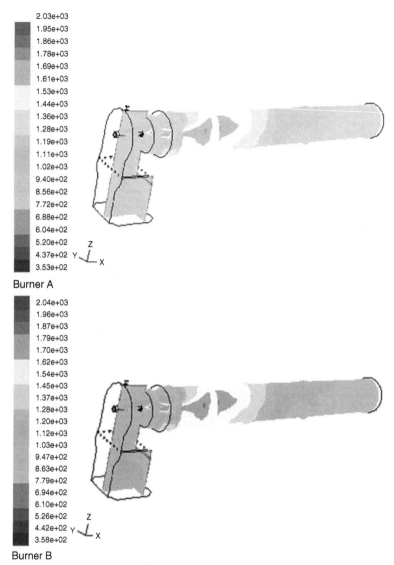

Figure 6.14 CFD predictions for three-dimensional temperature distribution mapping the combustion and flame shape.

compared to 1.34 for Burner A. As we have already discussed, a nozzle with a higher Craya–Curtet parameter is expected to provide a more intense flame and this flame will be drawn closer to the burner tip as the CFD results show (Figures 6.14 and 6.15). As indicated in the theoretical section, combustion efficiency depends on the ratio of the diffusion time to that of the chemical reaction. Since diffusion has the lowest

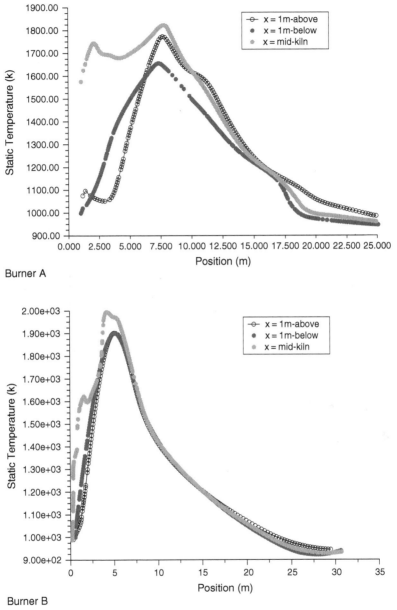

Burner A

Burner B

Figure 6.15 CFD predicted axial temperature profiles.

timescale, it determines the combustion outcome. The bad news is that due to the intensity of the flame associated with the flame stabilizer, a threefold increase in the NO_x emission is predicted based on the equations shown earlier. As a result of the more intense recirculation eddies (vortex shedding) a higher rate of combustion products

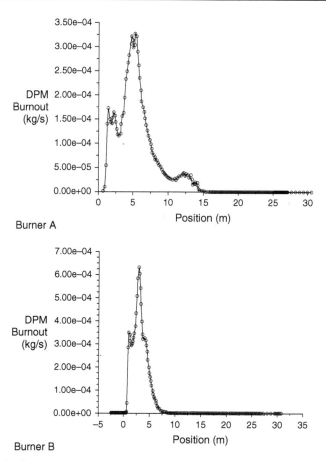

Figure 6.16 Predicted coal particle burnout.

including unburned char particles return to the flame front for a prolonged residence time. The burnout of coal particles for Burner B completes at almost half of the axial distance of burnout in Burner A (Figure 6.16).

It has been demonstrated that CFD has the capacity to predict the flame character and the temperature profiles observed by experience. Although intrinsic details of the combustion and flame can be obtained from such models, large computational demands are required as shown by the large number of calculation cells in Figure 6.12. Even with a high-speed computer, it is not uncommon to expect 3–4 days of real time running before results are obtained for a medium-sized rotary kiln. The good news is that qualitative results by flow visualization and partially quantitative estimates (e.g., the use of similarity parameters such as that of Craya–Curtet) can suffice with some experience in making interim judgments on rotary kiln combustion performance.

References

Field, M.A., Gill, D.W., Morgan, B.B., Hawksley, P.G.W., 1967. Combustion of Pulverized Coal. British Coal Utilization Research Association (BCURA), Leatherhead, UK.

Jenkins, B.G., Moles, F.D., 1981. Modelling of heat transfer from a large enclosed flame in a rotary kiln. Trans. IChemE 59, 17–25.

Law, C.K., 1982. Recent advances in droplet vaporization and combustion. J. Prog. Energy Combust. Sci. 8, 171–201.

Moles, F.D., Watson, D., Lain, P.B., December 1973. The aerodynamics of the rotary cement kiln. J. Inst. Fuel 353–362.

Mullinger, P.J., Chigier, N.A., 1974. The design and performance of internal mixing multijet twin fluid atomizers. J. Inst. Fuel 47, 251–261.

Musto, R.L., 1978. Coal Firing of Cement Kilns. Presented at Cement Manufacturing Technology Seminar, Pennsylvania State University.

Polak, S.L., 1991. Alcan Report #A-rr-1475-71-08. Arvida, Canada.

Ruhland, W., 1967. Investigation of flames in the cement rotary kiln. J. Inst. Fuel 40, 69–75.

Scott, D., 1995. Coal Pulverizers—Performance and Safety. IEA Coal Research.

Shaddix, C.R., Hardesty, D.R., 1999. Combustion properties of biomass flash pyrolysis oils. In: Final Project Report, Sandia National Laboratories Report SAND99–8238.

Vichnevetsky, R., 1981. Computer Methods for Partial Differential Equations. Prentice Hall, Englewood Cliffs, NJ.

Visser B.M., 1991. Mathematical Modeling of Swirling Pulverized Coal Flames (Ph.D. dissertation), IFRF.

Watson, K.C., 1992. Energy Conversion. West Publishing Co, New York.

Wang, S., Lu, J., Li, W., Li, J., Hu, Z., 2006. Modeling of pulverized coal combustion in cement rotary kiln. Energy Fuels 20, 2350–2356.

Westenberg, A.A., 1971. Kinetics of NO and CO in lean, premixed hydrocarbon-air flames (Reaction kinetics of NO and CO formation in lean premixed hydrocarbon-air flames). Combust. Sci. Technol. 4, 59–64.

Appendix 6A: Calculation of Primary Requirements for 10,000 lb/h Coal Combustion

CALCULATIONS FOR ESTIMATING COMBUSTION AIR USING COAL/COKE PULVERIZER

		SI			SI
Feed rate [FR]:	10000 lb/hr	4535.9 kg/hr	Specific heat capacity of gas, c_p	0.24 Btu/lb. °F	1004.9 J/kg K
Initial moisture content [IM]:	15%		Specific heat capacity of solid, c_p	0.3 Btu/lb. °F	1256.1 J/kg K
Final moisture content [FM]:	2%		Latent Heat of Evap.	996.3 Btu/lb	553.5 cal/g
Initial gas temperature [IT]:	600F	315.6 C	Specific volume of dry air @ 170°F	15.88 ft³/lb	
Vent gas temperature [VT]:	170F	76.7 C	Specific volume of water vapor @ 170°F	25.3 ft³/lb	
Product temperature [PT]	130F	54.4 C	Infiltration [INFIL]:	10%	10%
Mill horsepower [BHP]	100	981 kW	Ambient temperature [AT]:	60°F	15.6°C
			Entering gas, E	lb/hr	

EVAPORATION CALCULATION

			GROSS HEAT INPUT FROM GAS		
Evaporation =			GROSS-HEAT = E × C_p × (IT - AT)	1993325 Btu/hr	584.0 kW
FR × [(IM-FM)/(1-FM)]	1326.5306 lb/hr	601.7 kg/hr			
Product Rate [PR] =	8673.469 lb/hr	3934.2 kg/hr	VENT QUANTITY		
[FR - EVAP].			PRIMARY AIR + EVAPORATION		
Dry Solids Rate [DSR] =	8500 lb/hr	3855.5 kg/hr	**PRIMARY AIR:**		
PR × (1-FM)			Entering gas (E) + infiltration (10% of E)	16918.7 lb/hr	7674.2 kg/hr
Moisture remaining in coal = [PR - DSR]	173.469 lb/hr	78.7 kg/hr	TOTAL VENT QUANTITY	18245.18 lb/hr	8275.9 kg/hr
CALCULATION OF ENTERING GAS [E]			VENT VOLUME (AT TEMP)		
			Primary (Dry Air)	268668.22 ft³/hr	
System heat input:			Water Vapor	33561.224 ft³/hr	
			Total	302229.44 ft³/hr	
Contribution from mill motor BHP × 0.75 × 2545	190875 Btu/hr	55.9 kW		5037.1574	
			Primary Air	ACFM	

				STOICHIOMETRIC AIR/FUEL RATIO	9 wt/wt	9 wt/wt	9 wt/wt
	Net heat input from gas [E]			(SAR)			
	$E \times C_p \times \Delta T = \Delta H \times E \Delta H$	103.2		Theoretical air requirement = SAR × DSR	76500 lb/hr		34699.9 kg/hr
	Net heat input from gas = 103.2E			For the kiln to operate at 5% excess air	5%		
	Total heat input =			Actual air =	80325 lb/hr		36434.9 kg/hr
A1	103.2E + 190875 Btu/hr			**SECONDARY AIR**			
	Heat absorbed and lost:						
B1	Heat absorbed in evaporation:			SA = ACTUAL AIR − PA	63406.347 lb/hr	28760.7 kg/hr	
	EVAP × [C_p(VT-AT) + Latent Heat of evap.]	1467540.8 Btu/hr	430.0 kW	Primary air percent (of total air)	21.1%		
C1	Heat absorbed in heating the product:			Primary air percent (of stoic. air)	22.1%		
	PR × C_p × ΔT						
	Solids:	178500					
	Moisture:	12142.857					
		Total 190642.86 Btu/hr	55.9 kW				
D1	Radiation heat loss:						
	5% of E = 0.05 × 103.2E = 5.16E	5.16					
E1	Infiltration heat loss:						
	10% of E = 0.1E × C_{pg} × ΔT = 2.64E	2.64					
	Estimate E by balance as:						
	Heat Input = Heat absorbed and lost						
	A1 + B1 = C1 + D1 + E1						
	103.2E + 190,875 = 1,467,541 + 190642 + 5.16E + 2.64E						
		E = 15380.6 Btu/hr	4.5 kW				

Freeboard Heat Transfer

7

We saw in Chapter 6 that the energy required to drive the direct-fired rotary kiln process is produced as a result of combustion, typically from fossil or waste fuel, in the freeboard. We also saw that the heat source to sustain the flame must come from heat transfer, hence the need to include enthalpy in the conservation equations. Heat transfer in the freeboard is more than just to sustain combustion at the combustion zone; it involves the exchange of energy from the freeboard to the bed to carry out the material process operation. For a process to be efficient, we must get most of the energy into the material and later exhaust the rest. In addition to driving the process, heat transfer is also important for its control. Since the rates at which chemical reactions proceed are strong functions of temperature, controlling the temperature profiles in the freeboard is tantamount to controlling the bed process. Heat transfer in rotary kilns encompasses all the modes of transport mechanisms, that is, conduction, convection, and radiation. In the freeboard, radiation is believed to be the dominant mode of heat transfer constituting over 90%, primarily due to the large flame and curvature of the combustion chamber. Convection in the freeboard occurs as a function of the turbulent flow of gases and participates in the transfer of heat to the bed's free surface and the refractory wall in a manner similar to flow over heated plates. Convection also occurs in the interparticle interstices within the particulate bed. Heat is conducted from the freeboard to the outside environment through the refractory wall and must overcome the resistances to heat flow through the composite walls of refractory materials to reach the outer kiln shell. Within the bed, heat transfer is accomplished by interparticle conduction, which, together with interparticle convection and radiation, leads to an effective conductance of heat through the particulate medium. In this chapter we will review classic heat transfer mechanisms and point out where the phenomenon comes into play in the freeboard of the rotary kiln. This will be followed, in Chapter 8, by similar treatment for the bed.

7.1 Overview of Heat Transfer Mechanisms

As we have mentioned, virtually all the modes of heat transfer, that is, conduction, convection, and radiation, occur in the kiln, although the contribution of each may vary depending on the temperature and for that matter on the axial location or zone. Conduction and radiation are fundamental physical transfer mechanisms, whereas convection is really conduction as effected by fluid flow. Classical definitions have it that conduction is an exchange of energy by direct interaction between molecules of a substance subjected to temperature difference. Conduction can occur in gases, liquids, or solids but it normally refers to heat transfer within solid materials. Radiation, on the other hand, is the transfer of thermal energy in the form of electromagnetic waves emitted by atomic agitation at the surface of a body. Like all electromagnetic waves, for example, light, X-rays, and microwaves, thermal radiation travels at the speed of light passing most easily through vacuum or nearly transparent gases such as oxygen or nitrogen. CO_2 and H_2O transmit a portion of thermal radiation. Radiation normally occurs between solid surfaces but does not require any contact or intervening medium. Radiation may also occur within some fluids or between these fluids and solid surfaces. Generally, this mechanism is important only at elevated temperatures. Convection may be described as conduction in a fluid as enhanced by the motion of the fluid. It normally refers to heat transfer between a fluid and a solid surface.

Although the division of heat transfer into these modes oversimplifies the situation, it is adequate provided it is realized that all three modes possess some common features. Despite the fact that each mode is represented by different types of equations they all reduce to the common form defined by the heat flux as

$$\text{Heat Flux} = \text{Driving Force} \div \text{Resistance, or}$$
$$\text{Conductance} \times \text{Driving Force} \quad (7.1)$$

where the driving force is the temperature difference and the resistance is a measure of the material's ability to transport energy. We might note that the units for heat flux are

$$\frac{\text{Energy}}{\text{Area} \times \text{Time}}; \quad \text{for example,} \quad \frac{\text{kJ}}{\text{m}^2\text{s}} = \frac{\text{kW}}{\text{m}^2} \quad (7.2)$$

Since energy is the property of systems, that is, of mass or, more specifically, molecules, we might correctly expect similarities with other transport processes that involve the transport of molecules, for example, mass transfer and momentum transfer. Hence many process problems involve the transfer and conservation of mass, momentum, and heat as we saw in the description of reactive flow equations in the computational fluid dynamics (CFD) approach.

7.2 Conduction Heat Transfer

In solid materials such as refractory walls or the shell of a rotary kiln, the molecules form a rigid lattice structure. Although molecules do not migrate within the lattice, they possess internal energy in various forms, such as vibrational and rotational kinetic energy, which, in turn, are in proportion to the local temperature. If the material is not uniform in temperature, that is, if there is a temperature difference within the material, energy will be transferred by molecular interaction from the more energetic molecules in the region with higher temperature to the less energetic molecules in the region with lower temperature. The rate at which this energy transfer occurs depends both on the temperature gradient in the direction of transfer and on the modes of molecular interaction present in that particular material (thermal conductivity), which is Fourier's law

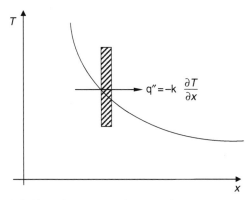

Figure 7.1 Fourier's definition of conduction heat transfer.

of heat conduction. Fourier's law states that the flux of energy or the heat flux by conduction across a plane normal to, for example, the x-axis (Figure 7.1) is proportional to the temperature gradient, that is,

$$q'' = -k\frac{\partial T}{\partial x} \left[\frac{\text{kW}}{\text{m}^2}\right] \tag{7.3}$$

By combining Fourier's law and the requirements of energy conservation, one can derive the governing equation for conduction when the temperature varies in only one coordinate, for example, the x-axis. We do that by considering a control volume of infinitesimal size in the solid lattice (Figure 7.2).

For a surface with area $dy \cdot dz$ at coordinate $x - dx/2$,

$$\begin{aligned} q_{x-dx/2,t} &= q_{x,t} + \frac{\partial}{\partial x}(q)_{x,t}\left(-\frac{dx}{2}\right) \\ &= -\left[k\,dy\,dz\frac{\partial T}{\partial x}\right]_{x,t} + \frac{\partial}{\partial x}\left[-k\,dy\,dz\frac{\partial T}{\partial x}\right]_{x,t}\left(-\frac{dx}{2}\right) \end{aligned} \tag{7.4}$$

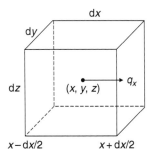

Figure 7.2 Control volume.

Similarly for the surface with area $dy \cdot dz$ at coordinate $x + dx/2$,

$$q_{x+dx/2,t} = -\left[k dy dz \frac{\partial T}{\partial x}\right]_{x,t} + \frac{\partial}{\partial x}\left[-k dy dz \frac{\partial T}{\partial x}\right]_{x,t}\left(+\frac{dx}{2}\right) \tag{7.5}$$

Adding Equations (7.4) and (7.5) gives the rate at which energy crosses into the control volume over the surfaces $x \pm dx/2$

$$q_{CS,t} = \frac{\partial}{\partial x}\left[k dy dz \frac{\partial T}{\partial x}\right]_{x,t}\left(\frac{dx}{2} + \frac{dx}{2}\right)$$

$$q_{CS,t} = k dx dy dz \frac{\partial^2 T}{\partial x^2} \tag{7.6}$$

For constant-pressure processes, the specific enthalpy h is the measure of the internal energy possessed by the material. The specific heat is related to the internal energy as

$$c_p = \left(\frac{\partial h}{\partial T}\right)_{p,T} \Rightarrow dh = c_p dT \tag{7.7}$$

Equation (7.7) can be equated to the rate of energy accumulation in the control volume as

$$\frac{\partial}{\partial t}(h\rho dV) = \rho dV \frac{\partial h}{\partial t} = \rho c_p dx dy dz \frac{\partial T}{\partial t} \tag{7.8}$$

If there is no source term generating energy within the control volume, then the first law of thermodynamics requires energy conservation for which the rate of heat accumulation must equal the rate at which heat crosses into the control surfaces, that is, Equations (7.6) and (7.8) are equal

$$\frac{c_p \rho}{k} \frac{\partial T}{\partial t} = \frac{\partial^2 T}{\partial x^2} \tag{7.9}$$

If there is a source of heat within the control volume, then the heat of accumulation within the control volume must equal the rate at which heat crosses the control surfaces plus the rate at which heat is evolved within the control volume (source of heat within, $\dot{q}$), that is,

$$\frac{c_p \rho}{k} \frac{\partial T}{\partial t} = \frac{\partial^2 T}{\partial x^2} + \dot{q} \tag{7.10}$$

Equation (7.10) is the representation of a transient (unsteady-state), one-dimensional heat conduction that must be satisfied at all points within the material. The combination of properties, $\alpha = k/c_p\rho$, which has units of square meters per second, is known as the thermal diffusivity, and is an important parameter in transient conduction problems. α is a measure of the efficiency of energy transfer relative to thermal inertia. For a given time under similar heating conditions, thermal effects will penetrate farther through a material of relatively high diffusivity. That is to say that heat will penetrate faster in metals, for example, aluminum with $\alpha \approx 1 \times 10^{-4}$ m²/s, than through a fireclay refractory brick with $\alpha \approx 1 \times 10^{-7}$ m²/s, but variations in temperature during heating will be less in aluminum.

Following the same analysis, the transient conduction for a multidimensional temperature field would be

$$\frac{c_p \rho}{k} \frac{\partial T}{\partial t} = \frac{\partial^2 T}{\partial x^2} + \frac{\partial^2 T}{\partial y^2} + \frac{\partial^2 T}{\partial z^2} + \dot{q} \tag{7.11}$$

and for steady-state conduction with no heat source, the equation reduces to what is known as the Laplace equation, that is,

$$\frac{\partial^2 T}{\partial x^2} + \frac{\partial^2 T}{\partial y^2} + \frac{\partial^2 T}{\partial z^2} = 0 \tag{7.12}$$

Equation (7.12) is the form that is required to solve for the temperature distribution in any rectilinear block of material, as those encountered in the refractory linings of rotary kilns. Consider a section of a plane wall that is some distance from any edge or corner, for example, a refractory section in the freeboard with certain thickness, receiving heat from the freeboard gas (Figure 7.3).

For this case, the Laplace equation will take on the very simple form, $\frac{\partial^2 T}{\partial x^2} = 0$, which has a general solution, $T(x) = C_1 x + C_2$. If we have temperature measurements for the boundaries such that

at $x = x_1$, $T = T_1$

at $x = x_2$, $T = T_2$

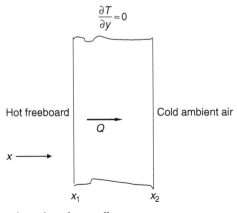

Figure 7.3 Conduction through a plane wall.

then the general solution yields a linear temperature distribution within the refractory lining at the location of interest as

$$T(x) = \frac{(T_2 - T_1)}{(x_2 - x_1)} x + T_1 \qquad (7.13)$$

Equation (7.13) is the analytical solution yielding the distribution of temperature from which heat transfer by conduction through the wall can be established. The rate of heat transfer through the lining can be estimated by invoking the Fourier law as

$$Q = Aq'' = -kA \frac{(T_2 - T_1)}{(x_2 - x_1)} = \text{constant} \qquad (7.14)$$

Although the local heat loss through a lining can be approximated by a plane slab, the global heat loss through a refractory lining in a rotary kiln is through a cylindrical surface. Nonetheless, the Laplace equation for a rectilinear control volume can be extended to a cylindrical control volume as well (Figure 7.4).

From Figure 7.4, conservation of energy requires that

$$\rho c_p \frac{\partial T}{\partial t} = \frac{1}{r} \frac{\partial}{\partial r}\left(rk \frac{\partial T}{\partial r}\right) + \frac{1}{r^2} \frac{\partial}{\partial \theta}\left(k \frac{\partial T}{\partial \theta}\right) + \frac{\partial}{\partial z}\left(k \frac{\partial T}{\partial z}\right) + \dot{q} \qquad (7.15)$$

and for steady-state conduction within a material of uniform conductivity, the equation reduces to the Laplace equation in cylindrical coordinates

$$\frac{\partial^2 T}{\partial r^2} + \frac{1}{r^2} \frac{\partial^2 T}{\partial \theta^2} + \frac{\partial^2 T}{\partial z^2} = 0 \qquad (7.16)$$

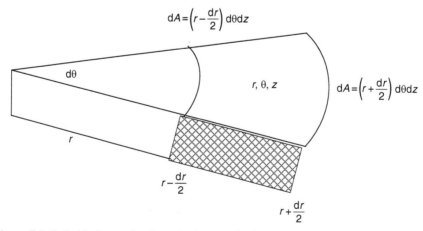

Figure 7.4 Cylindrical control volume for heat conduction.

For the specific case of one-dimensional steady-state heat transfer through a cylindrical wall, the Laplace equation reduces to

$$\frac{d^2T}{dr^2} + \frac{1}{r}\frac{dT}{dr} = 0 \qquad (7.17)$$

which, for the case of a refractory lining section a distance away from edges, presented earlier, has a specific solution,

$$T(r) = T_1 + (T_2 - T_1)\frac{\ln(r/r_1)}{\ln(r_2/r_1)}$$

or, with some rearrangement, gives the temperature distribution,

$$\frac{T(r) - T_1}{T_2 - T_1} = \frac{\ln(r/r_1)}{\ln(r_2/r_1)} \qquad (7.18)$$

Equation (7.18) shows that unlike the rectilinear plane wall, the temperature distribution is not linear. The heat flux declines as r increases due to increasing area $rd\theta dz$. Nonetheless, we can apply Fourier's law to this result to obtain an expression for the heat transfer rate through the wall of a pipe with length L or for that matter a rotary drum cylinder of length L as

$$Q = \frac{T_1 - T_2}{\frac{\ln(r_2/r_1)}{2\pi kL}} \qquad (7.19)$$

7.3 Convection Heat Transfer

As mentioned earlier, convective heat transfer occurs between a solid surface and an adjacent moving fluid. The rate of heat transfer between a surface at temperature T_w and a fluid at T_f can be calculated from Newton's law of cooling, which is mathematically stated as

$$Q = hA(T_w - T_f) \qquad (7.20)$$

The simplicity of Eqution (7.20) should not obscure the complexity of determining h, the heat transfer coefficient, which is a complicated function of both flow and the thermophysical properties of the fluid, that is, viscosity and density, and their relationships such as the Reynolds number $\rho uL/\mu$; Prandtl number $\mu c_p/k$; and the ratio of the momentum diffusivity (μ/ρ) to thermal diffusivity $(k/c_p\rho)$. Convection problems are classified as (1) forced convection, where the fluid is induced by a fan, pump, and so on or (2) free convection, where the flow is driven by temperature-induced density variations in the fluid, that is, buoyancy or a combination thereof. The study of convection heat transfer concerns the ways and means of calculating values for the convective coefficient, h. This involves solution of the Navier–Stokes equations given previously in Chapter 6.

Some analytical solutions of boundary layer flows have been used to deduce the relationships between the heat transfer coefficient and the flow properties. For example, fluid flow over heated surfaces develops two boundary layer thicknesses, that is, the hydrodynamic boundary layer and the thermal boundary layer. When these two layers coincide, then $Pr = 1$. For flow over a flat surface the condition of "no slip" at the wall suggests that the heat transfer in the fluid directly contacting the wall, that is, in the thin layer adjacent to the wall, occurs by pure conduction. Therefore the heat flux is given by

$$q'' = k_f \left(\frac{\partial T}{\partial y}\right)_{y=0} = h(T_\infty - T_w) \tag{7.21}$$

where

$$h = k_f \left(\frac{\partial T}{\partial y}\right)_{y=0} \left(\frac{1}{T_\infty - T_w}\right) \tag{7.22}$$

Like the dimensionless Reynolds number, the nondimensional form of the heat transfer coefficient is defined by the Nusselt number, $Nu_x = hx/k_f$, over the boundary layer. One correlation relating heat and flow over heated surfaces is

$$Nu_x = 0.322 Re_x^{1/2} Pr^{1/3} \tag{7.23}$$

One can also integrate the local values of the Nusselt number to obtain the average value for a plate of length L as

$$Nu_x = 0.664 Re_x^{1/2} Pr^{1/3} \tag{7.24}$$

There is also flow around heated bluff bodies and other complex surfaces for which modified forms of the Nusselt correlations have been deduced. For rotary kiln applications, the convective component of the heat transfer is small and in most cases Equation (7.23) may suffice in estimating the convective heat transfer coefficient, h.

7.4 Conduction–Convection Problems

Conduction usually occurs in conjunction with convection, and if the temperatures are high, they also occur with radiation. In some practical situations where radiation cannot be readily estimated, convection heat transfer coefficients can be enhanced to include the effect of radiation. Combined conduction and convection led to the concept of thermal resistances, analogous to electrical resistances, which can be solved similarly.

Freeboard Heat Transfer

We will now consider the problem of calculating the heat transfer from a hot fluid to a composite plane of refractory wall and through an outer steel shell.

1. Convection from the freeboard to the inside refractory surface will follow Newton's law of cooling as

$$q''_g = h_g(T_g - T_{iR}) \tag{7.25}$$

2. This is followed by one-dimensional steady-state conduction through the refractory lining

$$q''_R = k_R \left(\frac{T_{iR} - T_{Rs}}{L_R} \right) \tag{7.26}$$

3. The next step is one-dimensional steady-state conduction through the steel shell

$$q''_s = k_s \left(\frac{T_{Rs} - T_{sa}}{L_s} \right) \tag{7.27}$$

4. After leaving the shell, heat is transferred to the atmosphere by convection to the air at ambient temperature, that is,

$$q''_a = h_a(T_{sa} - T_a) \tag{7.28}$$

For steady-state conditions with no accumulation of energy within the wall or the gas

$$q''_g = q''_R = q''_s = q''_a = q'' \tag{7.29}$$

Assuming that T_g and T_a are known and we want an expression for determining q in terms of these temperatures and h, k, and L, we can rearrange the equations and sum them up as follows:

$$q'' \frac{1}{h_g} = T_g - T_{iR}$$

$$+ q''_R \frac{L_R}{k_R} = T_{iR} - T_{Rs}$$

$$+ q''_s \frac{L_s}{k_s} = T_{Rs} - T_{sa} \tag{7.30}$$

$$+ q''_a \frac{1}{h_a} = T_{sa} - T_a$$

$$\overline{q'' \left(\frac{1}{h_g} + \frac{L_R}{k_R} + \frac{L_s}{k_s} + \frac{1}{h_a} \right) = T_g - T_a}$$

The heat flux between the gas and ambient air can be calculated from their temperatures only using an expression from electrical resistances, that is, $I = E/R$, as

$$q'' = \frac{T_g - T_a}{\frac{1}{h_g} + \frac{L_R}{k_R} + \frac{L_s}{k_s} + \frac{1}{h_a}} = \frac{T_g - T_a}{\sum R} \tag{7.31}$$

where $\sum R$ is the thermal resistance for heat transfer between the gas and the ambient air. A similar expression for a composite cylinder shown in Figure 7.5 can be deduced using Equation (7.19) as follows:

$$q'' = \frac{T_g - T_a}{\frac{1}{h_g 2\pi r_1} + \frac{\ln(r_2/r_1)}{2\pi L k_1} + \frac{\ln(r_3/r_2)}{2\pi L k_2} + \frac{\ln(r_4/r_3)}{2\pi L k_3} + \frac{1}{h_a 2\pi r_4 L}} \tag{7.32}$$

From these we can also define the thermal conductance through the composite wall as $U = 1/RA$, that is, the rate of heat transfer per degree of temperature drop per square meter, from which follows the heat transfer

$$Q = UA\Delta T \tag{7.33}$$

7.5 Shell Losses

Shell heat loss plays a major role in rotary kiln operation. For some applications such as cement kilns, the shell temperature scans can be one way of controlling the process. Heat losses through the shell can give an indication of what is happening inside the kiln. Even if the kiln is equipped with cameras, refractory failure can only be detected by increased shell temperature at that particular location of failure. Conditions such as

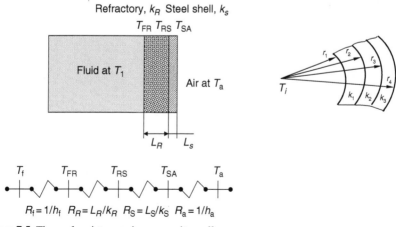

Figure 7.5 Thermal resistances in composite walls.

these are normally evidenced by hot spots. Some modern cement kiln operations are equipped with temperature scanners that continuously scan the shell and record real-time temperature profiles. Any causes that effect temperature changes inside the kiln, for example, increased feed rate, low fuel rate, and changes in airflow rate, will change the shell temperature profile. The drawback of using the shell wall for kiln control is that the refractory lining has a large thermal mass and its response time to temperature change inside the kiln is slow. That is the reason why some operators prefer to use the inside wall temperature as a control point by pointing cameras equipped with temperature sensors (e.g., Micron or Ircon).

7.6 Refractory Lining Materials

The type and thickness of the refractory material used for kiln lining are critical to heat losses through the kiln shell and play a major role in the design and maintenance of the process. The mechanics of wall–shell heat transfer are treated later. The critical thickness required for a lining to insulate the kiln is a heat transfer problem because beyond a certain threshold it no longer performs as an insulator. The thermal properties of the refractory materials, particularly their thermal conductivity, do control the rate of shell heat loss. Considering that about 25% of the total heat loss is through the shell, such losses can dramatically influence process efficiency, fuel efficiency, and lining life.

Materials for refractory lining are chosen based on several factors, including their stability and durability. While the main intention is to protect the steel shell, they can serve as an insulation material necessary to retain the heat within the kiln interior. At the sections near the combustion zone, conductive materials are preferred for lining so as to dissipate some heat away and prevent excessive temperature buildup and thereby avoid thermal stresses and associated refractory damage. Lining materials come in all forms including bricks and mortars of all consistency that can be spread or sprayed on surfaces. The material composition and thermophysical properties of some lining materials are presented in Table 7.1.

Figures 7.6 and 7.7 show the calculated shell losses based on refractory type and thickness. Although the actual mechanism of heat loss through a rotary kiln wall is more involved, the obvious result is that heat loss increases with conductivity and also with reduced refractory thickness. The shell heat losses also depend on the difference between the freeboard temperature and the ambient temperature, as is demonstrated in Figure 7.7.

7.7 Heat Conduction in Rotary Kiln Wall

Heat transfer in a rotary kiln wall is composed of heat loss through the freeboard gas or flame to the exposed wall and heat loss from the material being processed to the wall covered with material. Because the wall is rotating, the problem is not that of a steady state heat transfer but a cyclic equilibrium including transient responses. As shown in

Table 7.1 Composition and Thermal Properties of Some Refractory Materials Used in Rotary Kilns

	Refractory Type		
Chemical Analysis (Calcined Basis), %	High-Alumina Brick	Low-Alumina Brick	Burnt Magnesite (Basic Brick)
Silica (SiO_2)	24.6	15.2	2.7
Alumina (Al_2O_3)	70.8	78.7	10.9
Titania (TiO_2)	3.0	3.4	0.3
Iron oxide (Fe_2O_3)	1.3	1.8	1.1
Lime (CaO)	0.1	0.3	1.9
Magnesia (MgO)	0.1	0.2	83.1
Alkalies ($Na_2O + K_2O$)	0.1	0.4	—
Properties			
Bulk density (kg/m^3)	2579	2707	2835
Apparent porosity (%)	17	19.8	19.4
Crushing strength at 21 °C (N/mm^2)	45	56	22
Modulus of rupture	10	9	5
Linear change at temperature (%)	+2.0 at 1600 °C	+1.2 at 1599 °C	0 at 1482 °C
Thermal Conductivity at Temperature (W/m K)			
At 1000 °C	1.78	1.79	2.56
At 1200 °C	1.77	1.82	2.47
At 1400 °C	1.78	1.85	2.37

Figure 7.8, the kiln wall can be divided into two regions: (1) the active layer at the inner surface that undergoes a regular cyclic temperature change as the wall rotates through the freeboard and beneath the bed burden, and (2) a steady-state layer extending from the active region to the outer surface, which does not experience any temperature variation as a function of kiln rotation.

The governing equation for heat conduction in the cross section of a rotating wall can be approximated by that of transient heat conduction in spherical coordinates as (Gorog et al., 1982)

$$\frac{\partial}{\partial r}\left(k_w \frac{\partial T_W}{\partial r}\right) + \frac{k_w}{r}\frac{\partial T_W}{\partial r} + \frac{1}{r^2}\frac{\partial}{\partial \phi}\left(k_w \frac{\partial T_W}{\partial \phi}\right) + \frac{\partial}{\partial z}\left(k_w \frac{\partial T_W}{\partial z}\right) = \rho_w c_{p_w}\frac{\partial T_W}{\partial t}$$

(7.34)

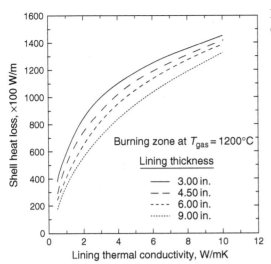

Figure 7.6 Refractory thermal conductivity and shell losses.

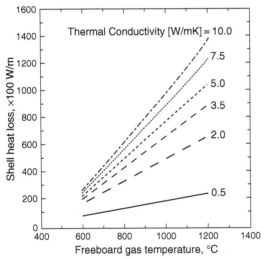

Figure 7.7 Flame temperature and shell losses.

It can be assumed that heat conduction in the longitudinal and the circumferential directions are negligible, that is,

$$\frac{\partial T_W}{\partial z} = \frac{\partial T_W}{\partial \phi} = 0 \tag{7.35}$$

For $R_i \leq r \leq R_f$, that is, in the active region

$$\frac{\partial^2 T_W}{\partial r^2} + \frac{1}{r}\frac{\partial T_W}{\partial r} = \frac{1}{\alpha_W}\frac{\partial T_W}{\partial t} \tag{7.36}$$

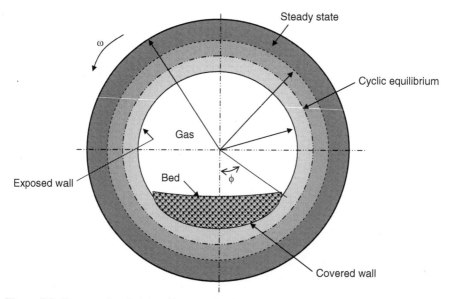

Figure 7.8 Cross-sectional view of heat conduction through rotary kiln wall.

with the following boundary conditions

1. At the inner surface of the exposed wall, for $0 \leq \phi \leq 2(\pi - \phi_L)$ and $t > 0$,

$$-k_w \frac{\partial T_W}{\partial r}\bigg|_{r=R_i} = h_{g \to ew}(T_g - T_w)$$

2. At the covered wall, for $2(\pi - \phi_L) < \phi_L < 2\pi$ and $t > 0$,

$$-k_w \frac{\partial T_W}{\partial r}\bigg|_{r=R_i} = h_{ch \to cw}(T_b - T_w)$$

3. At the interface between the two regions, for $r = R_f$, $0 \leq \phi \leq 2\pi$, and $t > 0$,

$$T_w = T_f$$

For the steady-state region, that is, $R_f \leq r \leq R_0$, Equation (7.34) reduces to

$$\frac{\partial^2 T_W}{\partial r^2} + \frac{1}{r}\frac{\partial T_W}{\partial r} = 0 \qquad (7.37)$$

With a boundary condition for $0 \leq \phi \leq 2\pi$,

$$-k_w \frac{\partial T_W}{\partial r}\bigg|_{r=R_0} = h_a(T_{sh} - T_a)$$

The solution of Equations (7.36) and (7.37) with their respective boundary conditions gives temperature distribution for the steady-state region as

$$T_{(r)} = T_f - \frac{T_f - T_a}{\ln\left(\frac{R_0}{R_f}\right) + \frac{k_w}{h_a R_0}} - \ln\left(\frac{r}{R_f}\right) \tag{7.38}$$

For the transient region the reader is referred to any heat transfer text for Heisler charts. Alternately, the transient region can be solved numerically by first guessing the location of the interface and iterating the procedure until the heat loss between the cyclic equilibrium region equals that of the steady-state region. The heat transfer coefficients include that of convection and radiation, which we will evaluate after treating radiative heat transfer.

7.8 Radiation Heat Transfer

Heat transfer by thermal radiation requires no intervening medium. Thermal radiation is the energy emitted by a body solely due to the temperature of the body and at a frequency that falls within a small portion of the electromagnetic wave spectrum, as shown in Figure 7.9.

7.8.1 The Concept of Blackbody

When radiation is incident on a homogeneous body, some of it is reflected and the remainder penetrates into the body. The radiation may then be absorbed as it travels through the medium. When the dimensions, that is, the thickness allows transfer of energy such that some of the radiation is transmitted through the body, then the transmitted radiation will emerge unchanged. If the material is a strong internal absorber, then the radiation that is not reflected from the body will be converted into internal energy within the layer near the surface. To be a good absorber for incident energy, a material must have a low surface reflectivity and sufficiently high internal absorption to prevent the radiation from passing through it. A surface is said to be radiatively "black" when it has zero surface reflection and complete internal absorption. A hypothetical blackbody is defined as an ideal body that allows all the incident radiation to pass into it (no reflected energy) and absorbs internally all the incident radiation (no transmitted energy). This is true for radiation for all wavelengths and for all angles of incidence. Hence a blackbody is a perfect absorber of incident radiation and in radiation heat transfer serves as the standard with which real absorbers can be compared. The Stefan Boltzmann law states that the total (hemispherical) emissive power of a blackbody is proportional to the fourth power of the absolute temperature

$$E_b = \sigma T^4 \left[\frac{W}{m^2}\right] \tag{7.39}$$

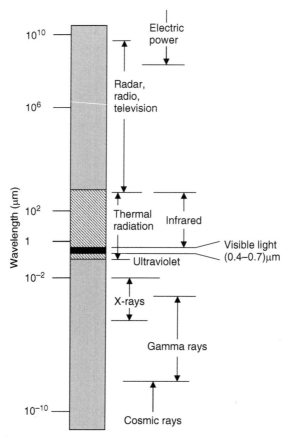

Figure 7.9 The electromagnetic wave spectrum.

where a hemispherical surface is defined as one that embodies all directions of the half space. Therefore, the total emissive power of a real surface can only be a fraction of that of the blackbody. This ratio is defined by the emissivity $\varepsilon = E/E_b$ of the body

$$E = \varepsilon E_b = \varepsilon \sigma T^4 \left[\frac{W}{m^2}\right] \quad (7.40)$$

This total emissive power includes all wavelengths within the thermal energy band. However, even a blackbody does not emit equally at all wavelengths. The blackbody emissive power at a particular wavelength is called the monochromatic emissive power and is related to the wavelength and temperature according to Plank's law as

$$E_{b,\lambda} = \frac{C_1 \lambda^{-5}}{e^{C_2/\lambda T} - 1} = f(\lambda, T) \left[\frac{W}{m^2 m}\right] \quad (7.41)$$

where $C_1 = 3.742 \times 10^{-16}$ W/m² and $C_2 = 1.439 \times 10^{-2}$ mK. Therefore plots of $E_{b,\lambda}$ indicate that as the temperature of the blackbody increases the maximum monochromatic emissive power shifts to shorter wavelengths, leading to Wein's displacement law, which states that the product of the temperature and the wavelength must be a constant, that is, $\lambda_{max} T = 2.898 \times 10^{-3}$. The relationship was derived by solving Plank's law for $\left(\frac{dE_{b,\lambda}}{d\lambda}\right)_T \to 0$. Similarly, real surfaces do not emit with equal efficiency at all wavelength hence one can define a monochromatic emissivity as $\varepsilon_\lambda = E_\lambda/E_{b,\lambda}$ leading to an idealized "gray" surface for which $\varepsilon_\lambda = \varepsilon = $ const.

7.9 Radiation Shape Factors

We have said earlier that the cylindrical enclosure of the rotary kiln maximizes radiation heat transfer in the freeboard. This is due to the fact that radiation exchange is geometry dependent. A fundamental problem in radiation heat transfer is estimating the net radiant exchange between surfaces. One can define the total energy streaming away from a surface as the *radiosity* and the energy incident on a surface as the *irradiation*. Figure 7.10 illustrates the concept whereby the surface is emitting its own energy at a rate E and is also reflecting energy incident on the surface or part of the irradiation G at a rate, ρG, where ρ is the reflectivity. The total energy leaving the surface, that is, the radiosity, J, is

$$J = E + \rho G = \varepsilon E_b + \rho G \tag{7.42}$$

where ε is the surface emissivity. If a number of surfaces i are involved in radiation exchange, each surface has a radiosity J_i. For nondiffuse surfaces, J_i will depend on the direction of the radiation as well.

One can derive the net radiative exchange between two surfaces separated by a nonabsorbing medium by considering Figure 7.11, which shows elemental areas dA_1 and dA_2 that form portions of the finite areas A_1 and A_2. If we define I_1 as the net intensity of radiation leaving surface 1 in the direction θ_1, then the total energy intercepted by surface dA_2 is

$$dE_{1 \to 2} = I_1 dA_1 \cos\theta_1 d\omega_1, \quad d\omega_1 = dA_2 \cos\theta_2/r^2 \tag{7.43}$$

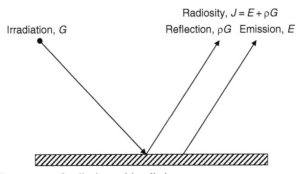

Figure 7.10 The concept of radiosity and irradiation.

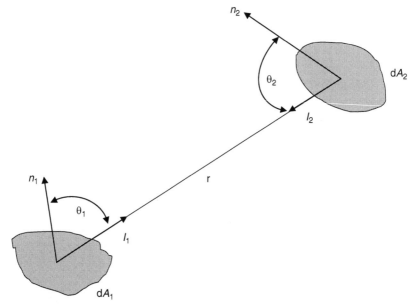

Figure 7.11 Radiation view factor.

The total energy leaving surface I is

$$dE_1 = dA_1 \left[2\pi \int_0^{\pi/2} I_1(\theta) \sin\theta (\cos\theta) \, d\theta \right] \quad (7.44)$$

The ratio of Equations (7.43) and (7.44) is what is known as the shape factor. It is also known as the geometric view factor, configuration factor, or angle factor $dF_{1 \to 2}$, and is the fraction of the total radiation leaving surface 1 that is intercepted by surface 2

$$dF_{1 \to 2} = dE_{1 \to 2}/dE_1 \quad (7.45)$$

View factor analysis is greatly simplified by assuming that both surfaces are diffuse, that is, both radiant intensities I_1 and I_2 are independent of θ_1 and θ_2, which implies that $I_1 = J_1/\pi$ and $dE_1 = J_1 dA_1$. Then Equation (7.45) becomes

$$dF_{dA_1 \to dA_2} = \frac{\cos\theta_1 \cos\theta_2}{\pi r^2} dA_2 \quad (7.46)$$

$$dF_{dA_2 \to dA_1} = \frac{\cos\theta_2 \cos\theta_1}{\pi r^2} dA_1 \quad (7.47)$$

The reader is referred to heat transfer texts for evaluation of the view factors for specific surfaces and enclosures (Siegel and Howell, 1981; White, 1988).

7.10 Radiation Exchange between Multiple Gray Surfaces

As Equation (7.45) indicates, knowledge of the view factor allows for the calculation of the net radiation between any two blackbodies as

$$Q(\text{net})_{1 \to 2} = A_1 F_{1 \to 2} \sigma \left(T_1^4 - T_2^4 \right) \tag{7.48}$$

For nonblack surfaces, the net interchange must account for the reflected radiant energies from other surfaces. For refractory material surfaces encountered in rotary kilns, perhaps one can assume that all surfaces are gray, diffuse, and opaque, that is,

$$\tau = 0, \quad \alpha = \varepsilon, \quad \rho = 1 - \varepsilon \tag{7.49}$$

Hence, knowledge of the emissivity completely characterizes the surface. Of course, nongray and nondiffuse surfaces bring several degrees of complication into the calculations. By employing the concept of radiosity and irradiation illustrated in Figure 7.10, the net thermal radiation heat exchange between surfaces i can be computed by

$$Q_i = A_i (J_i - G_i) \tag{7.50}$$

Substituting J_i from Equation (7.42) and ρ_i from Equation (7.49) gives

$$Q_i = \frac{\varepsilon_i A_i}{1 - \varepsilon_i} (E_{bi} - J_i) \tag{7.51}$$

Equation (7.51) can be interpreted either algebraically or electrically as was done for Newton's law of cooling in convection or for Fourier's law of conduction through composite materials. The electrical analogy for Equation (7.51) can be made by comparing $(E_{bi} - J_i)$ to the driving potential difference and Q_i to an electrical current.

$$Q_i = \frac{E_{bi} - J_i}{R_i} \quad R_i = (1 - \varepsilon_i)/(\varepsilon_i A_i) \tag{7.52}$$

where R_i is the radiative resistance of the gray surfaces. Similarly, the shape factor relation between surfaces in Equation (7.48) can be interpreted as

$$Q(\text{net})_{1 \to 2} = \frac{\left(T_1^4 - T_2^4 \right)}{R_{12}} \quad R_{12} = \frac{1}{A_1 F_{1 \to 2} \sigma} \tag{7.53}$$

where R_{12} is the "shape" or "view" resistance between the two surfaces. With the help of the electrical analogy, all the three heat transfer modes (conduction, convection, and radiation) can be linked into an electrical network to calculate the net heat transfer in an enclosure.

7.11 Radiative Effect of Combustion Gases

The term emissivity relates the radiant ability of a surface to the actual temperature of the surface. Because of the formation of radiatively emitting gaseous products of combustion, especially CO_2 and H_2O, all hydrocarbon flames produce significant levels of radiative heat transfer. Also, the presence of particles in a flame can enhance radiative transfer on account of emission from these particles. Flame luminosity is primarily associated with soot particle emission. Flames that do not contain particulates or do not generate significant levels of carbon by pyrolysis, for example, natural gas flames, are invisible, whereas those with high levels of carbon loading, for example, coal and oil, are clearly visible. The flame shape, color, emissivity, and temperature together with bed movement will determine the effectiveness of heat transfer from the flame to the bed.

It was mentioned earlier that over 90% of the heat transfer in the rotary kiln is by radiation. It is not surprising, therefore, that natural gas, despite its high energy content (about 44–48 MJ/kg), does not sustain heat as effectively as coal (with 23–28 MJ/kg) in kilns. Most of the constituents in liquid waste fuels used in kilns are solvents and alcohols that contain a low carbon–hydrogen ratio. This therefore produces barely visible flames, which have low radiant energy emission. It is not uncommon for the kiln operator to face a situation where the measured energy content of such fuels would be high but cannot produce a quality product even when it is possible to increase fuel flow rate within compliance limits.

In modeling the radiative heat transfer from flames, the emissivity and absorptivity of combustion gases are usually represented by the weighted gray gas mixture model first developed by Hotel (Guruz and Bac, 1981; Jenkins and Moles, 1981). This stems from the fact that gases having a dipole moment (e.g., CO_2, CO, H_2O, and hydrocarbons) are selective absorbers and emitters of radiation, that is, they absorb and emit radiation only within certain wavelengths, and this selectivity varies with the gas temperature, pressure, and the geometric shape. Gray gases absorb and emit the same fraction of energy at each wavelength. Hottel and Sarofim (1967) have shown that the emissivity and absorptivity of an equimolar CO_2–H_2O mixture, each with a partial pressure of 11.65 kPa, can be represented as a series of one "clear," where the gray gas absorption coefficient is zero ($K_1 = 0$), and three nonzero gray gases (K_2, K_3, K_4). The absorptivity and emissivity are related to the temperature and wavelength as

$$\varepsilon_g(T_g) = \sum_{n=1}^{4} a_n(T_g)[1 - \exp(-K_n pL)] \tag{7.54}$$

$$\alpha_g(T_g) = \sum_{n=1}^{4} b_n(T_s)[1 - \exp(-K_n pL)] \tag{7.55}$$

where the gray gas absorption coefficients are treated as independent of temperature but the weighting factors are temperature dependent.

7.12 Heat Transfer Coefficients for Radiation in the Freeboard of a Rotary Kiln

Several models exist for estimating the radiative heat transfer in kiln enclosures, based, in part, on some of the theoretical foundations presented herein (Barr et al., 1989; Boateng and Barr, 1996; Gorog et al., 1981, 1983; Guruz and Bac, 1981; Jenkins and Moles, 1981; and others). Almost all of these follow the zone model of Hotel and Cohen (1958) whereby models for calculating radiative exchange within enclosures are constructed by subdividing the enclosure, including the gas contained within, into numerous zones and then formulating the expressions for radiative exchange among the zones. What is different among the several radiation models presented by the various authors is the tools used for the solution of the large number of algebraic equations that result from the large number of zones involved in practical combustion chambers, for example, large rotary kilns. While the early researchers used a statistical solution approach such as the Monte Carlo method to simulate the radiative interchange (e.g., Guruz and Bac, 1981), the advent of high-speed computers has made it possible to apply such models through their inclusion in CFD modeling. We will follow a model by Barr et al. (1989) and Boateng (1993) to illustrate the estimation of the radiative heat transfer coefficients in the rotary kiln enclosure.

The radiant heat transfer between the freeboard gas and the kiln enclosure occurs by the exchange mechanism shown in Figure 7.12.

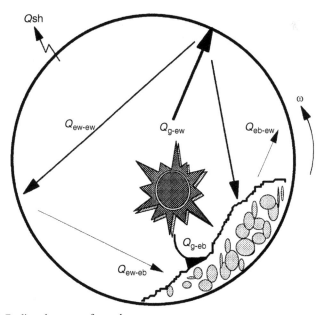

Figure 7.12 Radiant heat transfer paths.

Radiant heat is transferred either from the gas volume composed of emission products of combustion, that is, CO_2 and H_2O, or by particulate emission in the freeboard region to the bed surface, $Q_{g\text{-eb}}$, or to the exposed wall, $Q_{g\text{-ew}}$. The wall (refractory lining) interacts radiatively with the bed with radiative heat transfer, $Q_{eb\text{-ew}}$. In addition, some sections of the wall do exchange radiant energy with other sections of the wall once the enclosure is such that they see each other. Only a portion of the energy absorbed by the inside wall (refractory) surface during exposure to freeboard gas is lost through the outer shell (by conduction) of the refractory lining. The remainder is transferred to the bottom of the bed (regenerative heat transfer).

Calculations of radiative exchanges are carried out by tracing individual beams of radiation from their source to the surface of interest. The circumferential strip of the kiln freeboard (Figure 7.13) is isolated and subdivided into N small area elements. The portion of the kiln freeboard, approximately three kiln inside diameters in each axial direction viewed by these area elements, is subdivided into volume zones such as V_j composed of the freeboard gas, and the surface zones such as A_j that form the exposed bed and wall surfaces. The exchange areas between each of the zones and the elements of the circumferential strip are then evaluated and used to calculate the radiation streaming to these elements from the various volumes and surface zones, either directly or after undergoing one or two reflections.

7.13 Radiative Exchange from the Freeboard Gas to Exposed Bed and Wall Surfaces

Radiative exchange between elements A_i (exposed bed) or refractory wall and the gas volume V_j is calculated by using the expression for gray surfaces.

$$Q_{V_j \leftrightarrow A_i} \approx \frac{\varepsilon_i + 1}{2} \sum_{n=1}^{4} (g_j s_i)_n (E_j - E_i)_n \qquad (7.56)$$

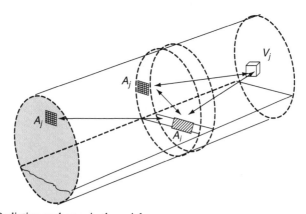

Figure 7.13 Radiative exchange in the axial zones.

Equation (7.56) accounts for the irradiation of the surface elements by the contribution of each component of the gas emissivity. Barr et al. (1989) found that for a rotary pilot kiln of about 40 cm diameter and 3.5 m length, more that 80% of reflected gas radiation leaving a surface was reabsorbed by the freeboard gas without impinging on another surface. Hence the effect of reflections could be ignored to simplify the calculations. The freeboard viewed by elements A_i can be considered as two symmetric regions, one upstream and the other downstream. The geometric exchange areas $g_j s_i$ can be calculated for one region and advantage of the symmetry can be taken to further simplify the model. The net radiative heat transfer computed by summing up Equation (7.56) over the subdivided volume zones is

$$Q_{g \leftrightarrow A_i} = \frac{\varepsilon_i + 1}{2} \sum_{j=1}^{N} \sum_{n=1}^{4} (g_j s_i)_n (E_j - E_i)_n \qquad (7.57)$$

where N is the number of zones or volume elements typically greater than 300. Like any computational domain, the larger the number of divisions N, the better the accuracy of the computations.

7.14 Radiative Heat Transfer among Exposed Freeboard Surfaces

The direct radiation exchange between a freeboard surface zone such as A_j and an area element A_i, that is, exposed wall to exposed bed, on the circumferential strip may be calculated using the expression

$$Q^0_{A_j \leftrightarrow A} = \sum_{n=1}^{4} (s_j s_i)_n (\varepsilon_j \varepsilon_i E_j - \varepsilon_i \varepsilon_j E_i)_n \qquad (7.58)$$

The energy streaming between A_i and A_j must traverse the intervening freeboard gas, hence the view factors $(s_j s_i)$ must be computed first. For surface emissivities less than 0.9, which is the case for refractory and bed surfaces, reflectivity can be significant. To account for the reflected energy in the interchange one can simplify the calculation by assuming a single reflection from the intermediate surfaces or multiple reflections depending on the accuracy of interest. For a single reflection depicted in Figure 7.14, the net radiative exchange between the element A_i and zone A_j involving a single reflection from an intermediate surface A_k can be expressed as

$$Q_{A_j \leftrightarrow A_k \leftrightarrow A_i} = \sum_{n=1}^{4} \varepsilon_i \varepsilon_j \rho_k (s_j s_k)_n \frac{(s_k s_i)_n}{A_k} (E_j - E_i)_n \qquad (7.59)$$

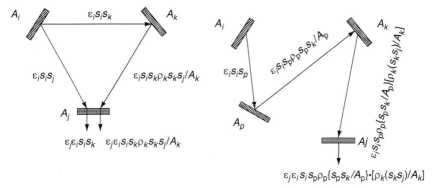

Figure 7.14 Radiative exchange with reflectivity.

By summing over all the reflecting surfaces, an expression for the radiative exchange between A_i and A_j involving a single reflection is obtained as

$$Q^1_{A_j \leftrightarrow A_i} = \sum_{\text{all}-k} Q_{A_j \leftrightarrow A_k \leftrightarrow A_i} \qquad (7.60)$$

Extending the analysis to include an additional reflection will yield a net radiative exchange between A_i and A_j as

$$Q_{A_j \leftrightarrow A_i} = Q^0_{A_j \leftrightarrow A} + Q^1_{A_j \leftrightarrow A_i} + Q^2_{A_j \leftrightarrow A_i} \qquad (7.61)$$

Barr et al. (1989) found that the error introduced by accounting for only two reflections was only 3% in pilot rotary kilns except that the lining material reduces the emissivity to below 0.5. For large rotary kilns, an emissivity of 0.7 is typical for the refractory walls.

For the process or kiln engineer perhaps the important application is to develop a catalog of radiative heat transfer coefficients, which can be combined with convection heat transfer coefficients to estimate the kiln heat balance. Cognizant of the fact that the radiative exchange between an area element and any zone in the freeboard is a function of temperature, pressure, and geometry, the radiative heat transfer coefficient can be cast into the form

$$h_{V_j - A_i} = \frac{Q_{V_j \leftrightarrow A}}{A_i(T_j - T_i)} \qquad (7.62)$$

Barr et al. (1989) employed the equations presented herein to assemble the radiative heat transfer coefficients for the entire freeboard gas, that is, $h_{g \to ewi}$ and $h_{g \to ebi}$; the entire exposed bed surface, $h_{eb \to ewi}$; and the entire exposed wall, $h_{ew \to ewi}$. Some radiative heat transfer coefficients from the freeboard gas to the participating surfaces are presented in Figures 7.15–7.17 for a 41-cm-diameter pilot kiln (Barr et al., 1989).

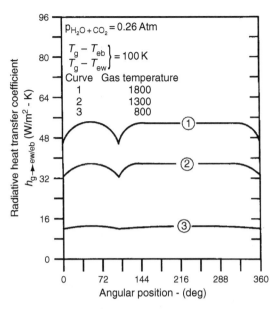

Figure 7.15 Predicted radiation heat transfer coefficients from freeboard gas to the exposed wall for a 41-cm-diameter pilot kiln (Barr et al., 1989).

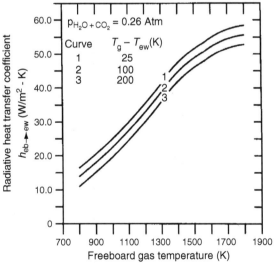

Figure 7.16 Radiative heat transfer coefficient from exposed bed to exposed wall as a function of gas-to-wall temperature difference (Barr et al., 1989).

These show that except near the corners formed by the exposed bed and wall, both $h_{g \to ewi}$ and $h_{g \to ebi}$ are nearly uniform and for all practical reasons can be replaced by one averaged value for both coefficients. As implied by Equations (7.54) and (7.55), $h_{eb \to ew}$ varies with the partial pressure and with the temperature difference between the gas and the participating surface, that is, the driving potential as per the electrical

Figure 7.17 Radiative heat transfer coefficient from exposed bed to exposed wall as a function of emitting gas partial pressure (Barr et al., 1989).

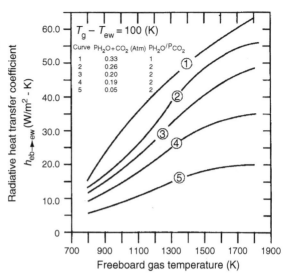

analogy. Calculations show that for larger kilns, for example, 4-m-diameter kilns at the same operating conditions, these coefficients can be an order of magnitude greater than for the pilot kilns essentially due to the geometric view factors. What can also be deduced from the plots is the effect of freeboard gas temperature, T_g, on the magnitude of the exposed wall heat transfer coefficient, that is, h_{ew}, increases with increased freeboard gas temperature. The regenerative action of the wall may see a temperature variation at the inside face showing that as the kiln rotates, the wall will pick up energy from the freeboard gas and will give it to the bed at the covered wall. Aside from the regenerative effect, the thickness of the wall active region depends on the freeboard gas temperature. The modeling of the bed heat transfer will be covered in detail in Chapter 8.

References

Barr, P.V., Brimacombe, J.K., Watkinson, A.P., 1989. A heat-transfer model for the rotary kiln: Part II. Development of the cross-section model. Met. Trans. B 20, 403–419.

Boateng, A.A., 1993. Rotary Kiln Transport Phenomena: Study of the Bed Motion and Heat Transfer (Ph.D. Dissertation). The University of British Columbia, Vancouver.

Boateng, A.A., Barr, P.V., 1996. A thermal model for the rotary kiln including heat transfer within the bed. Int. J. Heat Mass Transfer 39 (10), 2131–2147.

Gorog, J.P., Brimacombe, J.K., Adams, T.N., 1981. Radiative heat transfer in rotary kilns. Met. Trans. B 12, 55–69.

Gorog, J.P., Adams, T.N., Brimacombe, J.K., 1982. Regenerative heat transfer in rotary kilns. Met. Trans. B 13 (2), 153–163.

Gorog, J.P., Adams, T.N., Brimacombe, J.K., 1983. Heat transfer from flames in a rotary kiln. Met. Trans. B 14 (3), 411–423.

Guruz, H.K., Bac, N., 1981. Mathematical modelling of rotary cement kilns by the zone method. Can. J. Chem. Eng. 59, 540–548.

Hottel, H.C., Sarofim, A., 1967. Radiative Heat Transfer. McGraw-Hill, New York.

Hottel, H.C., Cohen, E.S., 1958. Radiant heat exchange in a gas-filled enclosure: allowance for nonuniformity of gas temperature. AIChE J. 4 (1), 3–14.

Jenkins, B.G., Moles, F.D., 1981. Modelling of heat transfer from a large enclosed flame in a rotary kiln. Trans. IChemE 59, 17–25.

Siegel, R., Howell, J.R., 1981. Thermal Radiation Heat Transfer. McGraw-Hill, New York.

White, F., 1988. Heat and Mass Transfer. Addison-Wesley, New York.

Heat Transfer Processes in the Rotary Kiln Bed

Thus far we have evaluated the freeboard heat transfer and quantified the radiation heat transfer to the walls and the exposed bed. The combined radiation and convection heat transfer is the total heat input into the bed. Perhaps the most important concern is what happens to the energy after it is directed to the bed. A truly complete evaluation must include the subsequent distribution of energy within the bed where the temperature-dependent process that we intend to make happen takes place, at least for materials processing. We saw from the computational fluid dynamics (CFD) modeling of pulverized fuel combustion that the temperature in the flame region, 1 m around the centerline was not dramatically different from the centerline temperature (cf. Figure 6.15). The assumption that the freeboard is well mixed has previously been applied to rotary kilns and has allowed the use of one-dimensional modeling of the axial temperature there (Sass, 1967; Tscheng and Watkinson, 1979; and others). In fact the International Flame Research Foundation (IFRF) cement flame (CEMFLAM) modeling for rotary kiln flames that drew on the work of Jenkins and Moles (1981) was based on one-dimensional representation. By assuming that the bed is also well mixed and axially moves in plug flow, the axial gradients of bed temperature and gas temperature could be related to the local rates of gas-to-exposed bed and wall-to-covered bed heat transfer by ordinary differential equations (Sass, 1967). This allows one to establish a representative "averaged" bed temperature at each axial location. One of the first early representations of one-dimensional modeling was by Sass (1967).

$$\frac{dT_s}{dz} = \frac{1}{C_{ps}G_s}[\alpha_2(T_g - T_s) + \alpha_3(T_w - T_s)] \tag{8.1}$$

$$\frac{dT_g}{dz} = \frac{1}{C_{pg}G_g}[\alpha_2(T_g - T_s) + \alpha_1(T_g - T_w)] \tag{8.2}$$

where $\alpha_i = h_i A_i$, the product of the heat transfer coefficient h of the interface with area A; C_{pg} and C_{ps} are the specific heats of gas and solid, respectively; T is the local thermodynamic temperature; and G is the mass flow rate.

These equations formed the basis for the various one-dimensional kiln models that have appeared in the literature (Brimacombe and Watkinson, 1979; Wes et al., 1976; and others). In these models an energy balance on the wall must be included, as well as the kinetic expressions for any reactions. The latter led to a set of mass conservation expressions that must be solved along with the energy equation. For example, if the evaporation of free moisture is controlled by heat transfer, rather than by mass transfer, an additional thermal balance on the moisture can be included as

$$\frac{dW}{dz} = \frac{1}{G_1\lambda}[\alpha_2(T_g - T_s) + \alpha_3(T_w - T_s)] \tag{8.3}$$

$$\lambda = \lambda_0 + C_v(T_g - T_s) \tag{8.4}$$

where W is the mass of water, λ_0 is the latent heat of liquid water, and C_v is the specific heat of water vapor.

Such models did not take into account the distribution of energy and, for that matter, temperature differences within the bed as a result of the process chemistry, bed segregation, and so on that become manifest in product quality issues. To address these one has to consider how heat is transferred to the bed, distributed, and later extracted. We saw in Chapter 7 how the heat transfer to the bed can be evaluated and how the wall takes heat from the freeboard and dumps it to the bottom of the bed (cf. Figures 7.16 and 7.17). In this chapter we will evaluate the heat transfer mechanisms between the wall and the particle bed and the redistribution of this heat and the heat from the free surface within the particle bed. These heat transfer mechanisms are the same as that used for packed beds except that there is an effect of flow induced by the rotating wall and the flow of granules within the bed.

8.1 Heat Transfer between the Covered Wall and the Bed

Various estimated heat transfer coefficients have been employed to calculate the covered wall–bed exchange. Evaluation of this coefficient has generally been either by pure guesswork or by adopting some type of surface renewal or penetration model. Typical guess-estimated values for the wall–solid heat transfer coefficient lie within the range of 50–100 W/m² K (Gorog et al., 1982) which might, perhaps, depend on the kiln conditions. However, the problem at hand can be equated to that of heat transfer between a heated surface and a flowing granular material where the influence of gas within the bed might have an effect. In the absence of chemical reactions such a process might represent a kind of moving pebble–bed heat exchanger in which particles form an extended surface. One of the early models employed for the situation was based on the surface renewal-penetration theory (Wes et al., 1976) and modifications thereof (Sullivan and Sabersky, 1975). In such a case, heat conduction between a wall and granular material might be calculated using the equation for a one-dimensional unsteady-state conduction as follows:

$$\frac{\partial T}{\partial t} = \alpha \frac{\partial^2 T}{\partial y^2} \tag{8.5}$$

Implicit in Equation (8.5) is that the granular material might be treated as a continuum having average or effective thermophysical properties from which the heat transfer coefficient at the wall might be calculated as

$$h(x) = \sqrt{\frac{2nR}{x}(k\rho C_p)_{\text{eff}}} \tag{8.6}$$

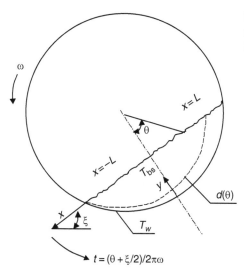

Figure 8.1 Wall-to-bed heat transfer—the modified penetration model (Ferron and Singh, 1991).

where $x = 2\pi nRt$ is the distance traveled by a particle from the lower edge along the circumference of the kiln, R is the kiln radius, and n is the kiln revolution. Since these models appeared, several modifications and variations have manifested, some very complex (Barr et al., 1989; Gorog et al., 1982; Lehmberg et al., 1977). We will follow the modified penetration model for the wall-to-bed heat transfer (Figure 8.1) developed by Ferron and Singh (1991) below.

8.2 Modified Penetration Model for Rotary Kiln Wall-to-Bed Heat Transfer

Figure 8.1 assumes that the kiln radius is large compared to the penetration depth, $d(\theta)$, that develops during contact of the heated wall coming from the exposed wall with the bed. We can set the initial conditions (ICs) and boundary conditions (BCs) for the solution of a two-dimensional transient heat transfer equation, similar to Equation (8.5), using the time variable in terms of the rotation as $t = (\theta + \xi/2)/2\pi\omega$, which is zero at the instant of initial wall-to-bed contact. The final time is $t = \xi/2\pi\omega$ when the wall exits the covered bed. Between the time of initial contact and the maximum time, $y = R - r$, where R is the kiln radius and r is the radius at the penetration depth, that is,

$$\text{for} \quad 0 \leq t \leq \xi/2\pi\omega, \quad y = R - r \tag{8.7}$$

By replacing the time variable with the subtended angle θ we can rewrite Equation (8.5) and its BCs to include the degree of fill, which is in essence a function of θ.

$$2\pi\omega \frac{\partial T_b}{\partial \theta} = \alpha_b \frac{\partial^2 T_b}{\partial y^2} \tag{8.8}$$

$$T_b(0, \theta) = T_w \tag{8.9a}$$

$$T_b\left[R - (R^2 - L^2)^{1/2}, \theta\right] = T_{bs} \tag{8.9b}$$

$$T_b[y, \theta_0(y)] = T^-\left[-(y^2 - 2Ry + L^2)^{1/2}, \infty\right] \tag{8.9c}$$

where $t \to \infty$ is the steady-state condition, that is, when $T_b \to T_{bs}$ and

$$\theta_0(y) = -\cos^{-1}\frac{(R^2 - L^2)^{1/2}}{(R - y)} \quad \text{for} \quad 0 \leq y \leq R - (R^2 - L^2)^{1/2} \tag{8.9d}$$

The first BC, Equation (8.9a), implies that we are assuming that the wall and bed temperatures are equal at the initial point of contact. Also, the mean free path for the gas at the contact wall is sufficiently large such that convective heat exchange by the gas is not in local equilibrium with the conduction through the bed. However, radiative heat transfer can play a vital role within the penetration layer (Ferron and Singh, 1991). As we did for the freeboard, the most practical approach is not only to solve the differential equation but to establish a heat transfer coefficient that can be used for practical calculations. The heat transfer coefficient per unit contact area may be written in terms of the overall heat balance using Newton's law of cooling,

$$h_{cw-cb} = \frac{-2LQ}{\xi R(T_w - T_{bs})} \tag{8.10}$$

Ferron and Singh (1991) solved the governing equations using the dimensionless temperature distribution

$$\Phi(y, \theta) = \frac{[T_b(y, \theta) - T_w]}{(T_{bs} - T_w)} \tag{8.11}$$

and found a solution to the transient equation after several mathematical manipulations as

$$\Phi(y, \theta) = erf\frac{\pi^{1/2}\zeta}{2\Delta} + 2S\sum_{j=1}^{\infty} B_j(\xi)\exp(-j^2\pi\zeta^2)\sin(j\pi\zeta) \tag{8.12}$$

for $R \gg d(\theta)$ and where

$$\zeta = \frac{y}{R - (R^2 - L^2)^{1/2}} \tag{8.13a}$$

$$\Delta(\theta) = \frac{d(\theta)}{R - (R^2 - L^2)^{1/2}} \tag{8.13b}$$

and

$$B_j(\xi) = \int_0^1 \sin(j\pi\zeta)\ln\left[\frac{L^*/L + \chi(\zeta)}{L^*/L - \chi(\zeta)}\right]d\zeta \tag{8.14a}$$

Here, we have used a case in which the heat capacity of the bed, mc, is constant and the heat delivered to the free surface is independent of position, then

$$\frac{L^*}{L} = \left(1 + 8.38\frac{\omega R}{V_m}\cos\frac{\xi}{2}\right)^{1/2} \tag{8.14b}$$

where V_m is the free surface velocity.

$$\chi(\zeta) = \frac{a_2(\zeta^2 - 2\zeta/a_2 + a_1/a_2)}{\sin(\xi/2)}, \tag{8.14c}$$

with $a_1 = 1 + \cos(\xi/2)$, and $a_1 = 1 - \cos(\xi/2)$.

By setting the heat balance as

$$Rk_b(T_w - T_{bs})\int_{-\xi/2}^{\xi/2}\left(\frac{\partial \Phi}{\partial y}\right)_{y=0} d\theta = h_{cw-cb}R\xi(T_w - T_{bs}) \tag{8.15}$$

recognizing that Nusselt number, $\text{Nu} = h_{cw-cb}R\xi/k_b$ and $\text{Pe} = P^2\xi\omega/\alpha_b$

$$\text{Nu} = \frac{2\sqrt{2\text{Pe}}}{1 + \sum_{j=1}^{\infty}\left[\frac{B_j(\xi)}{\pi a_1}\right]\left[1 - \exp\left(\frac{-j^2\pi\xi^2}{2a_2^2\text{Pe}}\right)\right]} \tag{8.16}$$

If the Peclet number is large, the sum of the denominator of Equation (8.16) does not contribute much and the heat transfer coefficient reduces to

$$\text{Nu} = 2\sqrt{2\text{Pe}} \tag{8.17}$$

Other relationships based on conventional penetration theory for packed bed have deduced that $\text{Nu} = 2^{3/2}\sqrt{\text{Pe}}$, and for low Peclet numbers to the order of 10^4, Tscheng and Watkinson (1979) deduced an empirical correlation, where $\text{Nu} = 11.6 \times \text{Pe}^{0.3}$. Any of these would suffice in estimating the wall-to-bed heat transfer coefficient as functions of the kiln's rotational speed, ω, and the dynamic angle of repose, ξ. The calculated values of Nusselt numbers using Ferron and Singh's analytical deduction (Ferron and Singh, 1991), the classical penetration theory, and the correlation equation of Tscheng and Watkinson (1979) have been compared in Figure 8.2 (Ferron and Singh, 1991).

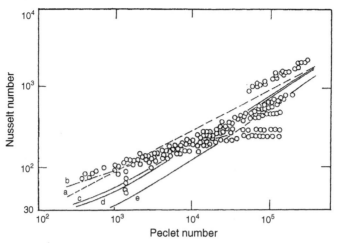

Figure 8.2 Wall-to-bed heat transfer (a) $Nu = 2^{3/2}\sqrt{Pe}$; (b) $Nu = 11.6 \times Pe^{0.3}$; (c) $Nu = 2\sqrt{2Pe}$ at $\omega R/V_m = 0.05$ and $\xi \leq 110°$; (d) $\omega R/V_m = 0.05$ and $\xi \leq 140°$; (e) $\omega R/V_m = 0.05$ and $\xi \leq 180°$ (Ferron and Singh, 1991).

8.3 Effective Thermal Conductivity of Packed Beds

Heat transfer in the bed of a rotary kiln is similar to heat transfer in packed beds except that in addition to the heat flow in the particle assemblage of the static structure (Figure 8.3), there is an additional contribution of energy transfer as a result of advection of the bed material itself. The effective thermal conductivity of packed beds can be modeled in terms of thermal resistances or conductance within the particle ensemble. As shown in Figure 8.3 almost all the modes of heat transfer occurs within the

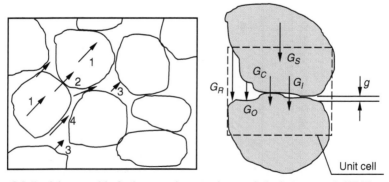

Figure 8.3 Particle ensemble for heat conductance in a static bed structure. 1: Within-particle conduction; 2: particle–particle conduction; 3: particle–particle and gas–particle radiation; 4: gas convection.

ensemble, that is, particle-to-particle conduction and radiation heat transfer as well as convection through the interstitial gas depending upon the size distribution of the material and process temperature. Several models are available in the literature for estimating the effective thermal conductivity of packed beds.

We saw in Chapter 7 that the heat flow between a temperature difference ΔT across a planar gap of Δx and area A is $Q = kA\Delta T/\Delta x = UA\Delta T = G \cdot \Delta T$, where G is the conductance. If we consider the unit cell in any spherical packing as shown in Figure 8.3, it is evident that the overall conductance can be expressed as a combination of series and parallel conductances including G_S, the conductance across the solid sphere; G_C, conductance through the points of direct contact between the two spheroids; G_R, the radiation component of the heat conductance; G_I, heat flow through the inner gap of average width, g; and the conductance, G_O, for the gas in the outer gap of width of the order of the radius of the spheroid. If T_1 and T_2 are the midpoint temperatures of the two spheres we can define the effective thermal conductivity over the entire cell using (Slavin et al., 2000; Siu and Lee, 2000)

$$Q = k_{\text{eff}} 4R^2 \frac{(T_1 - T_2)}{2\alpha_e R} = G(T_1 - T_2) \tag{8.18}$$

where R is the radius of the sphere and α_e is a geometrical correction factor for the conductances. Using the electrical conductance analogy, it is evident from the flow paths in Figure 8.3 that the conductance G_S is in series with the parallel combination of G_O, G_I, and G_C, which is in parallel with G_R, that is,

$$G = G_R + \frac{G_S(G_O + G_I + G_C)}{G_S + (G_O + G_I + G_C)} \tag{8.19}$$

It is worth mentioning that in a vacuum, G_O and G_I are 0 and since $G_S \gg G_C$, Equation (8.19) becomes $G = G_R + G_C$ (Slavin et al., 2000). The use of this equation is for all practical purposes sufficient. The radiant heat transfer coefficient can be expressed in terms of the average temperature, $\overline{T}$ (Wakao, 1973)

$$h_R = \frac{4\sigma \overline{T}^3}{2\left(\frac{1}{\varepsilon} - 1\right) + \frac{2}{1+F_{12}}} \tag{8.20}$$

where σ is the Stefan–Boltzmann constant, ε is the emissivity, and F_{12} is the view factor, which was calculated as a simple cubic array by Wakao (1973) as $F_{12} = 0.151$. If the void is filled with gray gas at uniform temperature, where several gas emissivities in the void can be replaced by an average emissivity value, ε_g, then the heat transfer coefficient becomes

$$h_R^* = \frac{4\sigma \overline{T}^3}{2\left(\frac{1}{\varepsilon} - 1\right) + \frac{2}{1+F_{12}(1-\varepsilon_g)}} \tag{8.21}$$

For solid particles larger than 0.5 mm, that is, nonradiating gas systems, the effective thermal conductivity at atmospheric pressure is little affected by solid–solid conductivity. The Nusselt number for radiation is defined as

$$\mathrm{Nu} = \frac{h_R d p}{k_s} \tag{8.22}$$

$$k_{\mathrm{eff}} = (k_e)_{\mathrm{COND}} + (k_e)_{\mathrm{RAD}} \tag{8.23}$$

from which the effective thermal conductivity of the unit cell can be established from Figure 8.4 knowing the thermal conductivity of the solid, k_s, and the fluid medium, k_f.

The radiative heat factor has been calculated using a cubic array packing with a void fraction $e = (0.5 - 0.35)$. The effective thermal conductivity involving the void fraction, which is frequently used in estimating the bed heat transfer, is expressed as (Schotte, 1960)

$$K_{\mathrm{eff}}^r = \frac{1-e}{\frac{1}{k_s} + \frac{1}{4\sigma\epsilon d_p T^3}} + e 4\sigma\epsilon d_p T^3 \tag{8.24}$$

For most applications, Equation (8.24) may be adequate since the advantages of more complex models have not been extensively verified. The porosities for

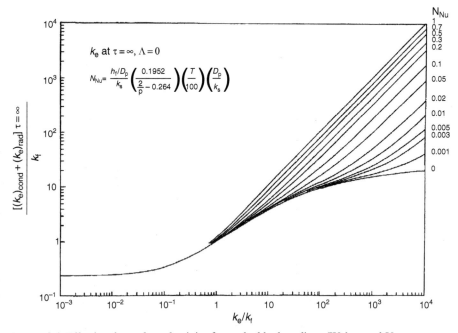

Figure 8.4 Effective thermal conductivity for packed bed medium (Wakao and Vortmeyer, 1971).

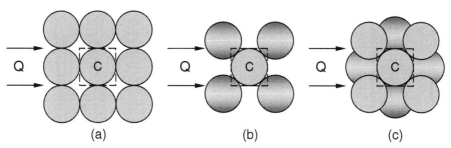

Figure 8.5 Packing arrangements for different porosity ranges. (a) Simple cubic, $N_C = 6 (e = 0.5 - 0.35)$. (b) Body-centered cubic, $N_C = 8 (e = 0.3 - 0.25)$. (c) Face-centered cubic, $N_C = 12 (e \leq 0.2)$.

body-centered cubic array and face-centered cubic array are in the range of 0.3–0.25 and 0.2, respectively (Figure 8.5).

8.4 Effective Thermal Conductivity in Rotating Bed Mode

While the effective thermal conductivity for the packed bed applies to the bulk of the bed mainly in the plug flow region, the effective conductance is enhanced by a self-diffusion coefficient component established by the bed motion. For good mixing, this component can increase the effective heat conductance by 10-fold (Boateng, 1998).

$$k'_{\text{eff}} = k_{\text{eff}} + \rho c_p \tilde{D}_y \quad \text{or} \quad \alpha_{\text{eff}} = \alpha_e + D_y \tag{8.25}$$

where $\alpha_e = f(T_b, v, d_p)$ and $D_y = f(\tilde{T}, v, d_p)$, and with the solid fraction $v = 1 - e$ and $D_y = \rho c_p \tilde{D}_y$. $\tilde{D}_y$ is the mass diffusion coefficient calculated in the flow model using the granular temperature. With these values, the modified effective thermal conductivity for a flowing active layer might be expressed as

$$K_{\text{eff}} = \frac{1-e}{\frac{1}{k_s} + \frac{1}{4\sigma e d_p T^3}} + \varepsilon 4\sigma \varepsilon d_p T^3 + \frac{\rho c_p}{9\sqrt{\pi}} \frac{d_p}{(1-e)g_0} \tilde{T}^{1/2} \tag{8.26}$$

It is evident from these relationships why it is preferable to feed a kiln with larger size particles as in the limestone calcination process so as to take advantage of the radiation effect enabled by the interparticle spacing. Through Equation (8.26), the effect of mixing on effective thermal conductance is achieved through two thermal diffusivities. For kiln control purposes, we can recast Equation (8.25) in the form

$$k_{\text{eff}} = \frac{k_e}{1 - \text{Le}'} \tag{8.27}$$

where $Le' = \rho c_p D_y / k_{\text{eff}}$ is a modified dimensionless Lewis number bounded by $0 \leq Le' \ll 1$. For the no-flow condition, that is, as $Le' \to 0$, the effective thermal conductivity is only by packed bed. As flow is induced by rotation, superimposed on the convective transport will be particle velocity fluctuations defined by granular temperature which, like thermodynamic temperature, will enhance granular conduction through small-scale particulate mixing (Boateng, 1998).

8.5 Thermal Modeling of Rotary Kiln Processes

In Chapter 6, we considered the combustion of fuel and discussed some results of combustion modeling using CFD. In Chapter 7 we discussed heat transfer in the freeboard and the type of radiation models that go into modeling combustion and establishing freeboard temperature profiles. Having now established the physical models for heat conduction through the bed, it is appropriate, at this time, to combine these models into a modeling process to establish the temperature and heat distribution in a rotary kiln system. Obviously, one can incorporate all these into a commercial CFD code through the help of some user-defined functions. Modeling the entire rotary kiln from the feed end to the discharge end will require the discretization of the entire freeboard space, bed, refractory walls, and so on, which not only is expensive and time-consuming but also unnecessary. We saw in Chapter 6 (Figure 6.15) that the temperature profiles for the centerline and the surroundings are almost the same especially when freeboard gas mixing is optimized. This result means that we can perhaps represent the freeboard temperature distribution by a single temperature profile with little error in the calculations. It is rather the bed temperature nonuniformity that determines product quality. So we can construct a combination of one-dimensional freeboard model and use that to estimate a two-dimensional bed temperature distribution so that the temperature-dependent chemical/physical reactions can be evaluated. Doing so will provide an opportunity to evaluate the effect of the flow behavior of the bed material as well as particulate mixing and segregation on the bed process. The objective of this section, therefore, is to describe the development of a thermal model for a transverse section of bed material and incorporate this two-dimensional representation of the bed into a conventional one-dimensional, plug flow-type thermal model for the rotary kiln. The resultant quasi-three-dimensional thermal model is used to examine the role of the various mechanisms for heat transfer within the cross section, for example, the regenerative action of the wall and the effect of the active layer of the bed on the redistribution of energy within the bed without resorting to CFD.

8.6 Description of the Thermal Model

A quasi-three-dimensional bed model is envisioned as one that comprises both an axial model (one-dimensional model) and a cross-sectional model (two-dimensional model). The former is used to independently determine the one-dimensional axial temperature profiles for the freeboard gas and the bulk bed. It is implicitly assumed that the details of the energy redistribution that occurs within the bed do not significantly influence heat transfer between the bed and the freeboard. As part of the procedure for calculating these axial temperature profiles, the surface heat flux to the bed is determined from the freeboard heat transfer and this becomes the thermal BC employed to drive the cross-sectional model. In doing so, the bed temperature gradient computed from the axial model is used as a sink term that represents the rate of energy removal due to the material flow in the axial direction. The one-dimensional bed temperature is also employed as a check on the mass-averaged temperature, which is estimated from the bed cross-sectional model. The two-dimensional model is employed to determine the thermal condition of the bed material and the kiln wall over successive transverse sections (or slices) of the kiln and incorporates both the flow and segregation models developed in the previous chapters. The interaction of the heat transfer model with other models in the global setup for heat and mass transport calculations in the rotary kiln can be envisioned as depicted in Figure 8.6.

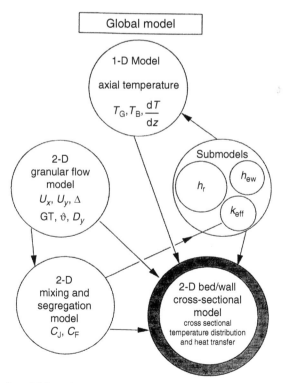

Figure 8.6 Global model layout.

8.7 One-Dimensional Thermal Model for Bed and Freeboard

Various models for the rotary kiln (Barr et al., 1989; Sass, 1967; Tscheng and Watkinson, 1979; and others) have the capability of predicting average conditions within both the bed and the freeboard as functions of axial position. The thermal component of these one-dimensional models can be derived by considering the transverse slice (Figure 8.7(a)), which divides the section into separate control volumes of freeboard gas and bed material. Under steady-state conditions energy conservation for any control volume requires that

$$\dot{Q}_{\text{NET}} = \sum (\dot{n}H)_{\text{out}} - \sum (\dot{n}H)_{\text{in}} \tag{8.28}$$

If conditions in the freeboard and bed are each assumed to be uniform in the transverse plane (the plug flow assumption), application of Equation (8.28) to the control volume of freeboard gas and bed material, respectively, yields a pair of ordinary

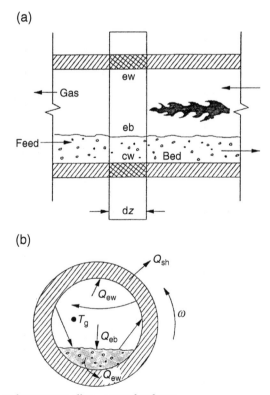

Figure 8.7 Axial and transverse slice—control volume.

differential equations relating axial gradients of temperature and composition to the net rates of heat transfer for each control volume

$$\dot{Q}_g = \sum_{i=1}^{N_g} \left[\dot{n}_{gi} c_{p_{gi}} \frac{dT_g}{dz} + H_i \frac{d\dot{n}_i}{dz} \right] \tag{8.29a}$$

$$\dot{Q}_b = \sum_{i=1}^{N_b} \left[\dot{n}_{bj} c_{p_{bj}} \frac{dT_b}{dz} + H_j \frac{d\dot{n}_j}{dz} \right] \tag{8.29b}$$

where N represents the total number of species in each region and T is the average or bulk temperature at that axial position.

In the absence of any chemical reaction or phase transformation, these equations simplify to

$$\sum \dot{m}_g c_{p_g} \frac{dT_g}{dz} = Q_{g \to ew} + Q_{g \to eb} \tag{8.30a}$$

$$\sum \dot{m}_b c_{p_b} \frac{dT_b}{dz} = Q_{g \to eb} + Q_{ew \to eb} + Q_{cw \to cb} \tag{8.30b}$$

where the various heat transfer paths are shown in Figure 8.7(b). One additional condition that must be met is that no net energy accumulation can occur within the wall. This yields an auxiliary condition

$$Q_{g \to ew} + Q_{eb \to ew} + Q_{cb \to cw} = Q_{shell} \tag{8.30c}$$

The system of equations, Equations (8.30a–c), can be solved for successive axial positions by any of a variety of techniques (e.g., Runge–Kutta) provided that the various heat transfer terms are characterized in terms of the local gas, bed, and wall temperatures. Thus, by starting at either end of the kiln, a complete solution of the thermal problem can be developed. It is chiefly the methodology employed in evaluating the heat transfer terms that distinguishes the various one-dimensional models.

Heat transfer at the interfacial surfaces is complex and involves radiation, convection, and at the covered bed–covered wall interface, conduction as well. Although a heat transfer coefficient can be allocated to each transport path shown in Figure 8.8

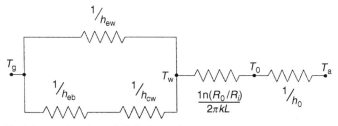

Figure 8.8 Heat transport paths and their thermal resistances.

(Gorog et al., 1983), this should not obscure the difficulty associated with realistic determination of values for these coefficients. As mentioned earlier, the one-dimensional model is required only to produce a framework from which to operate the two-dimensional thermal model for the bed and therefore existing models can be used to evaluate heat transfer at the interfaces. In the freeboard, the radiation model in Chapter 7 is utilized to develop coefficients for radiative heat transfer, that is, $h_{r,g \to ew}$, $h_{r,g \to eb}$, $h_{r,eb \to ew}$, and $h_{r,ew \to ew}$. Convection to the exposed wall and exposed bed may be calculated as per Gorog et al. (1983) as

$$h_{c,ew} = 0.036 \frac{k_g}{D_k} \text{Re}^{0.8} \text{Pr}^{0.33} \left(\frac{D_k}{L_k}\right)^{0.055} \tag{8.31}$$

$$h_{c,eb} = 0.4 G_g^{0.62}$$

$$G_g = \frac{\dot{m}_g \times 3600}{A_g} \; [\text{kg m}^2 \text{ h}^{-1}] \tag{8.32}$$

In applying these expressions the hydraulic diameter and transverse area of the freeboard must be utilized in evaluating the dimensionless groupings and mass flux terms. At the covered wall–covered bed interface the model described earlier is applied.

Since there is no intent to restrict the work to nonreactive conditions in the bed and freeboard, Equation (8.30) can be expanded to include the reactive terms, which are originally present in Equation (8.29), to yield the system

$$\left(\sum n_i c_{p_i}\right) \frac{dT_g}{dz} = h_{ew} A_{ew} (T_g - T_w) + h_{eb} A_{eb} (T_g - T_b) + \sum \gamma_i A_g \tag{8.33}$$

$$\left(\sum n_j c_{p_j}\right) \frac{dT_b}{dz} = h_{eb} A_{eb} (T_b - T_g) + h_{cw} A_{cw} (T_b - T_w) + \sum \gamma_j A_b \tag{8.34}$$

where T_b and T_w are the average temperatures over the interfacial surfaces and γ are the production rates for various species involved in either chemical reactions, for example, freeboard combustion or phase changes (such as evaporation of free moisture), each to be determined by the appropriate kinetic expressions. Since mass must also be conserved for the control volumes of the bed and freeboard, a system of ordinary differential equations representing the mass balance also results from these same kinetic expressions. For example, the calcination reaction for limestone, which is

$$\text{CaCO}_3 = \text{CaO} + \text{CO}_2 \tag{8.35}$$

generates the interrelated mass balance expressions

$$\frac{d\dot{n}_{\text{CaCO}_3}}{dz} = -\frac{d\dot{n}_{\text{CaO}}}{dz} = \left(\frac{d\dot{n}_{\text{CO}_2}}{dz}\right)_{\text{freeboard}} \tag{8.36}$$

where

$$\frac{dn_{CaCO_3}}{dz} = \gamma_{CaCO_3} A_b = A_0 \exp(E/\overline{R}T) \tag{8.37}$$

In developing the global solution for the kiln model the complete system of ordinary differential equations (i.e., the two energy balance equations, the mass balance equations, and the auxiliary energy condition for the wall) must be solved simultaneously.

8.8 Two-Dimensional Thermal Model for the Bed

Although useful results can be obtained from one-dimensional models, the assumption that conditions will be uniform across any transverse section of the bed material will hold only for a well-mixed bed. Since segregation is known to occur within the bed, a two-dimensional model provides the opportunity to examine the effects of "demixing" within the bed on kiln performance. As we saw in Chapter 5, segregation in the transverse plane is driven by the bed motion, which is established by the rotation of the kiln. However, it was not until an adequate granular flow model was developed that mixing and segregation effects on temperature distribution could be modeled. For example, during rotary kiln limestone calcination, product quality is tested by checking the extent of dissociation of fine particles at the discharge end rather than that of larger particles. One would think that smaller particles would have a quicker thermal response and reach dissociation temperature earlier than the larger ones. However, they segregate to the core and do not see higher freeboard temperatures, due to temperature gradient across the free surface and the core.

We had mentioned earlier that the rolling bed mode would be the preferred mode of operation in most kiln processes due to its potential to achieve adequate mixing. We will set out the bed model by applying a Cartesian coordinate to the active layer since the flow there is primarily parallel to the free surface. For simplicity, we begin by first modeling an inert bed and by assuming it to behave like a continuum. Energy conservation for a control volume in the active layer (Figure 8.9(a)) requires that

$$\frac{\partial}{\partial x}\left(k_{eff}\frac{\partial T}{\partial x}\right) - \rho c_p u_x \frac{\partial T}{\partial x} + \frac{\partial}{\partial y}\left(k_{eff}\frac{\partial T}{\partial y}\right) - \rho c_p u_y \frac{\partial T}{\partial y} + \dot{m}_b c_{p_b} \frac{dT_{ba}}{dz} = 0 \tag{8.38}$$

We will further assume that mixing is sufficient to ensure that, within the active layer, the temperature gradient in the axial direction of the kiln (i.e., dT_{ba}/dz) in Equation (8.38) is uniform. Since transverse mixing is at least two orders of magnitude more effective than axial mixing (Barr et al., 1989) this latter condition appears justified. The last term in Equation (8.38) includes the axial gradients of temperature in the

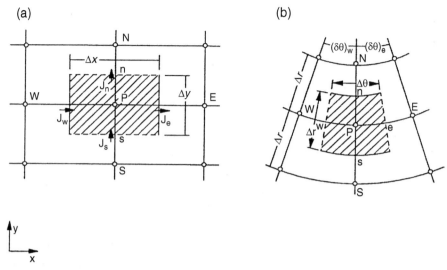

Figure 8.9 Control volumes for bed heat transfer calculations: (a) active layer; (b) plug flow region and refractory lining.

active layer, and accounts for the removal of energy from the control volume by axial bed flow. It is assumed that all particles within the active layer advance axially at the same rate and that, because the plug flow region behaves as a rigid body, this axial advance occurs only within the active layer. Thus the mass flux $\dot{m}_b$ in Equation (8.38) is set by the kiln feed rate and the transverse area of the active layer, $\dot{m} = \dot{M}/A_{AL}$.

Because the plug flow region rotates as a rigid body about the kiln axis, a cylindrical coordinate system is employed in this region and for the refractory wall. Energy conservation for any control volume in the plug flow region and the wall (Figure 8.9(b)) requires that

$$\frac{\partial}{\partial r}\left(k_{pf}\frac{\partial T}{\partial r}\right) + \frac{1}{r}\frac{\partial}{\partial \theta}\left(k_{pf}\frac{\partial T}{r\partial \theta}\right) - \rho c_{p_{pf}}\omega r\frac{\partial T}{r\partial \theta} = 0 \qquad (8.39a)$$

$$\frac{\partial}{\partial r}\left(k_{w}\frac{\partial T}{\partial r}\right) + \frac{1}{r}\frac{\partial}{\partial \theta}\left(k_{w}\frac{\partial T}{r\partial \theta}\right) - \rho c_{pw}\omega r\frac{\partial T}{r\partial \theta} = 0 \qquad (8.39b)$$

where the first two terms constitute the radial and transverse conduction, respectively, and the third term represents the movement of energy through the control volume due to the rotation of the kiln. Since the bed particles are assumed to advance axially only in the active layer, Equation (8.39a) does not include any term for energy transport in the axial direction, it being implicitly assumed that axial conduction in the plug flow region is negligible. The effective thermal conductivity is applied with the appropriate diffusional contribution depending on the flow field.

8.9 The Combined Axial and Cross-Sectional Model—The Quasi-Three-Dimensional Model for the Bed

In order to apply the understanding of the bed motion in the distribution of energy within the bed over the entire length of the kiln, we combine the two-dimensional model with a conventional one-dimensional axial model for the kiln to derive a quasi-three-dimensional model for the bed material. Although a rigorous three-dimensional formulation of the problem might be easily accomplished, the extreme aspect ratio of most kilns, $L/D \geq 10$, makes the solution of the resulting equations somewhat cumbersome. The impetus for the quasi-three-dimensional model is the need to significantly improve the capability of predicting conditions within the bed while maintaining relatively modest demands on computing capability. Two solution approaches are possible.

1. *Synchronous solution of the individual formulations*: Here, the heat transfer problem external to the bed might be determined using current values of the freeboard gas temperature and mean bed temperature. Once the net rate of heat transfer to the bed is determined (including the flux distribution to the covered wall surface), the axial gradients of the freeboard gas and mean bed temperatures become available to advance the temperatures to the next axial position by solving Equation (8.33). Before doing so the axial temperature gradient in the bed could be used to obtain a solution for Equations (8.38) and (8.39) and hence determine the temperature distribution within the bed at the current axial position.
2. *Asynchronous solution of the one-dimensional and two-dimensional problems*: The approach involves developing the one-dimensional solution over the entire kiln length before returning to the charge end to expand the axial bed temperature profile into the transverse plane. Implicit in doing so is the assumption that, at any axial position, heat transfer in the transverse plane (within the bed) will not significantly alter the freeboard, bed, and wall. The axial temperature profiles for the freeboard gas, bed, and wall are first developed using the one-dimensional model. The gas temperature and axial gradient of the bed temperature are then employed to drive the transverse model of the bed and kiln wall at a series of axial kiln positions and thereby determine, for each position, the temperature field in the bed and wall material. It should be pointed out at this stage that the two-dimensional model employs only the freeboard gas temperature and the axial bed temperature gradient at the given axial location in order to calculate the bed and wall temperatures. Both may also be supplied either from experimental data or by a one-dimensional model. Thus a check on the model can be made by comparing the computed bed and wall temperatures with either the measured values or, once the one-dimensional model has been verified, with values obtained from the latter.

8.10 Solution Procedure

The axial temperature profiles can be developed beginning from either end of the kiln by means of Equations (8.30) and (8.33). However, because direct-fired kilns operate in a countercurrent flow mode, we can employ a shooting method (Tscheng and

Watkinson, 1979) to arrive at a solution. We can start from the charge end where the material temperature is known but the gas temperature is unknown. We show an example of the calculation using an inert bed medium and a freeboard gas consisting of carbon dioxide and water vapor as the only combustion products. The thermal properties of the gas, as well as the calculation of the gray gas emissivities and absorptivities required for computing the radiation BCs, are based on these two gas constituents. The exit gas temperature and composition are needed to initiate the solution (e.g., the Runge–Kutta solution procedure) but the measured data may be used if available so as to avoid destabilizing the calculation as a result of the large number of unknowns involved.

At this point, the finite difference method described for convective diffusion equations (Patanker, 1980) might be used for the solution of the governing bed/wall equations derived for the kiln cross section

$$\rho u \frac{\partial T}{\partial x_j} - \frac{\partial}{\partial x}\left(\Gamma \frac{\partial T}{\partial x_j}\right) = S \tag{8.40}$$

where S is the source term and Γ is the effective mass flux, which can be given as

$$\Gamma = \frac{k_{\text{eff}}}{c_p} [\text{kg/m s}]$$

Equation (8.40) may be rearranged to give

$$\frac{\partial}{\partial x_j}\left(\rho u T - \Gamma \frac{\partial T}{\partial x_j}\right) = S \tag{8.41}$$

where the term in the parentheses, represented by J_i, is the sum of the convection and diffusion fluxes. By introducing the continuity equation into the above equations, the general algebraic form of the discretized differential equations for the bed and wall become

$$A_P T_{i,j} = A_E T_{i+1,j} + A_W T_{i-1,j} + A_N T_{i,j-1} + A_S T_{i,j+1} - \dot{m} c_p \frac{dT_b}{dz} \tag{8.42}$$

The coefficients, A, are expressed as

$$A_E = D_e A |P_e| + [-F_e, 0]$$
$$A_W = D_w A |P_w| + [F_w, 0]$$
$$A_N = D_n A |P_n| + [-F_n, 0] \tag{8.43}$$
$$A_S = D_s A |P_s| + [F_s, 0]$$
$$A_P = A_E + A_W + A_N + A_S$$

The heat diffusion flux, D_e; the convection flux, F_e; and the associated Peclet number are

$$D_e = \frac{\Gamma \Delta y}{\delta x_c}, \quad F_e = (\rho u_x)_{i+1,j} \Delta y \quad \text{and,} \quad P_e = \frac{F_e}{D_e}, \text{etc.}$$

Within a numerical model, the function $A|P|$ is represented by a power law scheme as (Patanker, 1980)

$$A(|P|) = \left[0, \left(1 - 0.1|P|^5 \right) \right] \tag{8.44}$$

The term in the square brackets represents the greater of the two quantities. Because of the inclusion of the source term, $\dot{m}c_p dT/dz$ in Equation (8.42), we multiply Equation (8.43), which represents the coefficients of the discretized equation, by c_p so as to obtain the desired units, that is, watts per meter kelvin.

For the plug flow region and the refractory wall, the discretized equations are the same as for the active region of the bed except that they are in cylindrical coordinates. For the radial direction, only the conduction component is considered and the coefficient of the algebraic (discretized) equation is equal to, for example,

$$A_S = \frac{k_w(r\theta)}{\Delta R}$$

with all units being consistent with that of the term, $\dot{m}c_p dT/dz$.

The mesh employed for the cross-sectional model is shown in Figure 8.10. Rectangular grids can be used for most of the nodes except at the interface between the plug flow and the active layer where triangular nodes automatically emerge as a result of the merging of the two coordinate systems; such nodes are considered as half of the rectangular nodes in the calculation of the nodal areas. Higher density mesh is required for the active layer of the bed and in the active region of the refractory wall. The latter is needed to capture temperature cycling that results due to kiln rotation. Although the specific depth of the wall active region is not known a priori, 10% is a good guess based on measurements. This value is only needed to establish the mesh because the actual depth is found from the results of the numerical calculation. Outside the active wall region, a steady state one-dimensional conduction can be applied

$$Q_{SS} = 2\pi k_w \frac{(T_{w_r} - T_{w_{r+\Delta r}})}{\ln|(r + \Delta r)/r|} \tag{8.45}$$

Doing this speeds up the numerical calculation, which otherwise is slow because of the slow thermal response of the wall. From here the Gauss–Seidel iteration method can be employed to solve the system of algebraic equations with a reasonable convergence criterion, for example,

$$\left| T_{ij}^n - T_{ij}^{n+1} \right|_{\max} < 10^{-5} \tag{8.46}$$

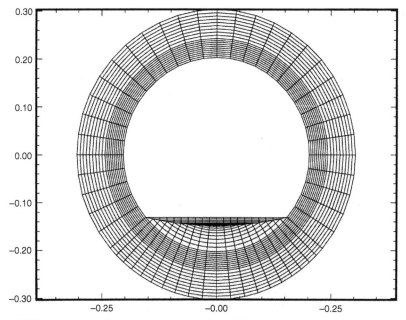

Figure 8.10 Mesh generation for cross-sectional modeling.

The solution requires an underrelaxation technique because of the numerous nonlinear temperature-dependent terms. To apply the segregation of binary mixture to the model, it must be recognized that the effective thermal conductivity is a function of particle size and that its magnitude depends upon the concentration gradient imposed by finer (jetsam) particles at each node i.e.,

$$k_{ef} = C_J(i,j) k_{\text{eff}|d_{p_s}} + [1 - C_J(i,j)] k_{\text{eff}|d_{p_L}} \tag{8.47}$$

8.11 Model Results and Application

The mathematical model presented here has been validated using a well-instrumented 0.41-m-diameter pilot kiln (Figure 8.11, Boateng and Barr, 1996). The predicted bed cross-sectional temperature distributions for uniformly sized (well-mixed) and a binary mixture (segregated bed) in the same pilot kiln are shown in Figures 8.12 and 8.13, respectively (Boateng and Barr, 1996).

As shown, radial segregation will tend to generate a cooler region that coincides with the segregated core of fine particles. This means that, for example, in limestone calcination, either the fine particles may not be fully calcined or the larger particles at the peripheral region may be overburned at a specified freeboard gas temperature. In other processes, for example, incineration of solid waste and cement clinkering, there is the possibility that the charge will contain materials of varied density. Density

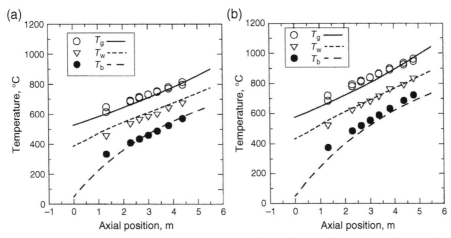

Figure 8.11 Predicted and measured temperature profiles, pilot kiln (Boateng and Barr, 1996): (a) and (b) are two different experimental measurements.

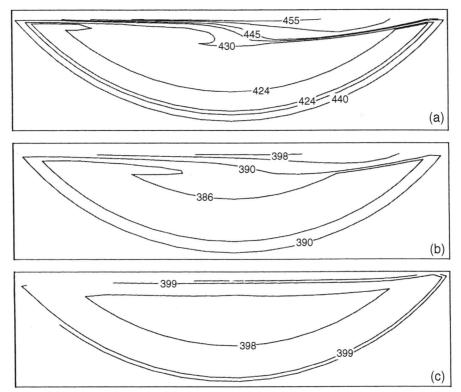

Figure 8.12 Contour plots for "mixed-bed" temperature distribution: $T_g = 631\ °C$, 12% fill: (a) 1.5 rpm, (b) 3 rpm, and (c) 5 rpm (Boateng and Barr, 1996).

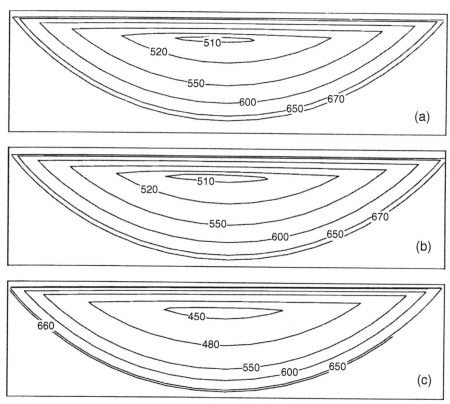

Figure 8.13 Contour plots for "segregated" temperature distribution: 12% fill, $T_g = 804$ °C: (a) 1.5 rpm, (b) 3 rpm, and (c) 5 rpm (Boateng and Barr, 1996).

differences also result in segregation. These are the areas where the thermal model can be used to control the temperature nonuniformities that accompany segregation. At the moment the only means of determining bed temperature in industrial kilns is through the use of a pyrometer, which usually measures bed surface temperature only and will not detect a lower core temperature, if any segregated core exists. Therefore the kiln operator will not know whether the core materials are fully calcined until the material is discharged. The usual practice is to sample the fines in the discharge by breaking them apart and visually inspecting the extent of calcination or by using a dissociation test in the laboratory. A typical case follows, where the thermal model may be applied to establish the bed temperature distribution at a particular axial position and thereby evaluate the quality and energy problems that may arise as a result of segregation. If the kiln operator, upon noticing undercalcined fine particles decides to increase the gas temperature (the obvious and usual practice), he or she will have to decide by how much the gas temperature should be increased. It is likely that increased gas temperature will result in an increase of the core temperature and overcook the larger particles at the periphery. In the case of density differences, for example, during solid waste incineration, the particle size of the feed can be controlled by using the model to decide jetsam loading so that the material with low heat capacity will be fed as fines in order to accomplish uniform burning.

The model was applied to the process of calcining limestone in a 3-m-diameter (10 ft) rotary kiln with 20 ton/h product capacity. Figure 8.14 shows the prediction of the axial temperature and concentration profiles. Here the temperatures for the surfaces and interfaces suggest heat exchange as a function of axial location. The bed temperature strictly follows the freeboard temperature. At about 35% along the axial distance, the bulk bed temperature supersedes the covered wall temperature showing that prior to that point the wall gives heat to the bed. The figure also shows that dissociation is complete by midlength at least for the conditions modeled. Figure 8.15 predicts the temperature distribution in the cross section including the bed and the refractory lining at an axial location with freeboard gas temperature of 2000 K.

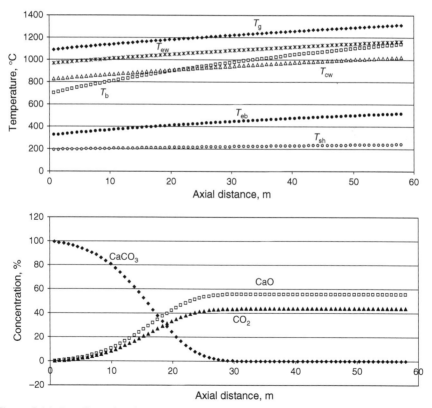

Figure 8.14 One-dimensional temperature and concentration profiles for 3.04-m-diameter (10 ft) preheater kiln for limestone calcination.

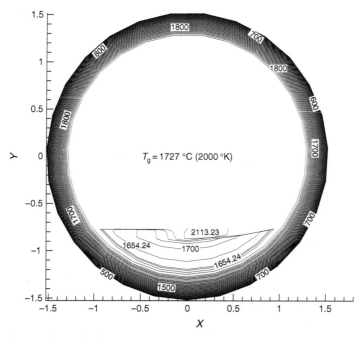

Figure 8.15 Two-dimensional temperature contours for 3.04 m-diameter (10 ft) preheater kiln for limestone calcination.

8.12 Single-Particle Heat Transfer Modeling for Expanded Shale Processing

We had mentioned previously that most of the chemical reactions in a rotary kiln take place in the active layer. The cross-sectional model shows that the extent of the reaction will depend on the exposure time. The modeling of the exposed bed heat transfer coefficients allows for the possibility of modeling the temperature response of a single particle that emerges from the plug flow region into the active layer and tumbles down the free surface. We will examine the ramification of the time–temperature history using an expanded shale process. This process is described later in Chapter 10 but for now we will look at the material transport in the cross section of such a kiln (Figure 8.16) and examine the classical transient heat transfer problem in the context of material processing and kiln control that might affect product quality. From the illustration in Figure 8.16 one can write balance equations in the Lagrangian framework and examine the temperature response and the expansion kinetics of an individual shale particle tumbling down the incline based on its velocity and the fluid–particle heat transfer rate. Let us assume that the particle emerges at the free surface at a specific temperature T_s and density ρ_0. Within the time the particle travels along the chord length, L_c, it will soak up heat and when the temperature reaches a certain threshold, called the bloating temperature, it will undergo an expansion. With little mass loss and

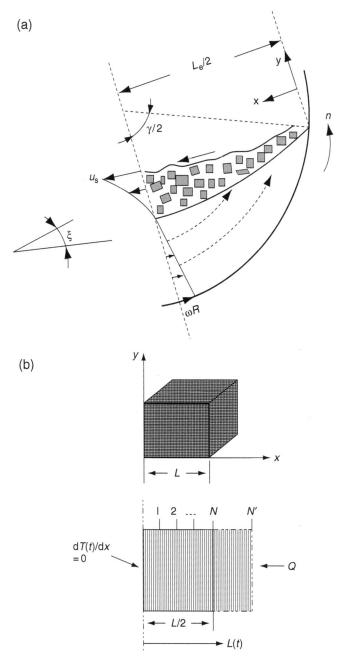

Figure 8.16 Modeling a single-particle temperature response using expanded shale kinetics: (a) cross section of kiln and (b) modeled particle (Boateng et al., 1997).

large expansion, the density will reduce to form a lightweight aggregate. The dependent property for the balance equation is the conversion, that is, $(\rho_0 - \rho)/(\rho_0 - \rho_f)$, which represents the extent of particle density or specific gravity change. This can be related to the Arrhenius equation as

$$\frac{\partial}{\partial t}\left(\frac{\rho_0 - \rho}{\rho_0 - \rho_f}\right) = -A_0 \exp\left(\frac{-E}{RT_s}\right) \tag{8.48}$$

where $\rho(x, 0) = \rho_0$ for $-L/2 \leq x \leq L/2$. Furthermore, it may be assumed that significant gradients exist within the particle due to the large size of the particles, typically about 2.5–5 cm, and therefore the dimensionless Biot number is large ($hL/k_s \gg 1$). The unidirectional energy balance on the particle (Figure 8.16(b)) is

$$\frac{\partial}{\partial t}(\rho c_p T) = \frac{\partial}{\partial x}\left(k_s \frac{\partial T}{\partial x}\right) - \Delta H A_0 \exp\left(\frac{-E}{RT}\right)(\rho - \rho_0) \tag{8.49}$$

IC

$$T(x, 0) = T_0 \quad \text{for} \quad -x_{L/2} \leq x \leq x_{L/2} \tag{8.50a}$$

BC

$$k_s \frac{\partial T}{\partial x}(\pm L/2, t) = h(T_{fb} - T_s) + \sigma\varepsilon(T_{fb}^4 - T_s^4) = Q \tag{8.50b}$$

For symmetry, if the particle center point is chosen for $x = 0$, the BC there is

$$\frac{\partial T}{\partial x}(0, t) = 0 \tag{8.50c}$$

The density or specific gravity change of the shale particle to three-dimensional diffusion will result in a three-dimensional expansion. The incremental change in each direction may be calculated independently using the unidirectional expression

$$\frac{1}{L_0}\frac{\partial L}{\partial t} = \beta_0 \varphi \frac{\partial T}{\partial t} + \beta_f(1 - \varphi)\frac{\partial T}{\partial t} + \frac{\eta}{\rho_0}\frac{\partial \rho}{\partial t} \tag{8.51}$$

where L_0 is the initial particle length; L is the instantaneous length; β_0 and β_f are, respectively, the initial and final coefficients of expansion; η is a decomposition or expansion factor; and φ is the dimensionless density. The activation energy for dehydroxylation of kaolin in a flash calciner has been reported as ranging between 27 and 70 kJ/mol (Meinhold et al., 1994). The initial density for shale is about 2785 kg/m^3, whereas the final density, achieved at 1400 °C, is about 442 kg/m^3 (Boateng et al., 1997).

The system of equations, Equations (8.49–8.51), can be solved analytically, graphically, or numerically using implicit finite difference methods utilizing the unidirectional nodal system shown in Figure 8.16(b). Since the problem is one-dimensional, it reduces to a system of algebraic equations, $[A][T] = [R']$, where $[A]$ is the coefficient tridiagonal matrix, $[T]$ is the temperature distribution, and $[R']$ is the set containing residual constant terms. A solution can be found by solving the matrix using the Thomas algorithm with the result shown in Figure 8.17. These results can be plotted using two dimensionless groupings for control of the operation (Figure 8.18). The first grouping

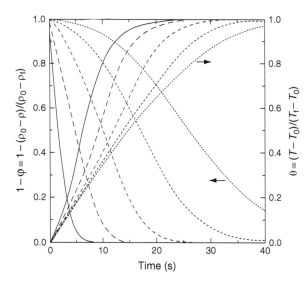

Figure 8.17 Predicted temperature and density changes at the exposure time for shale materials of various origins (Boateng et al., 1997).

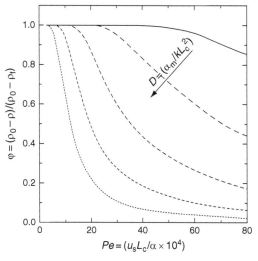

Figure 8.18 Kiln and product quality control curves for shale expansion in a 2.44 m diameter rotary kiln (Boateng et al., 1997).

is the Peclet number for heat transfer, $Pe(u_s L_c/\alpha)$, and representing the ratio of the heat absorption time to the time of traverse at the exposed bed surface. The second is the modified Damkohler group II grouping, $D(\alpha_m/\kappa L_0^2)$, where κ is the expansion rate constant in units 1/s, defined as the ratio of the timescales of particle self-diffusion in the active layer to the particle expansion time. This parameter is essentially material dependent and can be related to the activation energy that governs the kinetics of the expansion. Since the surface velocity depends on the rotational rate, Pe is similar to the rotational Froude number $Fr(\omega^2 R/g)$, which is more convenient and practical for an operator.

The graph in Figure 8.18 can be employed as a control procedure for quality assurance. For example, if it is desirable for the lightweight aggregate kiln operator to change the kiln feed rate and maintain the same degree of fill (i.e., constant chord length), then the kiln speed must be altered according to Pe. If the heat transfer rate is not changed after this action, then the quality φ will change for that particular material. If each curve represents a specific material, that is, easy to cook (low E) or hard to cook (high E), then the desired quality can be achieved by following the appropriate path on the curve either by reducing kiln speed or by increasing the heat flux.

References

Boateng, A.A., 1998. On flow-induced kinetic diffusion and rotary kiln bed burden heat transfer. Chem. Eng. Comm. 170, 51–66.

Boateng, A.A., Barr, P.V., 1996. A thermal model for the rotary kiln including heat transfer within the bed. Int. J. Heat Mass Transfer 39, 2131–2147.

Boateng, A.A., Thoen, E.R., Orthlieb, F.L., 1997. Modeling the pyroprocess kinetics of shale expansion in a rotary kiln. Trans. IChemE 75 (Part A1), 278–283.

Barr, P.V., Brimacombe, J.K., Watkinson, A.P., 1989. A heat-transfer model for the rotary kiln: Part I. Pilot kiln trials. Met. Trans. B 20B, 391–402.

Brimacombe, J.K., Watkinson, A.P., 1979. Heat transfer in a direct fired rotary kiln: Part I. Pilot plant and experimentation. Met. Trans. B 9B, 201–208.

Ferron, J.R., Singh, D.K., 1991. Rotary kiln transport processes. AIChE J. 37, 754–758.

Gorog, J.P., Adams, T.N., Brimacombe, J.K., 1982. Regenerate heat transfer in rotary kilns. Met. Trans. B 13B (2), 153–163.

Gorog, J.P., Adams, T.N., Brimacombe, J.K., 1983. Heat transfer from flames in a rotary kiln. Met. Trans. B 14B, 411–424.

Jenkins, B.G., Moles, F.D., 1981. Modelling of heat transfer from large enclosed flame in a rotary kiln. Trans. IChemE 59, 17–25.

Lehmberg, J., Hehl, M., Schugerl, K., 1977. Transverse mixing and heat transfer in horizontal rotary drum reactors. Powder Technol. 18, 149–163.

Meinhold, R.H., Salvador, S., Davies, T.W., Slade, R.C.T., 1994. A comparison of the kinetics of flash calcination of kaolinite in different calciners. Trans. IChemE 72 (A1), 105–113.

Patanker, S.V., 1980. Numerical Heat Transfer and Fluid Flow. Hemisphere. New York.

Sass, A., 1967. Simulation of the heat transfer phenomenon in a rotary kiln. I&EC Process. Des. Dev. 6 (4), 532–535.

Schotte, W., 1960. Thermal conductivity of packed beds. J. AIChE 6 (1), 63–67, 1960.

Siu, W.W.M., Lee, S.H.-K., 2000. Effective conductivity computation of a packed bed using constriction resistances and contact angle effects. Int. J. Heat Mass Transfer 43, 3917–3924.

Slavin, A.J., Londry, F.A., Harrison, J., 2000. A new model for the effective thermal conductivity of packed beds of solid spheroids: alumina in helium between 100 and 500 °C. Int. J. Heat Mass Transfer 43, 2059–2073.

Sullivan, W.N., Sabersky, R.H., 1975. Heat transfer to flowing granular media. Int. J. Heat Mass Transfer 18, 97–107.

Tscheng, S.H., Watkinson, A.P., 1979. Convective heat transfer in rotary kilns. Can. J. Chem. Eng. 57, 433–443.

Wakao, N., 1973. Effect of radiating gas on effective thermal conductivity of packed beds. Chem. Eng. Sci. 28, 1117–1118.

Wakao, N., Vortmeyer, D., 1971. Pressure dependency of effective thermal conductivity of packed beds. Chem. Eng. Sci. 26, 1753–1765.

Wes, G.W.J., Drinkenburg, A.A.H., Stemerding, S., 1976. Heat transfer in a horizontal rotary drum reactor. Powder Technol. 13, 185–192.

Mass and Energy Balance 9

Heat balance calculations are usually carried out when developing new rotary kiln chemical processes or when improving old ones. No thermal process would work if too much heat is released or if there is a lack of sufficient thermal energy to drive the process, in other words, to maintain the reaction temperature. Heat balance can only be calculated with given mass balances as the boundary conditions, hence a quantitative description of the chemical processes on the basis of physical or chemical thermodynamics is required. While chemical thermodynamics establishes the feasibility of a particular reaction under certain reactor conditions, chemical kinetics determines the rate at which the reaction will proceed. Before we establish the global rotary kiln mass and energy balance, it is important to examine some fundamental concepts of thermodynamics that provide the pertinent definitions essential for the design of new rotary kiln bed processes.

9.1 Chemical Thermodynamics

Chemical processes are normally carried out at constant pressure (*isobaric*). Under isobaric conditions, the total enthalpy, or the total heat content of a chemical species, consists of the following three thermodynamic properties: (1) the heat of formation, (2) sensible heat, and (3) heat of transformation. When chemical elements react to form a compound, a certain amount of heat is released to the environment in accordance with the first law of thermodynamics, which states that the internal energy, U, of a system changes as a result of the exchange of energy with the surroundings, in the form of heat, Q, or through some form of work, that is, $\Delta U = Q + \dot{W}$. Hence, consistent with energy conservation, the *heat of formation* of the reaction or for short *heat of reaction*, ΔH, is equal to the sum of the total enthalpies of the products minus the total enthalpies of the reactants, as we saw in stoichiometric balances. The *sensible heat* is literally the heat that can be sensed because it is associated with temperature change, that is,

$$Q_S = \int_0^T C_p dT \qquad (9.1)$$

where C_p is the molar heat capacity of the species in J mol^{-1} K^{-1}, that is, its specific heat times its molecular weight ($C_p M$). The heat of transformation is usually associated with phase change such as crystal transformation, melting, or evaporation and is expressed in J mol^{-1}. For convenience, the total enthalpies of all chemical species are tabulated with respect to a reference temperature, T_0, at a standard pressure, p_0, which are usually taken to be 25 °C (298.15 K) and 1 bar (0.9869 atm). Therefore, the enthalpy of a chemical species, i, might be expressed as (Themelis, 1995)

$$\Delta H_i = \Delta H_i^0 + \int_{T_0}^{T} C_p dT + \sum (\Delta H_{\text{trans}}) \tag{9.2}$$

where ΔH_i^0 is the total enthalpy of species at T_0. The last term in Equation (9.2) represents the heat that is absorbed or released in the phase transformations, if any. For chemical compounds, the term ΔH_i^0 is equal to the heat of formation of the compound at T_0, that is, $\Delta H_i^0 = \Delta H_f^0$.

9.2 Gibbs Free Energy and Entropy

The second law of thermodynamics, as was introduced by Clausius and Kelvin, is formulated on the basis of theoretical analyses of reversible cycles established by Carnot (Knacke et al., 1991). From the reversible Carnot cycle, it follows mathematically that the reversible exchange of heat, dQ_{rev} between systems and surroundings, divided by the absolute temperature T is a total differential dS, where S is the property of the state called entropy, that is,

$$dS = \frac{dQ}{T} \tag{9.3}$$

The second law can therefore be stated as: the total entropy change in a system resulting from any real processes in the system is positive and approaches a limiting value of zero for any process that approaches reversibility (Themelis, 1995). The entropy of a reversible process is equal to the heat absorbed during the process, divided by the temperature at which the heat is absorbed (Equation (9.3)). Reversibility denotes a process that is carried out under near-equilibrium conditions and therefore means that it is carried out most efficiently. Combining the first law and the second law gives the expression

$$dS = \frac{dU + PdV}{T} = \frac{dH}{T} \tag{9.4a}$$

from which it can be deduced that

$$\left(\frac{\partial S}{\partial U}\right)_V = \frac{1}{T} \tag{9.4b}$$

$$\left(\frac{\partial S}{\partial V}\right)_U = \frac{P}{T} \tag{9.4c}$$

$$S = S(U, V) \tag{9.4d}$$

With these expressions, the derivation from cyclic processes and reversible heat exchange could be abandoned in exchange for the state functions T, P, V, U, and S of the individual systems (Rogers and Mayhew, 1982).

Similar to the first law, the total entropy of a chemical species, i, at its equilibrium state; temperature, T, and standard pressure, p_0, can be expressed as

$$\Delta S_i = \Delta S_i^0 + \int_{T_0}^{T} \frac{C_p dT}{T} + \sum \left(\frac{\Delta H_{trans}}{T_{trans}}\right) \tag{9.5}$$

where ΔS_i^0 is the total entropy of the species at its equilibrium state and at T_0 (298.15 K) and standard pressure p_0 (1 bar or 1 atm). Equation (9.5) applies at absolute zero temperatures by setting $\Delta S_i^0 = 0$ and $T_0 = 0$. At 0 K, the entropy of all crystalline elements and compounds is zero and for all other species, may be zero, that is, $\Delta S_{f,0} = 0$, a statement that defines the third law of thermodynamics.

According to Gibbs, a phase is understood to be a chemically and physically homogeneous substance, irrespective of the amount and shape. Every phase has a limited region of existence with respect to composition, temperature, and pressure. Phases in equilibrium with each other are called coexisting phases. Single-phase systems are homogeneous phases, whereas multiphase systems are heterogeneous. Closed, homogeneous or heterogeneous equilibrium systems have maximum entropy, which is a function of U and V (Equation (9.4)). The Gibbs free energy of a species is a function of its enthalpy and entropy and is defined as:

$$G(T, P) = H - TS \tag{9.6}$$

G is a thermodynamic characteristic function that can be expressed in terms of its derivatives with respect to the corresponding independent variables. From the equilibrium requirement of maximum entropy follows the minimum Gibbs energy for chemical equilibrium, that is,

$$G = \min \text{ at constant } T, P \tag{9.7}$$

From the general definition of the Gibbs free energy, one can express the Gibbs free energy of a chemical species i at equilibrium in terms of its total enthalpy and total entropy at T and standard pressure, p_0, that is,

$$\Delta G_i^0 = \Delta H_i^0 - T \Delta S_i^0 \tag{9.8}$$

The heat capacities, enthalpies, entropies, and Gibbs free energies of chemical species at their equilibrium state and at standard temperature and pressure can be found in the public domain (e.g., Knacke et al., 1991). Some of these are presented in the Appendix. Usually, included in such tables is the dimensionless Planck's function B that allows a simple calculation of the equilibrium constant as

$$B_i^0 = -G_i^0/RT \log 10 \tag{9.9}$$

$$\log K_r = \sum n_i^r B_i^0 \tag{9.10}$$

where n_i are the stoichiometric mole fractions of the reaction (Knacke et al., 1991). A reaction will proceed at a given temperature when $\log K_r > 0$, similar to the situation that water boils when temperature reaches 100 °C. A useful tool for estimating chemical thermodynamic conditions of several processes is the HSC Chemical Reaction and Equilibrium software (Outokumpu, 2002). Several example chemical equilibrium calculations using the HSC software can be found in Themelis (1995).

9.3 Global Heat and Material Balance

Thus far we have covered all the transport phenomena and transport processes that determine the conditions within the rotary kiln. These include bed material flow, fluid flow in the freeboard, combustion, and associated heat transfer that result in temperature distribution at any section of the kiln. With these, the extent of temperature-dependent chemical reactions involved in the pyroprocessing of minerals and materials can be established and from that, product quality can be estimated. While all the models covered thus far can provide details of the intrinsic behavior within the kiln, including whether the reactions will or will not proceed at the reactor temperatures, it is important to step outside and look at the kiln as a thermodynamic system that interacts with the atmosphere and, like all thermodynamic systems, determine the thermal efficiency. The basic approach to doing this is to consider the kiln as a furnace system composed of a control volume with well-defined spatial boundaries into which mass and energy enter at a certain rate and leave at another rate. It is important to keep track of material and energy crossing the boundaries by doing some accounting. The bookkeeping relation for keeping track is the mass and energy balance, which is based on the steady-flow form of the first law of thermodynamics. If one mass flow rate exceeds the other, then mass is either accumulated or depleted. Also, if $\dot{Q}$ is the rate of heat flow into the furnace and $\dot{W}$ is the rate at which work is delivered to the surroundings, then the first law for conservation of energy requires that

$$\dot{Q} = \sum_o m_o h_o - \sum_i m_i h_i + \dot{W} = m_2 h_2 - m_1 h_1 + \dot{W} \tag{9.11}$$

where m and h are rates of mass and enthalpy crossing the system boundary. We know by now that rotary kilns are the equipment of choice when material processing requires high temperatures and long residence times. Because of these two important attributes, rotary kilns have also become the equipment of choice for waste destruction including hazardous waste fuels under the Environmental Protection Agency's Resource Conservation Recovery Act. Every kiln must therefore be permitted to operate within certain operational limits so as to ensure that harmful pollutants, whether they be dust, metals, or any other chemical compounds, are not discharged into the atmosphere. A mass and energy balance must, therefore, be carried out for the operation to show, at a minimum, mass flow rates, temperatures, and pressures of all process input and output streams at each discrete unit operation within the overall process such as the rotary kiln and all auxiliary devices including coal mills and air pollution control devices. A mass and energy balance is also required to ensure that process variables do not exceed the design or permitted limits. Additionally, it is necessary to establish the thermal efficiency of the process so that necessary action can be taken to optimize fuel use and conserve energy.

9.4 Thermal Module for Chemically Reactive System

The problem of the mass and energy balance is addressed by considering the kiln as a combustion system in which fuel and oxidizer flow into a control volume and combustion products flow out. Without the details of all the transport models that have been discussed in the previous chapters, one should be able to relate the combustion products to the input mass flow rates and energies and to account for any mass or energy accumulation or destruction that might take place and how they do. The steady-flow first law equation, Equation (9.11), may be written for the balance as

$$Q = H_p - H_r + \dot{W} \tag{9.12}$$

It is easy to envision that no shaft work is involved and that the work done on the surroundings is what transforms the raw material into usable material and which can be accounted for as energy of formation of the intended product. Therefore H includes chemical as well as thermal energy. The enthalpy H_p is the sum of the enthalpies of all product streams leaving the system and H_r is that for the entering reactant streams. The individual enthalpies may each be written as the product of the number of moles of the component in the reaction equation and the respective enthalpy per mole of the component. For example, for k number of products

$$H_p = n_1 h_1 + n_2 h_2 + \cdots + n_k h_k \text{ [kJ]} \tag{9.13}$$

where n is the stoichiometric coefficient of the chemical equation and the enthalpies are on a per mole basis. The enthalpy of any component of the reactants or products may be written as the sum of (1) its enthalpy of formation at the standard or reference temperature, T_0, and standard pressure, and (2) its enthalpy difference between the

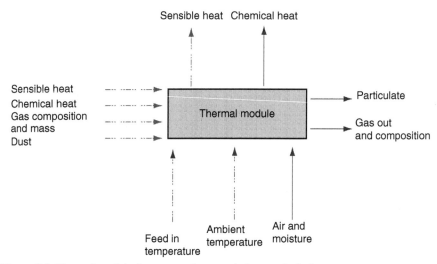

Figure 9.1 Thermal module for mass and energy balance calculations.

actual state and the standard state of the components. Assuming pressure independence for now, the enthalpy for each component becomes

$$h(T) = h_f(T_0) + [h(T) - h(T_0)] \; [\text{kJ/mol}] \tag{9.14}$$

where the term in the square brackets is the sensible enthalpy relative to the standard reference state. Hence the thermal module for the kiln component (Figure 9.1) can be balanced by balancing the streams and assigning enthalpies at the temperature state.

9.5 Mass Balance Inputs

As Figure 9.1 shows, a mass and energy balance requires knowledge of the conditions of the state, hence measurements of temperatures and pressures are required. For existing operation a complete audit is required to take measurements in addition to the data gathered by supervisory or monitoring instrumentation if any. The input to the mass balance module may consist of mass quantities, chemical composition, and any other pertinent qualitative information. These include feed rate or product rate, fuel input, and measured or calculated product rates.

9.6 Chemical Compositions

The input streams must include air composition in weight-%, fuel composition in weight-%, feed composition and moisture in weight-% on dry feed basis, effective volatile content in weight-% on dry feed basis; for solid fuel combustion, volatile composition in weight-% dry feed basis, and product composition by specification on product weight basis.

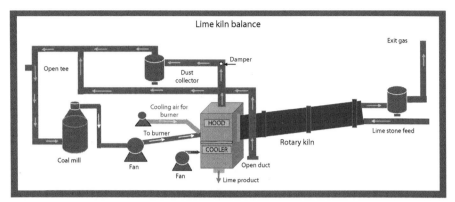

Figure 9.2 Process flow and component layout for dolomitic limestone calcination used for mass and energy balance calculation.

9.7 Energy Balance Inputs

Energy inputs include (1) thermodynamic data such as specific heats, heats of combustion, enthalpies of phase change, densities, and so on; (2) kinetic rate data; and (3) their state, that is, input and output temperatures.

The mass and energy balance for a 15 by 300-ft (4.57 by 92-m) kiln for the dolomitic limestone calcination process is presented here. The procedure starts with information on the kiln dimensions and layout (Figure 9.2) followed by process audit, which provide the necessary process data. The reference temperature used for enthalpy calculations is set at 25 °C (298 K). Mass and energy balance is achieved from which the thermal efficiency and any other process efficiency indicators can be estimated.

9.8 Site Survey—Measured Variables

The tables that follow go through a site survey. Table 9.1 covers the ambient air to the cooler. Table 9.2 covers the cooling air to the burner. Table 9.3 covers the coal conveying air, and Tables 9.4 and 9.5 cover the limestone and the dust. Table 9.6 covers the properties of the natural gas input and the kiln exit gas. Table 9.7 covers the kiln dimensions and shell temperatures.

9.9 Shell Heat Loss Calculations

Table 9.8 covers the calculated heat loss from the shell.

9.10 Calcination Module Calculation

Table 9.9 covers the calcination module calculation, including heat of formation, heat of calcinations, and heat of water evaporation.

Table 9.1 Ambient Air to the Cooler[a]

Measurement Number[b] (#)	Value (m/s)	Measurement Number (#)	Value (m/s)	Average Velocity (m/s)	Area (m²)	Volume (m³/s)
1	8.45	14	9.00	8.73	0.237	2.069
2	10.8	13	9.00	9.90	0.201	1.987
3	14	12	12.00	13.00	0.164	2.134
4	14.5	11	12.60	13.55	0.13	1.730
5	15.5	10	14.60	15.05	0.091	1.373
6	15.5	9	14.80	15.15	0.055	0.829
7	14	8	14.00	14.00	0.018	0.255
Total	–	–	–	–	0.894	10.4

[a]Cooler fan outside diameter is 42 in. (1.07 m). Temperature is 78 °F (25.6 °C). Total area is 0.894 m². Note that the velocity of air into the fan (given in units of m/s) is usually measured at variable points on the radius. The numbered pair points (1 and 14, 2 and 13, etc.) are opposite sides of same radius. Density = 1.18 kg/m³. Mass flow = 12.27 kg/s.
[b]Note that the distance between two measurement radial points is 0.08 m.

Table 9.2 Cooling Air to the Burner[a]

Fan	Side	Center	Volume (m³/s)
Velocity (m/s)	11	9	–
Area (m²)	0.02	0.01	–
Volume	0.2	0.1	0.3

[a]Internal diameter of the fan is 0.1905 m (7.5 in). Density = 1.19 kg/m³. Flow rate = 0.36 kg/s.

Table 9.3 Coal Conveying Air[a]

Measurement Number	Value (m/s)	Area (m²)	Volume (m³/s)
1	6	0.456	2.736
2	6	0.355	2.128
3	5	0.253	1.267
4	3.5	0.152	0.532
5	2	0.051	0.101
Total		1.267	6.76

[a]Temperature = 149 °C. Pipe outside diameter = 50 in (1.270 m); total area = 1.267 m². Distance between two measured points is 0.127 m. Density = 0.84 kg/m³. Mass flow = 5.66 kg/s. Flows are split before entering the coal mill. The mixing air going to the coal mill $m_0 = 5.66$, with a temperature $T_0 = 278$ °F. The air from the hood to the coal mill $m_1 = 0.28\%$, with a temperature $T_1 = 800$ °F. The tempering air to the duct $m_2 = 0.72\%$, at a temperature $T_2 = 78$ °F. The coal mill off-take $m_1 = 1.57$ kg/s through the hood duct. The tempering air $m_2 = 4.09$ kg/s ambient air.

Table 9.4 Limestone[a]

	High Calcium (HI-CAL) Limestone					
	MW	Weight-%	%C	Ca/Mg		
$CaCO_3$	100	95.61	11.47	42.96		
$MgCO_3$	84.3	1.7	0.24	—		
S	—	0.06568	—	—		
H_2O	18	1.96	—	—		
Inert	100	0.66	—	—		
Total	—	100.00				
	DOLOMITE					
	MW	Weight-%	%C	Ca/Mg		
$MgCO_3$	84.3	43.94	6.25	—	N_2	1.52
S	—	0.016	—	—	O_2	6.51
H_2O	18	1.94	—	—	C	79.28
Inert	100	1.46	—	—	H	5.02
Total	—	99.99			H_2O	3.00
					S	0.080
					Ash	4.58

MW, molecular weight.
[a] The temperature of both the high-calcium (HI-CAL) and dolomite inputs was 78 °C. For the HI-CAL, the %C in product was 0.08, and the flow rate was 47 short tons per hour (STPH) (11.48 kg/s). For the dolomite, the %C in product was 0.16 kg/s, and the flow rate was 52 STPH (13.1 kg/s).

Table 9.5 Dust and Coal[a] Composition (Weight-%)

% Dust	15							
Dust/product	Composition (lab)	CaCO$_3$%	CaO%	MgCO$_3$%	MgO%	S%	Ash%	
Temperature (°F)	1148	Weight-%	0.7	56.4	0.6	40	—	2.6
Coal	170.5	Composition	N%	O%	C%	H%	H$_2$O%	S%	Ash%
Temperature (°F)		Weight-%	1.52	6.51	79.28	5.02	3.0	0.08	4.58

[a]The flow rate for coal is 6.7 STPH (6.08 metric ton/h, or 1.69 kg/s).

Table 9.6 Natural Gas and Exit Gas Components[a]

Natural gas

Composition	N$_2$%	O$_2$%	CO$_2$%	CH$_4$%	C$_2$H$_6$%	C$_3$H$_8$%	C$_4$H$_{10}$%	C$_5$H$_{12}$%	C$_6$H$_{14}$%
	1.35	0	1.25	94.5	2.5	0.011	0.03	0.04	0.02
MW	28	16	44	16	30	44	58	72	86

MW of natural gas = 16.9

Kiln exit gas

Composition	N$_2$%	NO%	NO$_2$%	O$_2$%	CO%	CO$_2$%	H$_2$O%	SO$_2$%
	—	—	—	0.5	—	—	—	—

MW, molecular weight.
[a]Temperature = 1326 °F, flow rate = 15,000 standard cubic feet per hour (424.75 m^3/h, or 0.089 kg/s).

Table 9.7 Kiln Dimensions and Shell Temperatures[a]

Measured Temperature on the Surface of the Kiln (°F)								
655	605	617	595	593	605	638	608	424
550	570	581	581	588	550	547	517	506
535	530	533	528	508	535	508	535	540
518	308	669	635	885	886	890	774	810
668	495	428	435	524	700	640	786	675
763	766	766	739	742	735	712	660	708
635								

Measured Temperature on the Surface of the Kiln (°C)								
346	318	325	313	312	318	337	320	218
288	299	305	305	309	288	286	269	263
279	277	278	276	264	279	264	279	282
270	353	354	335	474	474	477	412	432
353	257	220	224	273	371	338	419	357
406	408	408	393	394	391	378	349	376
335								

Note: measurements run left to right and top to bottom.
[a] Kiln diameter, 15 ft (4.57 m); kiln length = 300 ft (92 m).

9.11 Combustion

Table 9.10 covers the heat of combustion reactions. Tables 9.11 and 9.12 cover the calculation of energy, of fuel in Table 9.11, and of natural gas in Table 9.12. Table 9.13 covers the calculation of the percent of water in air. Table 9.14 covers the heat capacity of the different components of fuel.

9.12 Energy Balance Module

Table 9.15 covers the formulas used to determine the sensible energy for input streams, as well as the sensible heat for the coal conveying air, the inlet air, and the limestone.

Table 9.8 Shell Heat Loss Calculated[a]

Shell Zone Temperature (°C)	Shell Zone Temperature (K)	Q1	Q2	Q3	Q5 (kcal/m²/h)	Length (m)	Area (m²)	Heat Loss kcal/h	Heat Loss kJ/h
346	619	0.51	4177.75	5628	7763	1.75	18.434	143,107	598,189
318	591	0.45	4235.58	4701	6595	1.75	18.434	121,580	508,203
325	598	0.46	4221.55	4923	6877	1.75	18.434	126,769	529,896
313	586	0.44	4245.42	4547	6398	1.75	18.434	117,950	493,031
312	585	0.43	4247.37	4516	6360	1.75	18.434	117,232	490,029
318	591	0.45	4235.58	4701	6595	1.75	18.434	121,580	508,203
337	610	0.49	4196.83	5319	7376	1.75	18.434	135,963	568,325
320	593	0.45	4231.60	4764	6675	1.75	18.434	123,050	514,347
218	491	0.25	4397.47	2189	3283	1.75	18.434	60,516	252,959
288	561	0.38	4292.27	3822	5465	1.75	18.434	100,748	421,129
299	572	0.41	4272.16	4131	5865	1.75	18.434	108,125	451,961
305	578	0.42	4260.85	4306	6091	1.75	18.434	112,275	469,309
305	578	0.42	4260.85	4306	6091	1.75	18.434	112,275	469,309
309	582	0.43	4253.19	4426	6243	1.75	18.434	115,092	481,085
288	561	0.38	4292.27	3822	5465	1.75	18.434	100,748	421,129
286	559	0.38	4295.84	3768	5394	1.75	18.434	99,439	415,657
269	542	0.34	4324.99	3323	4812	1.75	18.434	88,703	370,778
263	536	0.33	4334.76	3175	4615	1.75	18.434	85,078	355,627
279	552	0.36	4308.10	3581	5150	1.75	18.434	94,934	396,825

277	550	0.36	4311.54	3528	5081	1.75	18.434	93,669	391,536
278	550	0.36	4311.54	3528	5081	1.75	18.434	93,669	391,536
276	549	0.36	4313.25	3502	5047	1.75	18.434	93,040	388,906
264	537	0.33	4333.15	3199	4648	1.75	18.434	85,677	358,128
279	552	0.36	4308.10	3581	5150	1.75	18.434	94,934	396,825
264	537	0.33	4333.15	3199	4648	1.75	18.434	85,677	358,128
279	552	0.36	4308.10	3581	5150	1.75	18.434	94,934	396,825
282	555	0.37	4302.89	3660	5254	1.75	18.434	96,850	404,835
270	543	0.35	4323.34	3349	4845	1.75	18.434	89,315	373,338
353	626	0.53	4162.60	5877	8073	1.75	18.434	148,814	622,042
354	627	0.53	4160.41	5913	8118	1.75	18.434	149,640	625,494
335	608	0.49	4201.00	5251	7291	1.75	18.434	134,404	561,810
474	747	0.85	3859.52	11,325	14,604	1.75	18.434	269,218	1,125,330
474	747	0.85	3859.52	11,325	14,604	1.75	18.434	269,218	1,125,330
477	750	0.86	3850.99	11,490	14,796	1.75	18.434	272,745	1,140,075
412	685	0.68	4024.52	8252	10,973	1.75	18.434	202,274	845,503
432	705	0.73	3973.56	9176	12,077	1.75	18.434	222,637	930,624
353	626	0.53	4162.60	5877	8073	1.75	18.434	148,814	622,042
257	530	0.32	4344.23	3030	4423	1.75	18.434	81,538	340,829
220	493	0.25	4395.14	2228	3337	1.75	18.434	61,513	257,124
224	497	0.26	4390.33	2308	3446	1.75	18.434	63,532	265,563
273	546	0.35	4318.33	3425	4946	1.75	18.434	91,167	381,077

Continued

Table 9.8 Shell Heat Loss Calculated[a]—cont'd

Shell Zone Temperature (°C)	Shell Zone Temperature (K)	Q1	Q2	Q3	Q5 (kcal/m²/h)	Length (m)	Area (m²)	Heat Loss kcal/h	Heat Loss kJ/h
371	644	0.57	4122.42	6547	8902	1.75	18.434	164,098	685,928
338	611	0.49	4194.73	5352	7418	1.75	18.434	136,746	571,599
419	692	0.69	4006.92	8568	11,352	1.75	18.434	209,269	874,745
357	630	0.54	4153.82	6022	8253	1.75	18.434	152,134	635,920
406	679	0.66	4039.40	7987	10,654	1.75	18.434	196,389	820,908
408	681	0.67	4034.46	8074	10,759	1.75	18.434	198,339	829,059
393	666	0.63	4070.99	7431	9981	1.75	18.434	183,993	769,090
394	667	0.63	4068.59	7472	10,032	1.75	18.434	184,929	773,005
391	664	0.62	4075.77	7347	9880	1.75	18.434	182,128	761,296
378	651	0.59	4106.33	6821	9237	1.75	18.434	170,281	711,777
349	622	0.52	4171.29	5734	7895	1.75	18.434	145,537	608,344
376	649	0.58	4110.96	6742	9141	1.75	18.434	168,501	704,333
335	608	0.49	4201.00	5251	7291	1.75	18.434	134,404	561,810
Total	—	—	—	—	—	—	—	7,255,192 kcal/h (8424 kJ/s)	30,326,702 kJ/h (28,507,100 Btu/h)

[a]Kiln shell diameter = 3.353 m, ambient air temperature = 301 K.

Table 9.9 Heat of Formation (for Calcination Calculation)[a]

Heat of Formation at 25 °C	\multicolumn{4}{c	}{Heat of Formation[a]}	
	kcal/kmol	kJ/kmol	kJ/kg	MJ/kg	Btu/lb
H_f—$CaCO_3$	289,500	1,211,246	12,100	12.10	5.22
H_f—CaO	151,700	634,701	11,314	11.31	4.88
H_f—$MgCO_3$	261,700	1,094,933	12,985	12.99	5.60
H_f—MgO	143,840	601,816	14,926	14.93	6.44
H_f—CO_2	−94,052	−393,507	−8943	−8.94	−3.86

Heat of Reaction at 25 °C	\multicolumn{4}{c	}{Heat of Calcination[b]}	
	kcal/kmol	kJ/kmol	kJ/kg	MJ/kg	Btu/lb
$CaCO_3 \rightarrow CaO + CO_2$	43,748	183,038	1829	1.829	0.79
$MgCO_3 \rightarrow MgO + CO_2$	23,808	99,611	1181	1.181	0.51

Heat of Evaporation	\multicolumn{4}{c	}{Heat of Water Evaporation[c]}	
	kcal/kmol	kJ/kmol	kJ/kg	MJ/kg	Btu/lb
$H_2O(l) \rightarrow H_2O(g)$ 25 °C	—	—	2452	2.45	1.06
$H_2O(l) \rightarrow H_2O(g)$ 57 °C	—	—	2365	2.37	1.02

[a]Boynton (1980) and Perry (1984).
[b]Note: $CaCO_3 = CaO + CO_2$; 770 kcal/kg CaO ⇔ 180,734 kJ/kmol $CaCO_3$ or CaO.
[c]Note: $H_2O \rightarrow H_2 + 0.5O_2 \rightarrow$ 970 Btu/lb of water at 212 °F.

Table 9.10 Heat of Combustion Reactions

Heat of Reaction	kcal/kmol	kJ/kmol	kJ/kg	MJ/kg	Btu/lb
$N + O_2 = NO$	21,600	90,373	6455	6.46	2783
$C + O_2 = CO_2$	−94,052	−393,507	−32,792	−32.79	−14,139
$C + O_2 = CO$	−26,416	−110,523	−9210	−9.21	−3971
$H + O_2 = H_2O$	−57,798	−241,822	−120,911	−120.91	−52,135
$S + O_2 = SO_2$	−70,940	−296,808	−9275	−9.28	−3999

Table 9.11 Calculation of Energy of Fuel

Elements	Fuel Composition		Heat of Combustion		
	wt%	Coefficient	kJ/kg Fuel	MJ/kg Fuel	Btu/lb
C	70	36.2	25,340	25.3	10,926
H	4.4	90	3960	4.0	1707
S	1.6	10.6	170	0.2	73
N_2	0.9	8	72	0.1	31
O_2	1.8	−10.6	−191	−0.2	−82
H_2O	0	0	0	0.0	0
Ash	15.3	0	0	0.0	0
Total	94	—	29,351	29.4	12,656

9.13 Sensible Energy for Output Streams

Table 9.16 covers the sensible heat of the components of the output gas. Table 9.17 covers the lime product output sensible heat. Table 9.18 covers the sensible heat from the dust. Table 9.19 covers the heat releases from the combustion reaction. Table 9.20 covers the heat consumed by calcinations. Table 9.21 covers the material and energy balance for the dolomitic lime process. Table 9.22 gives a summary and analysis, including the material balance and the heat balance.

Table 9.12 Calculation of Energy of Natural Gas

	Stoichiometric Combustion			Gross (kcal/mol)	Net (kcal/mol)
	O_2 Required	CO_2 Produced	H_2O Produced		
CH_4	2	1	2	212.798	191.759
C_2H_6	3.5	2	3	372.820	341.261
C_3H_8	5	3	4	530.605	488.527
iso-C_4H_{10}	6.5	4	5	686.342	633.744
n-C_4H_{10}	6.5	4	5	687.982	635.384
iso-C_5H_{12}	8	5	6	843.240	780.120
n-C_5H_{12}	8	5	6	845.160	782.040
C_6H_{14}	9.5	6	7	1002.570	928.930
O_2	−1	0	0	0	0

Table 9.13 Calculation of Percent of Water in Air

Ambient Air Conditions					Calculation Factors
Temperature		20	°C		0.00856
Relative humidity		60	%		0.99977
Humidity		0.00856 kg H_2O/kg dry air			
Humidity		0.01378 mol H_2O/mol dry air			
×H_2O		0.01359 mol H_2O/mol H_2O + mol dry air			
Humidity		1.38%			
	N_2%	O_2%	CO_2%	H_2O%	Total%
	77.43	20.71	0.49	1.38	100.02

Table 9.14 **Heat Capacity of Constituents**

Component	T (K)	Heat Capacity (C_P) Formula	Kcal/kmol K	kJ/kmol K	kJ/kg	Btu/lb
$CaCO_3$	400	$19.68 + 0.01189T - 307,600/T^2$	22.51	94.2	0.94	0.41
CaO	700	$10.0 + 0.004484T - 108,000/T^2$	13.17	55.1	0.98	0.42
$MgCO_3$	692	16.90	16.90	70.7	0.84	0.36
MgO	692	$10.86 + 0.001197T - 208,700/T^2$	11.25	47.1	1.17	0.50
C	662	$2.673 + 0.002617T - 116,900/T^2$	4.14	17.3	1.44	0.62
CO	662	$6.6 + 0.00122T$	7.39	30.9	1.10	0.48
CO_2	962	$10.34 + 0.00274T - 195,500/T^2$	12.76	53.4	1.21	0.52
N_2	962	$6.5 + 0.001T$	7.46	31.2	1.12	0.48
NO	962	$8.05 + 0.000233T - 156,300/T^2$	8.11	33.9	1.13	0.49
NO_2	963	$36.07 + 0.0397 \times T - 0.0000288T^2 + 7.87 \times (10^{-9}) \times T^3$		47.59	1.03	0.45
O_2	962	$8.27 + 0.000258T - 187,700/T^2$	8.07	33.8	1.05	0.45
S	389	$3.63 + 0.00064T$	6.12	25.6	0.80	0.34
SO_2	962	$7.7 + 0.0053 - 0.00000083T^2$	12.03	50.3	0.79	0.34
Cl_2	962	$8.28 + 0.000567T$	8.82	36.9	0.53	0.23
H	962	4.97	4.97	20.8	20.79	8.97
H_2	962	$6.62 + 0.00081T$	7.40	31.0	15.48	6.67
H_2O(l)	283	4.18	4.18	17.5	0.97	0.42
H_2O(g)	283	$8.22 + 0.00015T + 0.00000134T^2$	8.37	35.0	1.95	0.84
Coal	283	$0.205 + (0.205 \times 10^{-3}) \times \%VM + (3.104 \times 10^{-4}) \times T + (2.77 \times 10^{-6}) \times \%VM$				

VM, volatile matter.

Table 9.15 Sensible Energy for Input Streams

Coal		kJ/s
H_2O	$M_{H_2O} \times 0.97 \text{ kJ/kg K} \times (T - 298)$	2.6
Dry coal	$M_{coal} \times (0.205 + 0.205 \times 10^{-3} \times (\%VM) + 3.104 \times 10^{-4} \times (T°C) + 2.77 \times 10^6 \times (\%VM) \times (T°C) \times (25 - T°C)$	85
	Total	87
Natural Gas		
N_2	$M_{O_2} \times (8.27 + 0.000258T - 187{,}700/T^2) \times (1/0.23901) \times (T - 298)$	0
O_2	$M_{O_2} \times (8.27 + 0.000258T - 187{,}700/T^2) \times (1/0.23901) \times (T - 298)$	0
CO_2	$M_{CO_2} \times (10.34 + 0.00274T - 195{,}500/T^2) \times (1/0.23901) \times (T - 298)$	0
CH_4	$((M_{CH_4}/16) \times ((33{,}300 + 79{,}930 \times ((2086.9/T)/\sinh(2068.9/T))^2 + 41{,}600 \times ((991.96/T)/\cosh(991.96/T))^2) \times (T - 298))/1000$	0.2
C_2H_6	$(M_{C_2H_6}/16) \times ((40{,}330 + 134{,}220 \times ((1655.5/T)/\sinh(1655.5/T))^2 + 73{,}220 \times ((752.87/T)/\cosh(752.87/T))^2) \times (T - 298))/1000$	0
C_3H_8	$((M_{C_3H_8}/13) \times ((51{,}920 + 192{,}450 \times ((1626.5/T)/\sinh(1626.5/T))^2 + 116{,}800 \times ((723.6/T)/\cosh(723.6/T))^2) \times (T - 298))/1000$	0
C_4H_{10}	$((M_{C_4H_8}/13) \times ((71{,}340 + 243{,}000 \times ((1630/T)/\sinh(1630/T))^2 + 150{,}330 \times ((730.42/T)/\cosh(730.42/T))^2) \times (T - 298))/1000$	0
C_5H_{12}	$((M_{C_5H_{12}}/13) \times ((88{,}050 + 301{,}100 \times ((1650.2/T)/\sinh(1650.2/T))^2 + 189{,}200 \times ((747.6/T)/\cosh(747.6/T))^2) \times (T - 298))/1000$	0
C_6H_{14}	$((M_{C_6H_{14}}/13) \times ((10{,}440 + 3.523 \times ((1694.6/T)/\sinh(1694.6/T))^2 + 236{,}900 \times ((761.6/T)/\cosh(761.6/T))^2) \times (T - 298))/1000$	0
Coal Conveying Air		
N_2	$M_{O_2} \times (8.27 + 0.000258T - 187{,}700/T^2) \times (1/0.23901) \times (T - 298)$	114
O_2	$M_{O_2} \times (8.27 + 0.000258T - 187{,}700/T^2) \times (1/0.23901) \times (T - 298)$	27
CO_2	$M_{CO_2} \times (10.34 + 0.00274T - 195{,}500/T^2) \times (1/0.23901) \times (T - 298)$	1
H_2O	$M_{H_2O} \times (8.22 + 0.00015T + 0.00000134/T^2) \times (1/0.23901) \times (T - 298)$	6

Continued

Table 9.15 Sensible Energy for Input Streams—cont'd

Coal		kJ/s
Inlet Air		
N_2	$M_{O_2} \times (8.27 + 0.000258T - 187{,}700/T^2) \times (11/0.23901) \times (T - 298)$	4213.4
O_2	$M_{O_2} \times (8.27 + 0.000258T - 187{,}700/T^2) \times (1/0.23901) \times (T - 298)$	1157.7
CO_2	$M_{CO_2} \times (10.34 + 0.00274T - 195{,}500/T^2) \times (1/0.23901) \times (T - 298)$	29.2
H_2O	$M_{H_2O} \times (8.22 + 0.00015T + 0.00000134/T^2) \times (1/0.23901) \times (T - 298)$	227
Limestone		
$CaCO_3$	$M_{CaCO_3} \times (19.68 + 0.01189T - 307{,}600/T^2) \times (1/0.23901) \times (T - 298)$	5.8
$MgCO_3$	$M_{MgCO_3} \times 16.9 \times (1/0.23901) \times (T - 298)$	4.8
H_2O	$M_{H_2O} \times (4.1812 \text{ kJ/kg K}) \times (T - 298)$	1.1
S	$M_S \times (3.63 + 0.0064T) \times (1/0.23901) \times (T - 298)$	0.0
Inerts	$M_{inert} \times (10.0 + 0.00484T - 108{,}000/T^2) \times (1/0.23901) \times (T - 298)$	0.0

Table 9.16 Sensible Heat for Flue Gas Constituents

Flue Gas T (K)	Formula	kJ/s
N_2	$M_{N_2} \times (6.5 + 0.001T) \times (1/0.23901) \times (T - 298)$	10,750
NO	$M_{NO} \times (8.05 + 0.000233T - 156,300/T^2) \times (T - 298)$	0
O_2	$M_{O_2} \times (8.27 + 0.000258T - 187,700/T^2) \times (0.23901) \times (T - 298)$	57
CO	$M_{CO} \times (6.6 + 0.0012T) \times (1/0.23901) \times (T - 298)$	0
CO_2	$M_{CO_2} \times (10.34 + 0.00274T - 195,500/T^2) \times (1/0.23901) \times (T - 298)$	8017
H_2O	$M_{H_2O} \times (8.22 + 0.00015T + 0.00000134T^2) \times (1/0.23901)$	2028
SO_2	$M_{SO_2} \times (7.7 - 0.0053 - 0.00000083T^2) \times (1/0.23901) \times (T - 298)$	1

Table 9.17 Sensible Heat for Lime Products

Lime Product	Formula	kJ/s
$CaCO_3$	$M_{CaCO_3} \times (19.68 + 0.01189T - 307,600/T^2) \times (1/0.23901) \times (T - 298)$	86
CaO	$M_{CaO} \times (10.0 + 0.00484T - 108,000/T^2) \times (T - 298)$	5659
$MgCO_3$	$M_{MgCO_3} \times 16.9 \times (1/0.23901) \times (T - 298)$	35.5
MgO	$M_{MgO} \times (10.86 + 0.001197T - 208,700/T^3) \times (1/0.23901) \times (T - 298)$	4013.9
S	$M_S \times (3.63 + 0.0064T) (1/0.23901) (T - 298)$	4.4
Inert	$M_{inert} \times (10.0 + 0.00484T - 108,000/T^2) \times (1/0.23901) \times (T - 298)$	146

Table 9.18 Sensible Heat for Constituents of Dust

Dust	Formula	kJ/s
$CaCO_3$	$M_{CaCO_3} \times (19.68 + 0.01189T - 307{,}600/T^2) \times (1/0.23901) \times (T - 298)$	1.7
CaO	$M_{CaO} \times (10.0 + 0.00484T - 108{,}000/T^2) \times (T - 298)$	121.9
$MgCO_3$	$M_{MgCO_3} \times 16.9 \times (1/0.23901) \times (T - 298)$	1.1
MgO	$M_{MgO} \times (10.86 + 0.001197T - 208{,}700/T^3) \times (1/0.23901) \times (T - 298)$	104.7
S	$M_S \times (3.63 + 0.00647T) \times (1/0.23901) \times (T - 298)$	0.1
Inert	$M_{inert} \times (10.0 + 0.00484T - 108{,}000/T^2) \times (1/0.23901) \times (T - 298)$	3.1

Table 9.19 Heat Releases from the Combustion Reaction

From C

$C + O_2 = CO_2 \rightarrow -94{,}052$ kcal/kmol $\leftrightarrow -393{,}507$ kJ/kmol
$\rightarrow (-393{,}507$ kJ/kmol °C$) \times$ kmol °C

From H

$H + O_2 = H_2O \rightarrow 57{,}798$ kcal/kmol $\leftrightarrow -241{,}822$ kJ/kmol
$\rightarrow (-241{,}822$ kJ/kmol H$) \times$ kmol H

From S

$S + O_2 = SO_2 \rightarrow -70{,}940$ kcal/kmol $\leftrightarrow -296{,}808$ kJ/kmol
$\rightarrow (-296{,}808$ kJ/kmol S$) \times$ kmol S

Heat of formation (J/mol)	CH_4	802,319	C_4H_{10}	2,658,446
	C_2H_6	1,427,836	C_5H_{12}	3,272,055
	C_3H_8	2,043,996	C_6H_{14}	3,886,643

Table 9.20 Heat Consumed by Calcination (H_r)

From $CaCO_3 = CaO + CO_2$

$H_r = H_f(CO_2) + H_f(CaO) - H_f(CaCO_3)$

$H_r = -94,052\ \text{kcal/kmol} - 151,700\ \text{kcal/kmol} + 298,500\ \text{kcal/kmol} = -43,798\ \text{kcal/kmol}$

Heat consumed by

$CaCO_3 = (183,038\ \text{kJ/kmol CaCO}_3) \times (\text{kmol CaCO}_3\ \text{reacted})$ | 12,668

From $MgCO_3 = MgO + CO_2$

$H_r = H_f(CO_2) + H_f(MgO) - H_f(MgCO_3)$

$H_r = -94,052\ \text{kcal/kmol} - 143,840\ \text{kcal/kmol} + 261,700\ \text{kcal/kmol} = -23,808\ \text{kcal/kmol}$

Heat consumed by

$MgCO_3 = (99,611\ \text{kJ/kmol MgCO}_3) \times (\text{kmol MgCO}_3\ \text{reacted})$ | 0

Heat of water evaporation (kJ/kg) 2365 × 0.27 | 629

Heat Loss (First Law of Thermodynamics)

Heat loss = inlet sensible heat + heat of combustion − outlet sensible heat − heat of calcination − heat of water evaporation | 10,994

Table 9.21 Material and Energy Balance for Dolomitic Lime Process

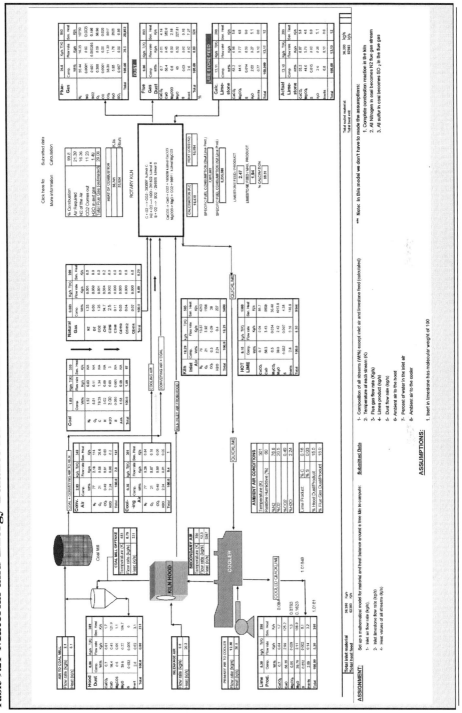

Table 9.22 Summary and Analysis for Material Balance and Heat Balance

Material Balance (kg/s)					
Feed Stream	**Actual**	**Balanced**	**Exit Stream**	**Actual**	**Balanced**
Limestone feed	13.1	13.1	Flue gas	–	29.50
Coal	1.69	–	Dust	–	1.59
Conveying air	2.8	–	Lime product	–	5.30
Natural gas	0.09	–	–	–	–
Cooling air	0.36	–	–	–	–
Inlet air	–	18.3	–	–	–
Material feed total	–	36.4	Outlet material total	–	36.4
Heat Balance (kJ/kg)					
Feed Stream	**Actual**	**Calculated**	**Exit Stream**	**Actual**	**Calculated**
Sensible heat	kJ/s	kJ/s	Sensible heat	kJ/s	kJ/s
Limestone	–	12	Flue gas	–	20,853
Coal	–	87	Flue gas dust	–	524
Conveying air	–	141	Hot lime	–	9944
Natural gas	–	0.2	Heat of calcination	–	19,439
Inlet air	–	5400	Heat of water evaporation	–	629
Cooling air	–	1	Total heat loss	–	10,994
Heat of combustion	–	56,741	Kiln heat loss	8424	–
Total	–	62,383	–	–	62,383

References

Boynton, R.S., 1980. Chemistry and Technology of Lime and Limestone. John Wiley & Sons, New York.

Knacke, O., Kubaschewski, O., Hesselmann, K. (Eds.), 1991. Thermochemical Properties of Inorganic Substances, vols. I and II. Springer-Verlag, New York.

Outokumpu Research Oy, 2002. Manual for Outokumpu HSC Chemistry. Chemical Reaction and Equilibrium Software, Finland.

Perry, R.H., 1984. Perry's Chemical Engineering Handbook. McGraw-Hill, NY.
Rogers, C.F.C., Mayhew, Y.R., 1982. Thermodynamic and Transport Properties of Fluids. Basil Blackwell Publishers, Oxford, UK.
Themelis, N.J., 1995. Transport and Chemical Rate Phenomena. Taylor & Francis, New York, 1995.

Rotary Kiln Minerals Process Applications

10

The processes that employ the rotary kiln as the primary reactor are many and they encompass all industries including food, dedicated dryers, and minerals processing. We will focus on minerals and materials processing and provide some of the key industrial processes that use a rotary kiln as the primary workhorse. As we saw from the history of cement in Chapter 1, rotary kiln technology as we know it today evolved from the struggles of the early engineers and inventors to produce cement and lime in an efficient, safe, and economic way. It is only prudent that we begin this chapter with these processes. We will first describe the cement and lime processes and then move on to describe some of the carbothermic reduction processes that employ the rotary kiln as the primary device. Here we will examine two processes, iron ore reduction (the SL/RN (acronym for steel-making companies Stelco, Lurgi, and Republic National) process) and ilmenite ore reduction (the Becher process). We will then examine the lightweight aggregate (LWA) process.

10.1 Lime Making

Lime manufacturing is one of the oldest processes known to man, having been carried out since prehistoric times. The remains of fossils, according to geologists, formed the limestone bed thousands of years ago. The abundance of limestone as well as high-magnesium limestone, called dolomite, in much of the world, has played an important roll in the development of many communities. Limestone is a rock composed of calcium, carbon, and oxygen. When it is heated, the carbon escapes as carbon dioxide, leaving calcium oxide, which is known as quicklime or, simply, lime. This process is called lime burning or limestone "calcination." In the United States, probably the first settlement to obtain lime from limestone was Rhode Island, followed by the Quakers when they settled in Philadelphia and the Dutch settlers of the Hudson River valley (Oates, 1998). It has been reported that cargoes of lime were shipped by boat from Maine to Boston and New York City in the early 1800s. The lime burning industry we see today came into prominence at the end of the 1890s. With so much suitable stone readily available and right at hand, it was natural that the first homes should have used lime burnt on the site. Hence, there were several lime kilns operating in many areas of the United States at the time. Surface stones were cracked for burning in kilns that were originally fired by wood. Often, large kilns were set in cliff faces as the sea winds assisted the burning process. However, much of the work was manual, including the crushing, until the crusher machine was designed, which was used primarily to crush limestone for road beds.

Lime was and is still used in soil stabilization, road building, chemical manufacture, tanning process, and for the purification of water, whitewashing buildings, treating animal hides and leather, and plaster and mortar. The US National Lime Association tends to think that lime is "the versatile chemical."

10.2 Limestone Dissociation (Calcination)

When limestone (calcium carbonate) is heated, it dissociates into quicklime and carbon dioxide in a process known as calcination. The calcination reaction is as follows:

$$CaCO_3(s) \Leftrightarrow CaO(s) + CO_2(g) \quad (10.1)$$

where s and g denote the solid and gaseous states, respectively. The equilibrium Gibbs free energy of the reaction at atmospheric pressure ($p_o = 1$) can be expressed as a function of temperature, that is,

$$\Delta G_{rx}^\circ = 182{,}837 + 13.402 T \ln T - 251.059 T \; [\text{J mol}^{-1}] \quad (10.2)$$

Recall that,

$$\Delta G_{rx}^\circ = -RT \ln K_e$$
$$\ln K_e = -\frac{21{,}990}{T} - 1.6119 \ln T + 30.196 \quad (10.3)$$

We assume that the activities of the solid phases are equal to unity. For the gaseous phase, the activity is a function of the dissociation pressure; therefore, the partial pressure of CO_2 at equilibrium with $CaCO_3$ is (Themelis, 1995)

$$P_{CO_2,e} \approx K_e p_o \quad (10.4)$$

HSC[1] calculations show that decomposition will proceed at about 900 °C. An alternative means of attaining decomposition at a temperature lower than 900 °C is to decrease the bulk concentration of CO_2 in the reactor by providing a flow of air, or any other gas with the exception of CO_2, over the decomposing limestone. The task is to select the appropriate reactor with the least heat transfer resistance so that the reactor temperature is well above the dissociation temperature necessary to accomplish the reaction. We will examine the appropriate heat transfer resistances later. The heat of reaction can be estimated from Table 10.1.

---

[1] HSC means enthalpy (H), entropy (S), and heat capacity (Cp).

Table 10.1 HSC Calculations

$CaCO_3 = CaO + CO_2(g)$					
T (°C)	ΔH (kcal)	ΔS (cal/K)	ΔG (kcal)	K	log(K)
0	42.604	38.345	32.13	1.95×10^{-26}	−25.709
100	42.475	37.954	28.313	2.61×10^{-17}	−16.584
200	42.258	37.442	24.543	4.60×10^{-12}	−11.337
300	41.99	36.928	20.824	1.15×10^{-08}	−7.941
400	41.68	36.432	17.156	2.69×10^{-06}	−5.571
500	41.334	35.952	13.537	1.49×10^{-04}	−3.827
600	40.949	35.485	9.965	3.20×10^{-03}	−2.495
700	40.525	35.026	6.44	3.58×10^{-02}	−1.446
800	40.062	34.573	2.96	2.50×10^{-01}	−0.603
900	39.556	34.123	−0.475	1.23	0.088
1000	39.007	33.674	−3.864	4.61	0.663
1100	38.416	33.227	−7.209	1.41×10^{1}	1.148
1200	37.781	32.78	−10.51	3.63×10^{1}	1.559
Formula	MW (g/mol)	Concentration (wt%)	Amount (mol)	Amount (g)	Volume
$CaCO_3$	100.089	100	1	100.089	36.933 mL
CaO	56.079	56.029	1	56.079	16.79 mL
$CO_2(g)$	44.01	43.971	1	44.01	22.414 L

MW, molecular weight.

$$CaCO_3 = CaO + CO_2 \quad \Delta H = -1768 \, \text{kJ/kg} (-760 \, \text{Btu/lb}) \, CaCO_3 \quad (10.5)$$

Magnesium carbonate, which is found in some quantities with limestone deposits, also dissociates the same way according to the following reaction:

$$MgCO_3 = MgO + CO_2 \quad \Delta H = -1298 \, \text{kJ/kg} (-558 \, \text{Btu/lb}) \, MgCO_3 \quad (10.6)$$

For the purpose of feedstock quality classification, mineral deposits that contain a significant amount of $MgCO_3$ (5−39%) are called dolomitic limestone, whereas those containing >40% are known as dolomite.

The dissociation of both calcium and magnesium carbonate is endothermic and needs elevated temperature to drive the reaction (see Figures 10.1 and 10.2). The reaction rate increases with increasing temperature. For magnesium carbonate,

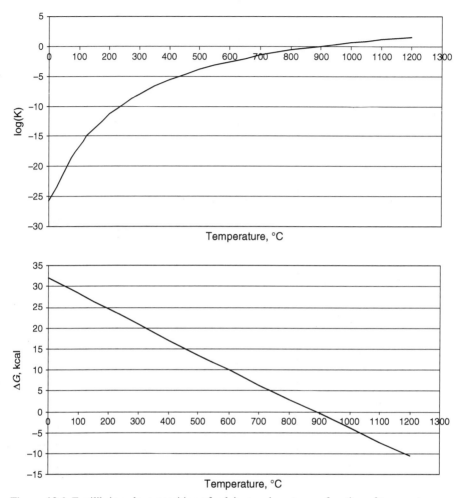

Figure 10.1 Equilibrium decomposition of calcium carbonate as a function of temperature.

the reaction commences at 250 °C (480 °F) and requires 410 °C (770 °F) to go to completion at atmospheric pressure (Figure 10.2), compared with calcium carbonate for which the reaction occurs between 500 and 805 °C (930–1480 °F). However, the decomposition of dolomites and magnesium/dolomitic limestone precursors is more complex than that of their pure compounds. It is still not clear whether the decomposition reactions proceed in a single path, two discrete paths, or a combination of both (Oates, 1998), that is,

$$CaCO_3 \cdot MgCO_3 + Heat = CaCO_3 \cdot MgO + CO_2 \tag{10.7}$$

$$CaCO_3 \cdot MgO + Heat = CaO \cdot MgO + CO_2 \tag{10.8}$$

$$CaCO_3 \cdot MgCO_3 + Heat = CaO \cdot MgO + CO_2 \tag{10.9}$$

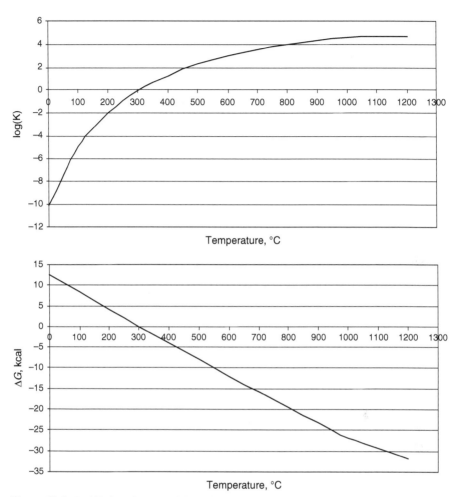

Figure 10.2 Equilibrium decomposition of magnesium carbonate as a function of temperature.

All dolomites and magnesium/dolomite limestone decompose at higher temperatures than magnesium carbonate, with the onset of dissociation varying between 510 and 750 °C. If the feedstock is heated to temperatures much greater than 900 °C, the quicklime product can become very dense with very low internal surface area and is described as overburnt. Overburning results in low-reactivity lime, which, although useful for certain applications such as refractory linings, is unacceptable for many chemical applications. The temperature at which overburning becomes serious is about 1400 °C for pure limestone and may be as low as 1200 °C for some of the less-pure materials. For many applications of quicklime, active lime is the preferred product, hence careful control of the dissociation (calcination) process is necessary, bearing in mind that the time required for complete calcination depends on factors such as kiln temperature, stone size, and porosity of feed material.

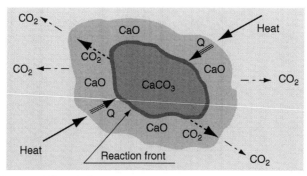

Figure 10.3 Limestone dissociation process steps.

The dissociation of limestone above the decomposition temperature is a heterogeneous reaction (Figure 10.3). This dissociation can be regarded as being made up of the following five process steps:

1. Heat is transferred from the furnace gases to the surface of the decomposing particle.
2. This is followed by heat conduction from the surface to the reaction front through the microporous lattice structure of lime.
3. Heat arrives at the reaction front and causes the dissociation of $CaCO_3$ into CaO and CO_2.
4. The CO_2 produced migrates from the reaction front, through the lime layer, to the particle surface.
5. CO_2 migrates away from the particle surface into the kiln's atmosphere.

Each of these events poses some degree of resistance to the overall process of achieving complete dissociation (calcination) and can be an impediment to achieving quality product. The step controlling the process will ultimately be the one with the highest degree of resistance. Since the decomposition reaction is fast, once the particle attains the appropriate temperature, perhaps the rate-controlling step will depend on how fast one can get the heat to the reaction front and get rid of the CO_2 accumulated at the front so as to prevent recarbonation of the quicklime already formed. Hence, process development and the evolution of lime kiln design over the years have been based on mixing that improves heat and mass diffusion. The selection of the appropriate kiln technology, kiln's dimensions, and internal mixing aid structures such as lifters, as well as operational procedures such as feed particle size distribution, are all factors that influence effective heat transfer to the material and the gas diffusion to the atmosphere.

10.3 The Rotary Lime Kiln

From the fuel requirement point of view, rotary kilns are the most flexible of all lime kilns (Oates, 1998). They are successfully fired with natural gas, fuel oil, and pulverized fuels of all types including coal, coke, and sawdust. According to Boynton (1980), the United States is by far the world's leader in rotary kiln lime production with about

Rotary Kiln Minerals Process Applications 237

Figure 10.4 Photo of preheater lime kiln.
Courtesy of Graymont, Inc., Cricket Mountain Plant.

88% of its commercial and about 70% of captive plant capacity provided by kilns. The conventional rotary lime kiln has a length-to-diameter (L/D) ratio in the 30–40 range with lengths of 75–500 ft (22.7–152.5 m) and diameter of 4–11 ft (1.2–3.3 m). Lime kilns are usually inclined at about 3°–5° slope with material charged at the elevated end and discharging at the lower end. The degree of fill is relatively deep, about 10–12%. Owing to its low thermal conductivity, limestone with a large diameter of about 2 in (5 cm) results in higher effective bed heat conduction than smaller stones. The larger feed material sizes tend to have larger pore volume in the bulk and thereby maximize the particle-to-particle heat transfer, which is usually dominated by radiation at the dissociation temperatures. The smaller feed stones tend to pack themselves upon rotation and render the bed a poor conductor of heat. For many years, most long kilns operated with deplorable fuel efficiencies because of poor or lack of heat recuperation such as coolers and preheaters (Figure 10.4) with thermal consumption as high as 12–15 million Btu/ton (3336–4170 kcal/kg) of lime. Thanks to ingenious heat recuperation systems such as coolers, preheaters, and lifters, today, thermal efficiencies of rotary lime kilns are in the 6–8 million Btu/ton range (1668–2224 kcal/kg), using fuel at about half the rate of early long kilns.

Some rotary lime kilns operate under reducing conditions by curtailing the combustion air to substoichiometric levels so as to volatilize any sulfur that may be in the limestone in order to meet the stringent sulfur specifications imposed by steel and chemical users. For most operations except for dead burnt dolomite, the burner tip velocities can range between a low of 25 m/s and a high of about 60 m/s. These are significantly lower than the velocities of cement kilns, which operate around 80–100 m/s. The momentum ratio and associated Craya–Curtet parameter is usually lower than 2, which means that the burner jet recirculation will have eddies and that fuel/air mixing is moderate and the flame is less intense than that in dead burnt dolomite kilns or cement kilns. A simple heat and mass balance for the kiln section of a lime-making process is shown in Figure 10.5.

10.4 The Cement-Making Process

Rotary kilns are synonymous with cement making, being the workhorses of this industry. There are many types of rotary kiln arrangements for producing cement clinker with each incremental design goal aimed at improving energy efficiency, ease of operation, and product quality and minimizing environmental pollutants. Rotary cement kilns can be classified into wet-process kilns, semidry kilns, dry kilns, preheater kilns, and precalciner kilns. All of these are described in the book by Peray (1986) and many others, hence we will not dwell upon them here. Rather, we will briefly show the pertinent process chemistry and the heat requirements that drive them, so as to be consistent with the transport phenomena theme.

10.5 The Cement Process Chemistry

The raw mixture of the cement kiln feedstock (or charge, raw meal) includes some formulations of limestone ($CaCO_3$), alumina (Al_2O_3), hematite (Fe_2O_3), and silica (SiO_3). As we discussed earlier, in its journey through the kiln, the charge undergoes all kinds of processes depending on the temperature, including drying, preheating, chemical reactions, a phase change, restructuring or sintering, and cooling. For kilns equipped with preheaters and precalciners (Figure 10.6), all the drying and some of the calcination reactions, for example, partial or full dissociation of limestone, take place there before entering the kiln proper. When the kiln feed enters the high-temperature zones in the rotary kiln, a series of chemical reactions occur in which the quicklime, alumina, ferric oxide, silica, and other metal oxides react to form four main compounds of cement (Wang et al., 2006), namely, $CaO \cdot SiO_2$ (C_3S), $2CaO \cdot SiO_2$ (C_2S), $3CaO \cdot Al_2O_3$ (C_3A), and $4CaO \cdot Al_2O_3 \cdot Fe_2O_3$ (C_4AF). The formation temperatures of these compounds differ, which therefore defines the axial zones in which each compound is formed. The kiln's axial temperature profile can be divided into three zones where all the reactions occur either independently or simultaneously. These delineations include (1) the decomposition zone (900 °C), (2) the transition zone (900–1300 °C), and (3) the sintering zone (1300–1400 °C).

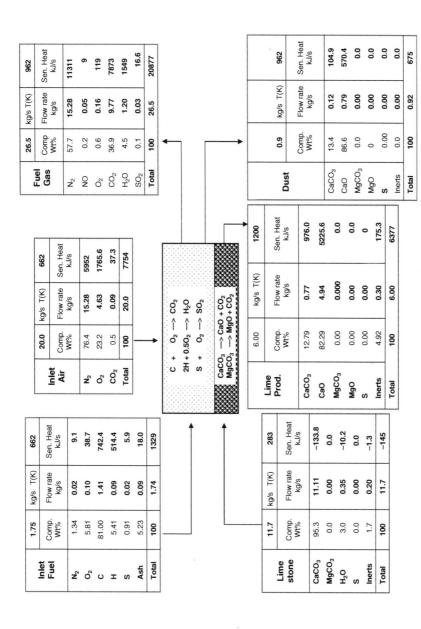

Figure 10.5 Lime kiln heat and material balance.
Comp. = composition; Sen. Heat = sensible heat; Lime Prod. = produced lime.

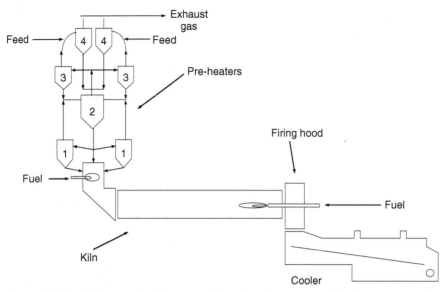

Figure 10.6 Typical schematic of a cement kiln with preheater, precalciner, and grate cooler.

10.5.1 Decomposition Zone

The amount of unreacted raw material in the outlet of the precalciner can be as much as 85–95% (Wang et al., 2006). Upon entering the decomposition zone, small amounts of $CaO \cdot Al_2O_3$ (CA), $CaO \cdot Fe_2O_3$ (CF), $2CaO \cdot Fe_2O_3$, and $5CaO \cdot 3Al_2O_3$ (C_5A_3) are formed following the reactions below:

$$CaCO_3 = CaO + CO_2 \ (600–900\ °C) \quad (10.10)$$

$$CaO + Al_2O_3 = CaO \cdot Al_2O_3 \ (800\ °C) \quad (10.11)$$

$$CaO + Fe_2O_3 = CaO \cdot Fe_2O_3 \ (800\ °C) \quad (10.12)$$

$$CaO + CaO \cdot Fe_2O_3 = 2CaO \cdot Fe_2O_3 \ (800\ °C) \quad (10.13)$$

$$3(CaO \cdot Al_2O_3) + 2CaO = 5CaO \cdot 3Al_2O_3 \ (900–950\ °C) \quad (10.14)$$

After the decomposition zone, that is, at the region with axial temperature greater than 900 °C, it can be assumed that the dissociation of calcium carbonate, Equation (10.10), an endothermic reaction with $\Delta H = -1660$ kJ/kg $CaCO_3$, is essentially complete.

10.5.2 Transition Zone

The key reactions in this zone are exothermic beginning with silica (C_2S), ($\Delta H = +603$ kJ/kg C_2S) followed by the formation of C_4AF ($\Delta H = +109$ kJ/kg C_4AF) and C_3A ($\Delta H = +37$ kJ/kg C_3A), that is,

$$2CaO + SiO_2 = 2CaO \cdot SiO_2 \ (1000\ °C) \quad (10.15)$$

$$3(2CaO \cdot Fe_2O_3) + 5CaO \cdot 3Al_2O_3 + CaO$$
$$= 3(4CaO \cdot Al_2O_3 \cdot Fe_2O_3) \, (1200-1300 \, °C) \tag{10.16}$$

$$5CaO \cdot 3Al_2O_3 + 4CaO = 3(3CaO \cdot Al_2O_3) \, (1200-1300 \, °C) \tag{10.17}$$

10.5.3 Sintering Zone

In this zone, liquid-phase reactions occur. The main component, C_3S ($\Delta H = +448$ kJ/kg C_3S), is formed by a reaction between C_2S formed earlier and any available free lime as

$$2CaO \cdot SiO_2 + CaO = 3CaO \cdot SiO_2 (1350-1450 \, °C) \tag{10.18}$$

The kinetic rate constants for the dissociation reactions, Equation (10.10) and Equations (10.15−10.18), assembled by Guruz and Bac (1981) are presented in Table 10.2. Similar rates were obtained by Mastorakos et al. (1999) by trial and error to give the expected (measured) composition of charge at the exit of cement kilns.

Obviously these equations are by no means the only reactions involved, since there are traces of other metal oxides that evolve or go into reaction as well. For example, magnesia, MgO, from the dissociation of $MgCO_3$, which usually accompanies $CaCO_3$ as dolomite or dolomitic magnesia, will, as we saw under lime kilns, undergo reactions similar to quicklime. It is said that the maximum permissible MgO in clinker should not exceed 6%. Alkaline metals such as potassium and sodium and also sulfur are present in the cement feedstock. K_2O generally enters the raw meal as a natural mineral in the form of $K \cdot AlSi_3O_6$ and exits with the clinker as K_2SO_4. It is often referred to as double alkali salt, since Na_2SO_4 also forms similarly, or in an even more complex form (Haspel, 1998). Sulfur also enters the kiln system through the raw material or fossil fuels used for combustion and forms in the clinker as SO_3.

Table 10.2 Parameters for Reaction Rate Constant, $k = A \, \exp(-E/RT)$[a]

Reaction	Frequency Factor, A (1/hr)	Activation Energy (kJ/kg mol)
$CaCO_3 = CaO + CO_2$	9.67×10^{24}	1,092,947
$2CaO + SiO_2 = C_2S \, (2CaO \cdot SiO_2)$	1.41×10^{15}	346,014
$CaO + C_2S = C_3S \, (3CaO \cdot SiO_2)$	4.18×10^{8}	461,352
$3CaO + Al_2O_3 = C_3A \, (CaO \cdot Al_2O_3)$	1.81×10^{9}	251,208
$4CaO + Al_2O_3 + Fe_2O_3 = C_4AF \, (4CaO \cdot Al_2O_3 \cdot Fe_2O_3)$	5.59×10^{11}	188,406

[a] Note that the following combinations are given in shortened form: $2CaO \cdot SiO_2$ (C_2S), $CaO \cdot SiO_2$ (C_3S), $3CaO \cdot Al_2O_3$ (C_3A), and $4CaO \cdot Al_2O_3 \cdot Fe_2O_3$ (C_4AF).

The sulfur that is exhausted from the kiln through the exhaust gas stream does so as SO_2. Various sulfur and alkali cycles exist in the rotary cement kiln and, when proper measures are not taken to have them bypass the heat transfer surfaces, they can condense on the heat transfer surfaces, causing them to foul. Obviously, there is more to cement making than described here, and readers interested in a more detailed process description, including the effect of each component on cement quality, should refer to the appropriate text.

10.6 Rotary Cement Kiln Energy Usage

The distribution of energy within the cement-making process is based on the theoretical minimum process heat of formation of the clinker (Table 10.3) (Haspel, 1998). From the table, the theoretical minimum process heat required to form 1 kg of clinker, which is defined as the difference between heat input and heat output of the process, is about 420 kcal/kg (about 1.5 MBtu/ton). This amount will decrease if the process is

Table 10.3 Theoretical Minimum Process Heat of Formation of Cement Clinker

Event/Process	Temperature Range (°C)	Energy (kcal/ kg Clinker)
Heat in		
Sensible heat to raw material at temperature	20–450	170
Dehydration of clay at temperature	450	40
Sensible heat into raw material at	450–900	195
Dissociation of $CaCO_3$ at	900	475
Sensible heat into material at	900–1400	125
Net heat of melting	1400	25
Subtotal	20–1400	1030
Heat out		
Exothermic crystallization of dehydrated clay	–	10
Exothermic formation of cement compounds	–	100
Cooling of clinker	1400–1420	360
Cooling of CO_2	900–920	120
Cooling and condensing of steam	450–470	20
Subtotal	1400–1420	610
Theoretical minimum process heat required to form 1 kg of clinker		420

Table 10.4 Typical Cement Kiln Heat Balance (Haspel, 1998)

Heat Loss/Transfer	Kcal/kg Clinker	Percent of Total (800 kcal/kg Basis)
Raw meal to clinker	417	52.1
Preheater exhaust	183	22.9
Cooler exhaust	78	9.8
Clinker discharge	17	2.1
Dust loss	17	2.1
Shell loss	88	14.75
Total	800	100

carried out in a rotary kiln equipped with preheater. Although this energy accounts for heat losses through the ancillary equipment (Table 10.4), the bulk of it (about 52%) goes to the transformation of the raw meal into clinker material. The energy usage or the specific heat consumption is about 800 kcal/kg clinker (2.88 MBtu/ton) compared to about 6 MBtu/ton for a lime-making kiln with preheater. This is because of the exothermic reactions associated with the clinker-making process and perhaps more efficient preheating systems, given the fact that the raw material is fed as powder with high particle surface area, thereby enhancing gas–solid heat exchange in the cyclone preheaters.

Cement kilns have been successfully modeled using the chemistry described above with the one-dimensional zone method for radiation heat transfer (Guruz and Bac, 1981) with plausible results. They have also been recently combined in commercial computational fluid dynamics (CFD) packages, for example, those offered by CINAR. Mastorakos et al. (1999) modeled the chemistry with an axisymmetric commercial CFD code (FLOW-3D) and obtained axial concentration of species similar to the one-dimensional concentration profiles reported by Guruz and Bac (1981).

10.7 Mineral Ore Reduction Processes in Rotary Kilns

In the field of extractive metallurgy, solid-state reduction of oxide ore is an important beneficiation process. Reduction of oxide minerals in the presence of suitable reducing compounds or elements (reductants) to metallic state or to lower oxide levels enables changes in the grain structure and the chemical activation in downstream pyrometallurgical metal extraction processes such as an electric arc furnace. The reductants for the direct reduction of iron ore used in the production of sponge iron and some important oxides such as rutile (TiO_2) can be either in the gaseous state (CO and H_2) or in the solid state (carbon). The latter route, known as carbothermic reduction process, employs the rotary kiln as the primary roasting (calcination) reactor. We will examine two carbothermic reduction processes, namely, the SL/RN process for direct reduction iron (DRI) and the Becher process for the carbothermic reduction of ilmenite ore, which employ rotary kilns as their primary reactor.

10.7.1 The Rotary Kiln SL/RN Process

The SL/RN process essentially involves a rotary kiln in which both solid reductant and the oxide ore are fed at one end and the DRI is collected at the opposite end. The ability of this process to use a wide range of carbonaceous matter comes from an important advantage that the rotary kiln offers. The kiln allows the coexistence of a reducing bed and an oxidizing freeboard at the same axial location, something unique to rotary kilns. The first DRI plant based on the SL/RN process was established in the 1960s by Lurgi Metallurgie, the predecessor of Outotec Oyj. Since then there have been a number of commercial plants commissioned worldwide operating with the widest range of iron-bearing materials including pellets, lump ores, beach sand, and ilmenite ore. The wide range of reductants used includes anthracite, bituminous and sub-bituminous coals, lignite, petroleum coke, and charcoal from biomass.

The kiln process can be described as per the schematic shown in Figure 10.7 (Venkateswaran and Brimacombe, 1977). The rotary kiln is charged with iron ore along with coal, recycled char, and dolomite. Some SR/LN kilns operate on a mixture of iron sand and char in a 10:3 ratio. The purpose of adding dolomite is to scrub the sulfur in the coal, if needed. Air is blown into a countercurrent freeboard with onboard (shell) fans located at several positions along the kiln's axial length. Because the bed is under reducing conditions, the onboard fans sustain combustion of evolved gas into the freeboard. Although the coal in the charge provides the energy needed to drive the process, provision is made for the supply of auxiliary fuel. Unlike cement, where

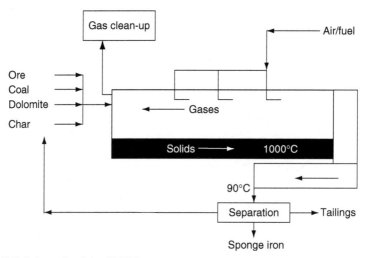

Figure 10.7 Schematic of the SL/RN process.

four reaction zones can be distinguished, in the SR/LN process there are two major zones of interest, that is, a preheat zone (about 40—50% of axial length) and a reduction zone. The span of the preheating zone is large, encompassing all the rotary kiln sequential processes beginning from drying, coal devolatilization, and the dissociation of dolomite, if present. Also, in the preheat zone, the hydrogen released during the devolatilization of coal initiates DRI, that is, "prereduction" of the iron ore. As the charge advances axially and reaches a temperature in excess of 900 °C, the reduction zone begins where the reduction reactions ensue. There is a two-step process that begins with the Boudouard reaction, in which the carbon in the char, a product of the coal devolatilization reaction in the preheat zone, reacts with CO_2 to yield CO. The CO, in turn, reacts with the iron sand or iron oxide pellet or metallic iron by the following reactions:

$$C + CO_2 = 2CO \tag{10.19}$$

$$Fe_xO_y + CO = Fe_xO_{y-1} + CO_2 \tag{10.20}$$

The overall reactions for Fe_2O_3 and FeO are

$$Fe_2O_3 + 3C = 2Fe + 3CO \tag{10.21a}$$

$$FeO + C = Fe + CO \tag{10.21b}$$

Because the Boudouard reaction is highly endothermic, char reactivity is an essential process consideration. Owing to the lack of external preheaters and precalciners as encountered in cement processing, the retention time for an SR/LN kiln with 4.1 m diameter and 65 m length processing about 40 metric tons/h iron sand and rotating at 0.5 rpm is in the 8—12 h range. As in all heterogeneous reactions, good granular mixing is essential to achieving an efficient operation. Mixing is even more important in carbothermic processes because the reactions involve several solid—gas interactions, which might be subject to mass transfer limitations associated with packed beds and slow rotational rates. With the density differences between the sand and char/coal, the process can also encounter severe axial and radial segregation patterns, which can manifest themselves into temperature and concentration gradients. These shortcomings can lead to product quality issues if mixing is inadequate. Venkateswaran and Brimacombe (1977) modeled the SR/LN process using the bed chemistry and the zone-type radiative heat transfer model described in Chapter 7 with considerable success. Such models have been extended to other carbothermic reduction processes such as ilmenite and laterite ore reduction in rotary kilns (Nicholson, 1995).

10.7.2 Roasting of Titaniferous Materials

The success of the carbonaceous reduction of iron oxide has opened the doors for the beneficiation of other minerals that coexist with oxides of iron also known as mineral sands. Examples of these ores are ilmenite, containing titanium dioxide, and laterite

Figure 10.8 The Becher process kiln with onboard fans.
Courtesy of Outotec Oyj.

ore, containing nickel oxide. Titanium dioxide (TiO_2) is one of the most important inorganic materials used as a pigment for paper, plastics, paints, textile, and so on. Titanium metal is also used to produce various types of surgical equipment and used extensively in the aerospace industry. Rutile, an almost pure TiO_2, is the preferred mineral precursor for titanium extraction. However, with its limited availability and recent instabilities encountered in rutile-rich deposit regions, efforts have intensified to exploit the more abundant titaniferous ilmenite ($Fe_2O_3 \cdot TiO_2$). In order to render it as a suitable feedstock, it is necessary to upgrade these mineral ores by selective removal of the iron and thereby enrich the TiO_2 to a grade known in the industry as synthetic rutile. Selective removal of iron from ilmenite minerals has been an area of extensive growth and a number of beneficiation processes have been suggested and tried. The most commonly practiced process of considerable importance is the thermal treatment (roasting) of titaniferous materials, for example, ilmenite ore in a rotary kiln, known as the Becher process (Becher, 1963). The Becher process involves several steps beginning with a reduction, in rotary kiln, of the iron oxides contained in the ilmenite feedstock largely to metallic iron using coal as the reductant as is done in the DRI process (Figure 10.8). This results in a mixture of metallic iron and titanium phases known as "reduced ilmenite." It has been reported (Welham, 1996) that the solid-state prereduction of ilmenite by carbon takes place at >400 °C, followed by a carbon monoxide reduction to elemental iron at >1000 °C. After cooling and discharging from the reduction kiln, the product is dry-separated to remove the reduced iron and the excess char remaining after the reaction. The reduced ilmenite is then subjected to aqueous oxidation (also know as aeration) to convert the metallic iron to iron oxide particles discrete from the TiO_2-rich mineral particles. Aeration is followed by a wet physical separation to remove the iron oxide, enriching the mix to synthetic rutile. Depending on the desired purity, further refinement might be carried out to remove the residual iron and traces of manganese and magnesium that naturally coexist with the ore using acid leaching followed by washing, dewatering, and drying.

The first commercial plant that employed the Becher process was built by Lurgi Metallurgie, the predecessor of Outotec, Oyj in 1968 following the SR/LN experience. It was for the reduction of Australian ilmenite at Western Titanium N.L. at Capel, Australia. It consisted of a 2.4-m-diameter by 30-m-long rotary kiln and is still in operation. It opened up many possibilities for the beneficiation of ilmenite to synthetic rutile without the consumption of acids and the generation of acidic iron effluents as associated with hydrometallurgical processes. Lurgi later built a slightly larger rotary kiln in 1969 (4-m-diameter by 75-m-long kiln) at New Zealand Steel Corporation at Glenbrook for processing TiO_2 containing iron sands ($FeTiO_3$). Instead of a long kiln, a moving grate preheater can be installed upstream of the rotary kiln where pre-hardened green pellets containing iron sands and lignite reductant can be heated to a prereduction temperature using waste gases from the rotary kiln. The Becher process has undergone a tremendous improvement thanks to research at Commonwealth Scientific and Industrial Research Organization (CSIRO) of Australia, which embarked on several mechanistic modeling and engineering endeavors aimed at understanding the chemistry and operation of the process (Nicholson, 1995). Figure 10.9 shows the 4.6 by 68-m rotary kiln for direct reduction of ilmenite concentrates with sub-bituminous coal at Iluka Resources I, Australia, built in 1984. It has a capacity of 900,000 metric tons per year ilmenite feed.

10.8 The Rotary Kiln Lightweight Aggregate-Making Process

A process that does not involve a purely chemical reaction but a combination of both physical and chemical ones is the processing of shale, clay, or slate at high temperatures (900–1200 °C) in a rotary kiln to produce a LWA. LWA belongs to the general body of aggregates used for mortar, plaster, concrete, and other masonry works. Naturally occurring LWAs are pumice, lava, slag, burned shale, slate, and clay. These LWAs are similar to other naturally occurring aggregates such as stone and sand in particle shape and gradation except for the aggregates' weight (some aggregates float on water). The aggregates' lightness is attributable to the multitude of tiny pores generated within each particle during formation. The origin of the LWA industry as it is known today can be traced to shipbuilding in the early 1900s. Studies at the time by marine engineers indicated that a concrete ship would be practical if the concrete used could meet strengths exceeding 35 MPa (5 psi) at a density less than 1760 kg/m^3 (110 lb/ft^3).

Figure 10.9 The 4.6 by 68-m rotary kiln for direct reduction of ilmenite concentrates with sub-bituminous coal at Iluka Resources I, Australia, built in 1984. Capacity: 900,000 metric tons per year ilmenite feed.
Courtesy of Outotec, Australasia, Pty.

Extensive investigations revealed that naturally occurring LWAs could not meet these specifications, whereas rotary kiln-produced expanded shale could. The American shipbuilding authority then commissioned Stephen J. Hayde to develop a strong, inert, and durable LWA from shale, slate, and clays in the rotary kiln. Shipbuilding during the great wars took advantage of this development and made extensive use of rotary kiln-produced LWA for constructing ship decks. Since then the LWA industry has employed the rotary kiln as the primary device for high-temperature processing of shale, slate, or clay achieving densities lower than 440 kg/m^3. For further reading on this role, the reader is referred to Holm (1995).

The thermal treatment of shale, slates, and clay to produce LWA involves the expansion of the material as a result of the evolution of gases at elevated temperatures. Upon heating, the shale, slate, or clay particle undergoes a sequential process beginning with drying, followed by preheating, and then expansion (also known as bloating). The last step occurs as the particle temperature approaches the 900–1000 °C (1650–1830 °F) range, and is believed to be caused by bloating agents such as CO_2 that evolve as the compounds in the raw material decompose. Because there is no appreciable mass loss associated with the bloating process, an increase in volume results in a decrease in specific gravity, a property that is important to structural

Figure 10.10 Replacement of concrete with LWA structural concrete on the roof of the US Capitol by Southern Lightweight Aggregate Corporation, later Solite Corporation of Richmond, sometime in the 1950s.
Courtesy of Northeast Solite Corporation.

concrete applications of LWA. In addition, the porous structure that results from bloating gives LWAs the added advantage of enhanced thermal and acoustic insulating properties (Holm and Bremner, 1990). Today, applications are found in many masonry works such as high-rise building construction, roofing, and bridge decks. In the 1950s, it was used to roof the House and Senate wings of the US Capitol (Figure 10.10) and to deck the Chesapeake Bay Bridge in Maryland. Mechanical properties, strength, and notable engineering applications of rotary kiln LWAs and concrete are widely published in the open literature (Holm, 1995; Holm and Bremner, 1990).

In the United States, today there are many LWA manufacturing companies some of which have been in operation for over 60 years. Many of these can be easily identified by the suffix "lite," which appears to be associated with the lightness of the material. The nature of the rotary kiln, which allows flame residence times on the order of 2–5 s and temperatures of over 2000 K (3100 F), also makes such kilns competitive alternatives to commercial incinerators of organic wastes and solvents. Today, there are a handful of LWA manufacturing facilities that use such wastes as supplemental fuels for combustion. Such kilns are regulated by the United States Environmental

Figure 10.11 Mining of shale feedstock for high-temperature rotary kiln LWA. Courtesy of Utelite Corporation, Coalville, UT.

Protection Agency as industrial furnaces and operate under the Boiler & Industrial Furnace compliance act. Like all rotary kiln processes, achieving product uniformity is the key attribute to achieving the early shipbuilder's dream of light, strong, and durable material. This, coupled with the responsibility of protecting the environment, makes LWA manufacturing a challenging task.

10.8.1 Lightweight Aggregate Raw Material Characterization

We mentioned that bloating or the expansion of slate, shale, or clay occurs at high temperatures in rotary kilns. In essence, the temperature window at which expansion occurs is quarry dependent. Even within the same quarry, this temperature window may change from strata to strata (or from bench to bench). In order to maintain a certain product quality, it is important to keep track of the mining operation to provide assays of the deposits so as to adjust the kiln temperature profiles to match their expansion rate. Figure 10.11 shows the mining of shale deposits strata by strata.

10.8.2 Lightweight Aggregate Feedstock Mineralogy

The mineralogy of shale, slates, and clays that bloat when subjected to elevated temperatures has been widely reported (Boateng et al., 1997; Epting, 1974). These minerals generally constitute, by volume, 80–97% mica clay, 0–10% quartz–feldspar, and

Table 10.5 Metal Oxide Composition of Expanded Shale, Clay, and Slate (Boateng et al., 1997)

Compound	Percent
SiO_2	61−73
Al_2O_3	11−26
Fe_2O_3	3.5−9
Oxides of Ca, Mg, Na, and K	<5
Loss on ignition	≈0.12

2−7% accessory minerals, for example, calcite and pyrite. X-ray analyses have indicated that the major mica−clay constituent of expandable materials is either kaolinite or montmorillonite in still clays, illite and chlorite in the slightly metamorphosed mudstones (shale), and muscovite and chlorite in the more strongly metamorphosed slates. Assays on the raw materials indicate that the oxides may be grouped as shown in Table 10.5.

There has been a considerable amount of discrepancy on the origin of the gas-producing agent that causes bloating. Ehlers' work (Ehlers, 1958) suggested that the bloating agents are calcite, $CaCO_3$, and ankerite, $Ca(Fe, Mg)(CO_3)$, all of which decompose at around 923 °C (1693 °F), and that the gases that cause bloating are CO_2 and traces of SO_2. Later, Phillips (1974) surveyed all the relevant literature and concluded that most of the mica−clay deposits can absorb various types of hydrocarbons and reasoned that as a result, they can retain some of their hydroxyl water beyond its major loss at the 500−700 °C (932−1292 °F) range, suggesting dehydroxylation as the cause of bloating. Notwithstanding, he concluded that CO_2 is the single gas common to the bloating of all the expandable materials. Differential thermal analysis (DTA) work (Epting, 1974) on these minerals indicate that there is an endothermic loss of bound water (dehydroxylation) at 600 °C (1112 °F) and another endothermic reaction at a relatively higher temperature around the 900−1200 °C (1652−2192 °F) range. The latter reaction may perhaps substantiate the claim that CO_2 is the bloating gas and that it evolves by the decomposition of calcite as claimed by Ehlers or, perhaps, by subsequent reduction reactions involving carbon.

Work on kinetics of the bloating of shale, slate, or clay has been lacking. However, there is little doubt that the process involves heterogeneous gas−solid chemical reactions coupled with a mass transfer phenomenon. Such reactions may proceed through the following three steps (Boateng et al., 1997) (1) chemical reaction involving the evolution of gas, (2) diffusion of the gaseous product through the core of the particle, and (3) diffusion of the gas through the outer boundary of the particle. Although it is not readily known which one of the three steps controls particle expansion, the internal voids created by the process suggest that the gases evolve at a rate faster than they can diffuse through the particle. Furthermore, like most ceramic materials, thermal conductivity tends to decrease at elevated temperatures (Zdaniewski et al., 1979) so that fusion of the periphery of the particle may lead to the formation of a pyroplastic exterior that is likely to increase the resistance to gas transport to the outer boundary layer. This phenomenon results in swelling with almost negligible mass loss. Since the particle is sealed on the periphery, the work done by internal gas movement goes to generate the internal porous structure. The low specific gravity of an LWA product is a result of volume increase with negligible mass loss.

To determine the extent of bloating or expansion in an industrial rotary kiln, one must carry out laboratory tests using bench-scale furnaces for the evolution kinetics and further correlation tests in a pilot rotary kiln for appropriate temperature profiles. The temporal events determined are, in turn, used to plan quarry operations for product quality control. The same data may also be useful in developing a mechanistic mathematical model that can predict temperature distribution and density changes in the raw material as it journeys through the kiln (Boateng et al., 1997). Such tools have been proved to be useful for the control of product quality as new mines are explored or even as different strata of the existing mine are explored for feedstock. Some of these time–temperature histories are discussed herein.

10.8.3 Lightweight Aggregate Thermal History

The physical events that occur as a shale, clay, or slate particle undergoes temperature changes can first be explored in the laboratory by various techniques. Thermogravimetric analysis (TGA) used for coal characterization is usually employed to determine the temperature at which mass changes occur in a sample. For instance, the temperature at which evaporation of free moisture or volatile evolution occurs can be established using this technique. Since there is no appreciable mass change during bloating itself, TGA is not appropriate for characterizing shale, slate, or clay expansion in the LWA-making process. TGA data on materials from a Virginia mine (Figure 10.12) show that there is a

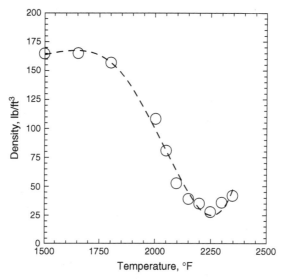

Figure 10.12 Density versus temperature curve for expanded shale precursor.

progressive decrease in sample density within the 1950—2150 °F (1066—1778 °C) temperature range but that within the same range the percent weight loss remains unchanged. Differential thermogravimetric analysis (DTA), on the other hand, can be used to establish the temperature ranges where any chemical reaction is likely to occur.

The density changes as a function of temperature and time can be established by following standard American Society for Testing Materials (ASTM) methods for measuring density. Usually, the particle is placed in a muffle furnace that is set at a predetermined temperature and is removed at a predetermined time. This is followed by a density test. Successive repetition of the procedure will map out density—temperature history. Since some expanded shale, clay, or slate may float on water, at times, some of these experiments are carried out using kerosene as the displacement agent. As seen from Figure 10.12, the density drop for this particular material begins at 1600 °F (870 °C) and the lowest density is achieved at 2250 °F (1232 °C) after which density rebound occurs as a result of pore collapse (vitrification). The expansion or coefficient of expansion, for that matter, can be determined using dilatometry. This is a technique employed to measure the change in length of a specimen as a function of temperature. The slope of the change in length per unit starting length is the coefficient of linear expansion. Highly metamorphosed shale (or slate) from quarries in the North Carolina area and the Hudson Valley in New York require a slightly higher temperature than the others to achieve complete expansion. From these it is possible to determine the activation energies for the expansion process, from which kinetic models can be developed for each quarry and used for quality control of the LWA. Density—temperature plots for several expanded shale and clay precursors from some quarries in the Eastern Seaboard of the United States and their likely kinetic data are shown in Figure 10.13 (Boateng et al., 1997). The activation energy for the expansion or bloating lies in the 30—70 kJ/mol range.

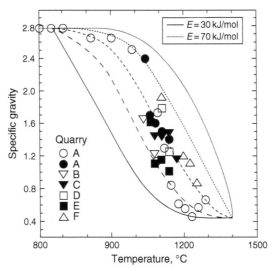

Figure 10.13 Density—temperature curves for expanded shale or clay raw materials from several quarries. Note: E lies between 30 and 70 kJ/mol.

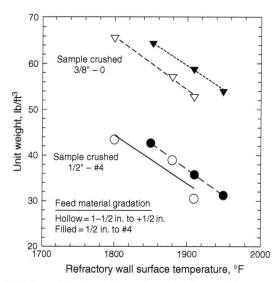

Figure 10.14 Unit LWA weight (density) as a function of rotary kiln temperature.

The tests described in the preceding section are carried out in a muffle furnace and are only useful as a first step in quantifying the behavior of the raw material. Specifically, they are only good for fundamental work on prospecting new quarries. For the purpose of designing large rotary kiln operations or in order to properly simulate the rock behavior in an industrial rotary kiln, pilot kiln studies are essential. It is a good engineering practice to collect yearly pilot kiln data on all samples so as to establish where to mine, what material requires extra heat, which one gives better weights, etc.

Figure 10.14 shows typical pilot kiln test results for intermediate strata material from two quarries. The plot also shows the effect of raw material gradation on product density. Most LWA correlation tests have shown that product quality improves with quarry depth. This is in part due to the fact that the rock from the top bench is likely to be withered by atmospheric conditions. As Figure 10.14 shows, there is a significant difference between the product qualities with regard to feeding coarse particles as compared to feeding fine particles. These tests reinforce that interparticle voids play a key role in bed heat transfer. More importantly, the variation in particle size impacts greatly on quality as smaller particles in a mix of larger ones will tend to segregate to the core and will not receive adequate heat. Depending on the raw material type and the operating conditions in the kiln, unit weights of the clinker can range between 45 and 65 lb/ft^3 for most LWA. Data also show that at a temperature of 2100 °F, the same material will expand to 28 lb/ft^3 if it is retained for an additional 5 min. This means that careful control of the kiln's temperature profile is essential due to the narrow temperature window within which one needs to operate to achieve the desired quality. Owing to the narrow temperature window of expansion for most of these materials, LWA kilns are prone to severe agglomeration, forming large balls at the burning zone, which have the tendency of obstructing the operation. Figure 10.15 shows an

Figure 10.15 Combustion zone region of an LWA kiln with material at incipient fusion. Courtesy of Utelite Corporation, Coalville, UT.

incipient fusion of expanded shale in the combustion zone of an industrial LWA kiln. A trained eye will notice that the product is well expanded and ready for discharge. However, further increases in temperature can increase the size and can cause severe agglomeration if not discharged.

10.9 The Rotary Kiln Pyrolysis/Carbonization

The making of charcoal or carbonization of organic material via the pyrolysis process has been around since the ancient times where the charcoal produced from wood has been the source of cooking fuel for many. The old art of making charcoal was carried out in mounds or piles of dirt, whereas the rotary retort kiln be it batch or continuous has become the main device for carbonization in the recent years. In this section, we will describe the rotary kiln pyrolysis process or carbonization process and present some case studies relating rotary kiln operation and biocharcoal quality.

10.9.1 Pyrolysis

Pyrolysis is the thermal conversion process in which the O_2 is completely devoid (or is present only for the purpose of generating the heat needed); it is a stand-alone process for the manufacture of biochar or charcoal. Pyrolysis is one of the most promising technologies for the production and sequestration of carbon (C) and, recently, for the production of pyrolysis oil (also known as bio-oil) that promises to be a feedstock for producing second-generation transportation fuels (Bridgwater and Peacocke, 2000; Woolf et al., 2010). When the aim of the pyrolysis process is to produce charcoal/biochar, the process is typically called carbonization. During pyrolysis, large complex hydrocarbon molecules of the biomass polymer break down into relatively smaller and simpler molecules of gas, liquid, and char (biochar). Pyrolysis has similarities to processes like cracking, devolatilization, carbonization, distillation, and thermolysis but is distinguished from gasification and combustion by the lack of oxygen. Pyrolysis is typically carried out at a relatively low-temperature range of 300–650 °C compared with 800–1000 °C for gasification and even higher temperatures for combustion. Pyrolysis of biomass can generally be represented by

$$C_nH_mO_p(\text{Biomass}) \xrightarrow{\text{Heat}} \underbrace{\sum C_xH_yO_z}_{\text{liquid}} + \underbrace{\sum C_aH_bO_c}_{\text{gas}} + H_2O + C(\text{char})$$

The product yield fractions and/or distribution depend on the temperature and the heat rate. The heat rate distinguishes slow pyrolysis (<100 C/s) from fast pyrolysis (>1000 C/s). Pyrolysis products are classified into three principal types: solid, mostly char (carbon and ash), and liquid formed from condensable gases (tars composed of heavier hydrocarbons and oxygenates and water) and noncondensable gases (CO, CO_2, C_xH_y). The relative amount of each fraction depends on the heating rate and the final temperature reached by the biomass. Pyrolysis maximizes biochar production if the heat rate is low (slow pyrolysis) and maximizes condensable gas (bio-oil) production if the heat rate is high (fast pyrolysis). Although pyrolysis and devolatilization are similar, the pyrolysis product should not be confused with the volatile matter of a solid fuel, e.g., coal as determined by its proximate analysis. In proximate analysis, the gases (condensable and noncondensable), for that matter liquid and gas, are lumped together as "volatile matter" and the char yield as "fixed carbon." Since the relative fraction of the pyrolysis yields depends on many operating factors, determination of the volatile matter of a fuel requires the use of standard conditions as specified in test codes such as ASTM D-3172 and D-3175. The latter standard, for example, follows a protocol of heating the specified sample of the fuel in a furnace at 950 °C for 7 min to measure its volatile matter.

10.9.2 Slow Pyrolysis Products

Carbonization is a result of slow pyrolysis and this process is slightly exothermic. The pyrolysis reaction can either be endothermic or exothermic depending on the temperature of the reactants, becoming increasingly exothermic as the reaction temperature decreases. The exothermicity of the slow pyrolysis reaction per unit of biochar yield is reported to range from 2.0 to 3.2 kJ/g biochar. Hence, because the fixed carbon content of biomass is high, biochar formation commences at low temperatures where autogeneous pyrolysis begins. The self-sustaining nature of the low-temperature reaction explains why traditional methods of making charcoal, where biomass was buried underground or in mounds (Figure 10.16), could carry on for days (Spokas et al., 2012).

Carbonization is as old as civilization itself, and for as long as human history has been recorded, heating or carbonization of wood for the purpose of manufacturing biochar or charcoal has been practiced (Lehmann and Joseph, 2015). The pyrolysis technologies whether they are used in producing tar/liquid or biochar/solid products occur simultaneously with one or the other always as a coproduct.

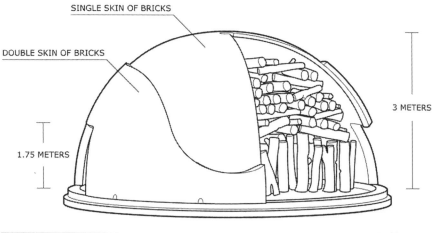

Figure 10.16 Mound kiln.
Adapted from Veitch (1907).

10.9.3 Fast Pyrolysis Products

The primary goal of fast pyrolysis is to maximize the production of the liquid fraction of the products also known as bio-oil (pyrolysis oil, biocrude, etc.), which has recently drawn attention as a renewable fuel precursor that can be upgraded to produce hydrocarbon fuels from biomass. Fast pyrolysis is achieved by heating the biomass very rapidly (1000–10,000 C/s) at a peak temperature in the 450–550 °C. Such high heat rates make the fast pyrolysis process heat transfer limited and therefore dictated by the reactor type and operational conditions that maximize heat transfer. The high heat rates make fluidized beds a more suitable reactor choice for fast pyrolysis but rotary kilns have been employed for fast pyrolysis in a limited way.

10.9.4 The Rotary Kiln Pyrolysis for Carbonization

Examples of large-scale slow pyrolyzers used by the carbonization industry include drum pyrolyzers and rotary drums or retorts that have been long used for the production of charcoal in batch. Rotary kilns can be employed as yet another form of continuous pyrolyzers (Lehmann and Joseph, 2015) for production of charcoal. The rotary kiln arrangements or design types used for this purpose are similar to drum pyrolyzers in the employment of an externally heated, cylindrical shell except that unlike retorts, they have all other features of a continuous rotary kiln whereby the shell is inclined at an angle to the horizon and rotated to allow gravity to move the biomass down the length of the kiln. The residence times of solids are similar and can range from 5 to 30 min depending on the product quality specification. The advantage of choosing rotary kilns over the drum pyrolyzer is the absence of moving parts in the interior. Rotary kilns for biomass pyrolysis have been investigated at low temperatures (350 °C) and at moderately high temperatures (600–900 °C). It has been shown that variations in biomass feed rate and operating temperatures for rotary kiln pyrolyzers may allow wider control on the relative yields of condensable vapors and noncondensable vapors, although biochar yield remained relatively constant in the range of 20–24%.

The main advantage of rotary kilns over others for producing biochar is its susceptibility to feedstock flexibility, scalability, and maturity of the technology. However, there are technical challenges to contend with when operating a rotary kiln including tighter sealing of the rotary drum to prevent air infiltration and/or escape of pyrolysis gases and the usual problems associated with heat penetration when large drums are externally heated.

The experimental rotary kiln unit at UK Biochar Research Centre (UKBRC), the University of Edinburgh (www.biochar.ac.uk), has been used to produce "standard biochar" for use in biochar applications research since 2013 (Figure 10.17) and exemplifies features of a rotary kiln for carbonization. We present herein some case studies carried out to provide insights into biochars; their yields, quality, properties; and their relationship with performance in soil and the environment when they are produced in a rotary kiln. The unit is composed of a variable-speed screw feeder with attached hopper, a sealed rotating drum (internal diameter, 0.244 m; heated length 2.8 m) heated by a set of electric heaters arranged in three heater banks of 16.67 kW each, a char handling screw conveyor that also serves as a heat exchanger to cool the char, a collection vessel, and an afterburner chamber. The unit can operate at temperatures of up to 850 °C and achieve mean residence times of solids between few minutes to over 40 min. Typical operation of the kiln consists of initial purging of the whole equipment with nitrogen and heating to the operating temperature. This is followed by introduction of biomass into the reactor at a preset rate by a horizontal screw feeder. Biomass then moves along the length of the reactor, gradually undergoing drying (moisture release), devolatilization, and charring. The residence time of material in the heated

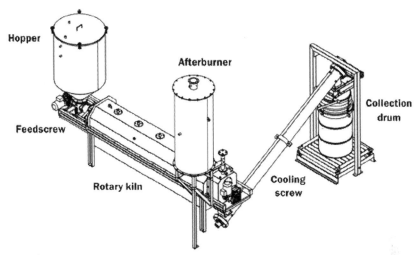

Figure 10.17 Rotary kiln pilot-scale pyrolysis unit (Stage III) at the UKBRC, University of Edinburgh.
Source: Ansac.

zone can be controlled by adjusting the speed of the kiln rotation. At the end of the rotary drum, biochar is separated from gases and vapors in an insulated discharge chamber, and transported by a screw conveyor into a sealed metal drum filled with nitrogen. The screw conveyor, with its water jacket through which cooling water was circulated, also acts as a heat exchanger, reducing the biochar temperature to well below 100 °C, without directly contacting it with water. As a result the produced biochar is dry and due to nitrogen blanketing, free of influence of air and air moisture. Gases and vapors separated in the discharge chamber are led through a heated tube into an afterburner chamber where they are combusted. Table 10.2 provides detailed information on the operating conditions used for all 10 standard biochar materials.

As reproducibility and consistency of a standard biochar is a priority, the production process is optimized to achieve this goal. The operation is started with purging the whole unit with nitrogen for approximately 20–30 min, while the electric heaters heat the unit to the preset operating temperature. When the reactor temperature is close to the desired temperature, biomass is introduced at a preset rate. The reactor is allowed to reach a steady state with the biomass flowing through, and further 30 min is allowed for material to flow through the kiln and screw conveyor before a new empty drum (purged with N_2) is installed and the collection of the target biochar is begun. The biochar produced up to this point, although a viable biochar on its own, is normally discarded and not included in the standard biochar set.

10.9.4.1 Temperature Control and Monitoring

One of the major advantages of externally heated rotary kilns is the superiority of temperature control as mentioned earlier in Chapter 1 and this is even more important

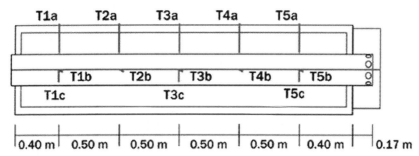

Figure 10.18 Position of thermocouples: external (T1a–T5a), kiln centerline (T1b–T5b), and kiln bed depth (T1c–T5c).
Source: Roy-Poirier, A. (UKBRC).

when it comes to the production of a standard biochar for research purposes. Although external drum heating can be accomplished by a gas flame or by electricity, the pilot-scale unit at UKBRC is heated electrically, to achieve a high degree of control and flexibility. Heating is provided by three independent heater banks with length of approximately 0.9 m each, distributed on one side, along the length of the kiln. The unit is equipped with an advanced system for monitoring and control of process temperatures, which allows careful adjustments of the reactor temperatures and monitoring of temperatures of the reactor, as well as material within the reactor. The arrangement of thermocouples around the rotary drum is shown in Figure 10.18.

The control of the heaters is achieved using a piping and instrument diagram/drawing controller for each bank, with input provided by thermocouples (T1a, T3a, and T5a) situated in the center of each heating zone (lengthwise) at the top of the kiln, i.e., 90° against the kiln rotation from the position of the heater banks, as shown in Figure 10.18. Besides these control thermocouples, there are two more thermocouples (T2a and T4a) situated between the heating zones. It is this set of five thermocouples that is used to control and monitor the temperature of the reactor.

To obtain data on temperature of the solids and gases within the reactor, a specially designed system was developed for inserting thermocouples into the rotary kiln. As a thermocouple system accessing the inside volume through the wall of the rotating drum, although not impossible, would be very complex, the access ports provided by hollow shafts of the screw feeder and the rotary kiln are used instead. Through these ports, it is possible to insert eight thermocouples (four from each side) into different parts of the kiln, as shown in Figure 10.18. The thermocouples are supported, and held in place by a high-grade stainless-steel wire, stretched through the axis of the kiln. To compensate for thermal expansion and prevent sagging of the wire and thermocouples, the wire is continuously tensioned by a counterweight. This array of thermocouples is distributed along the length of the kiln in such a way that temperatures of both the solids and gases could be continuously monitored. This is critical when assessing the operating conditions under which the standard biochar were produced. Positions of the thermocouples are selected so that they match the positions of the external thermocouples measuring the reactor wall temperatures, which therefore make it possible to establish the temperature profiles inside the reactor and compare these to the reactor wall temperature profile.

Observed results clearly show differences between the temperature profile of the reactor tube, the gas inside of the reactor, and the solids flowing through the reactor. In general, the temperature of the solids lagged behind the reactor wall temperature by several hundred degrees in the first third of the reactor and only equilibrated at the same temperature in the second half or last third of the reactor.

10.9.4.2 Material Flow and Particle Residence Time Distribution

Besides reactor and material temperatures, the residence time and particle residence time distribution in the reactor are critical parameters controlling the pyrolysis process and the degree of devolatilization and carbonization of the resulting biochar. Experimental determination of residence time distribution of biomass particles undergoing pyrolysis in a rotary kiln is not straightforward. Therefore, the UKBRC have adopted several approximations. First, they assume that there would be only small difference between the ways in which raw biomass (in the form of pellets) and charred pellets (still retaining their shape) would flow through the unit. This allowed for the use of a tracer method, where feedstock pellets colored to three different colors are used to study the residence time distribution, while running the rotary kiln at room temperature. To do this, the kiln is fed manually with small amounts (approximately 1 kg) of each colored pellet sequentially directly into the screw feeder, and how these come out at the end of the rotary kiln is observed. Pellets are collected every 30 s directly from the exit of the rotary drum, to avoid any mixing in the screw conveyor. Collected pellets are then spread in a monolayer on a tray with black background and photographed (Nikon D7000). The acquired images are then processed using image processing software to analyze the areas corresponding to each of the three colors used. This provides a comprehensive data set from which a typical residence time distribution curve can be reconstructed. Given the comparable size of pellets for all the feedstock used, with the exception of rice husk, the residence time distribution for all four pelleted feedstock can be assumed to be comparable. Results of such measurements show that the mean residence time of pellets in the rotary kiln, rotating at 4 rpm, is approximately 20 minutes for this experimental biochar kiln, with relative standard deviation of approximately 10%.

10.9.4.3 Energy Consumption

It is equally important to understand the energy requirements of the rotary kiln biochar-making process besides producing highly specified set of biochar materials. Here, however, we present (Table 10.6) the energy consumption of the UKBRC pilot-scale pyrolysis unit as determined by measuring electricity consumption of the whole unit and its individual processes. The table also shows the fraction used for heating the reactor (and maintaining temperature) during steady-state operation, while Table 10.7 provides charcoal yields as function of temperature.

Table 10.6 Steady-State Power Consumption of Pilot-Scale Pyrolysis Unit during Production of Standard Biochar from SWP and OSR at 550 °C and 700 °C

Biochar	Total Power Consumption (kWh/kg of Biochar)	Fraction Used for Heating (%)
SWP 550	3.07	81
SWP 700	5.98	89
OSR 550	2.72	85
OSR 700	4.53	89

OSR, oilseed rape straw pellets; SWP, softwood pellets.

Table 10.7 Yields of Biocharcoal as Function of Rotary Kiln Temperature

Material	Pyrolysis Temperature (°C)	Char Yield (wt% db)
OSR	550	28.87
OSR	700	22.62
SWP	550	21.80
SWP	700	17.34
MSP	550	22.81
MSP	700	21.07
RH	550	37.20
RH	700	32.77
WSP	550	24.11
WSP	700	23.54

MSP, miscanthus straw pellet; OSR, oilseed rape straw pellets; RH, rice husk or rice hull; SWP, softwood pellets; WSP, wheat straw pellets.

These results show that the production of biochar at 700 °C in a rotary kiln requires nearly twice the amount of energy than production at 550 °C. This is mainly due to increased heat losses from the reactor and reduced yield of biochar at elevated temperature. The energy required for maintaining the reactor at the desired temperature during steady-state operation contributes the major part of the overall consumption, i.e., between 80 and 90% in all cases. The energy consumption is expressed on biochar basis, and therefore the relative energy required per kilogram of biochar produced would be lower in units with lower heat losses, i.e., in larger scale units.

10.9.4.4 Effect of Pyrolysis Unit on Metal Content in Biochar

Detailed characterization of the rotary kiln-produced biochar shows an interesting trend that can have significant consequences for production of biochar for environmental applications. The UKBRC have found that the content of certain metals, some of which have prescribed environmental limits, increases as a result of the biomass being pyrolyzed in a high-grade stainless-steel reactor. They attribute this to gradual abrasion of the reactor tube. Although the exact mechanisms and extent of this problem will require further work, it is clear that in some cases biochar produced in this way may not comply with regulation or guideline values for heavy metal content in biochar (Buss et al., 2015).

10.9.4.5 Effect of Pyrolysis Unit Setup on Organic Contaminants in Biochar

Compared to fixed or moving bed reactors, rotary kilns reduce the degree of interaction between pyrolysis vapors and biochar. This can have several consequences. First, increased interaction between volatiles (including tars) and pyrolyzing biomass can increase the yield of biochar, as a result of secondary char formation (Huang et al., 2013). On the other hand, it could also lead to contamination of biochar with organic compounds that have been shown to be toxic to plants (Huang et al., 2013; Buss and Mašek, 2014; Buss et al., 2015). Such problems with biochar contamination can, however, be avoided if the vapor−biochar interaction is restricted to areas with sufficiently high temperatures, to avoid condensation of organic vapors on biochar.

References

Becher, R.G., 1963. Roasting of Titaniferous Materials, Australian Patent # 247110.

Boateng, A.A., Thoen, E.R., Orthlieb, F.L., 1997. Modeling the pyroprocess kinetics of shale expansion in a rotary kiln. Trans. IChemE 75, part A1, 278−283.

Boynton, R.S., 1980. Chemistry and Technology of Lime and Limestone. John Wiley & Sons, New York.

Bridgwater, A.V., Peacocke, G.V.C., 2000. Fast pyrolysis processes for biomass. Renewable Sustainable Energy Rev. 4, 1−73.

Buss, W., Masek, O., 2014. Mobile organic compounds in biochar − a potential source of contamination − phytotoxic effects on cress seed (*Lepidium sativum*) germination. J. Environ. Manage. 137, 111−119.

Buss, W., Masek, O., Graham, M., Wust, D., 2015. Inherent organic compounds in biochar−Their content, composition and potential toxic effects. J. Environ. Manage. 156, 150−157.

Ehlers, E.G., 1958. Mechanism of lightweight formation. Ceram. Bull. 37 (2), 95−99.

Epting, C.R., 1974. Mechanism of Vesiculation in Clays (MS thesis). Clemson University.

Guruz, H.K., Bac, N., 1981. Mathematical modelling of rotary cement kilns by the zone method. Can. J. Chem. Eng. 59, 540−548.

Haspel, D.W., 1998. Cement Handbook. Summary of Key Data. Prolink Ltd.

Holm, T.A., 1995. Lightweight Concrete and Aggregates. In Standard Technical Publication STP, 169C, American Society for Testing Materials (ASTM).

Holm, T.A., Bremner, T.W., 1990. 70 year performance record for high strength structural concrete. In: Proceedings, First Materials Engineering Congress, August (American Society of Civil Engineers), Denver, CO.

Huang, Y., Kudo, S., Masek, O., Norinaga, K., Hayashi, J., 2013. Simultaneous maximization of the char yield and volatility of oil from biomass pyrolysis. Energy Fuels 27, 247–254.

Lehmann, J., Joseph, S. (Eds.), 2015. Biochar for Environmental Management, Science, Technology and Implementation. Routledge Taylor & Francis Group, NY.

Mastorakos, E., Massias, A., Tsakiroglou, C., Goussis, D.A., Burganos, V.N., Payatakes, A.C., 1999. CFD predictions for cement kilns including flame modelling, heat transfer and clinker chemistry. Appl. Math. Model. 23 (1), 55–76.

Nicholson, T., 1995. Mathematical Modeling of the Ilmenite Reduction Process in Rotary Kilns (Ph.D. thesis). University of Queensland.

Oates, J.A.H., 1998. Lime and Limestone, Chemistry and Technology, Production and Uses. Wiley-VCH, Weinheim.

Peray, K.E., 1986. The Rotary Cement Kiln. Chemical Publishing Co., Inc., New York.

Phillips, E.L., 1974. Gas Producing Agents in Expansion of Shales, Slates, and Clays (Unpublished).

Spokas, K.A., Cantrell, K.B., Novak, J.M., Archer, D.W., Ippolito, J.A., Collins, H.P., Boateng, A.A., Lima, I.M., Lamb, M.C., McAloon, A.J., Lentz, R.D., Nichols, K.A., 2012. Biochar: a synthesis of its agronomic impact beyond carbon sequestration. J. Environ. Qual. 41, 973–979.

Themelis, N.J., 1995. Transport and Chemical Rate Phenomena. Taylor & Francis, New York.

Venkateswaran, V., Brimacombe, J.K., 1977. Mathematical model of SR/LN direct reduction process. Met. Trans. B 8, 387–398.

Veitch, F.P., 1907. Chemical Methods for Utilizing Wood, Including Destructive Distillation, Recovery of Turpentine, Rosin, and Pulp, and the Preparation of Alcohols and Oxalic Acid. Washington Government Printing Office.

Wang, S., Lu, J., Li, W., Li, J., Hu, Z., 2006. Modeling of pulverized coal combustion in cement rotary kiln. Energy Fuels 20, 2350–2356.

Welham, N.J., 1996. A parametric study of the mechanically activated carbothermic reduction of ilmenite. Mater. Eng. 9 (12), 1189–2000.

Woolf, D., Amonette, J.E., Street-Perrott, A., Lehmann, J., Joseph, S., 2010. Sustainable biochar to mitigate global climate change. Nat. Commun. 1, 1–56.

Zdaniewski, W., Hasselman, D.P.H., Knoch, H., Heinrich, J., 1979. Effect of oxidation on the thermal diffusivity of reaction-sintered silicon nitride. Ceram. Bull. 58 (5), 539–540.

Rotary Kiln Petroleum Coke Calcination Process: Some Design Considerations

11.1 Introduction

Petroleum coke, at times called "pet coke" for short, is a by-product of the coker process in the oil industry. In its raw form, it is also called "green coke" or green petroleum coke (GPC). Calcined petroleum coke (CPC) is an important industrial commodity that links the oil and the metallurgical industries as it provides a source of carbon for various metallurgical applications. In the aluminum industry, CPC is required for the manufacture of anodes for the aluminum potlines, and in the steel industry, it is used to manufacture carbon and graphite electrodes. In the chemical industry, pet coke finds an application as a feedstock for the manufacture of titanium dioxide via the chlorination process. In 2004 alone, calcined coke consumption was about 10 million metric tons and has since grown. The term "anode-grade coke" has been used broadly by the aluminum industry to describe delayed coke with a sponge structure containing relatively low levels of metals like vanadium (typically <400 ppm) and low to moderate levels of sulfur (0.5–4.0%). Delayed coke is classified as needle, anisotropic coke or shot coke, isotropic, and sponge coke, which is mixed anisotropic and isotropic, all varying in their optical textures and mineral/volatile composition (Edwards, 2015). Today, these classifications are less relevant because much wider ranges of green coke are used in anode blends (Edwards et al., 2012). The volatile combustible matter of green coke ranges from as low as 6 to as high as 20 wt%, so for all these applications, the raw petroleum coke should be thermally upgraded prior to its use in a process called calcination to obtain CPC. Pet-coke calcining is therefore the process whereby "green coke" or raw petroleum coke is thermally treated or upgraded to remove moisture and volatile combustion matter and improve physical properties suitable for the metallurgical applications mentioned. Such properties include electrical conductivity, real density, crystallinity, and oxidation characteristics. Pet-coke calcining may be carried out in a rotary kiln, rotary hearth kiln, or shaft kiln described in the early chapters. However, the rotary kiln has become the industrial workhorse for this purpose due to its flexibility of accepting a wide range of green coke as feedstock, large production capacity, low dust and harmful emissions, high product quality in terms of grain stability, and good operational experience historically associated with rotary kilns. The ability for the rotary kiln to utilize the volatile combustion matter to supplement fuel use (autogenetic) saves energy, a large part of the operating cost for any rotary kiln operation. In this chapter, we will describe some of the characteristics of petroleum coke calcination in the rotary kiln and use this as a design case for sizing a rotary kiln for the said application.

11.2 The Rotary Calcining Integrated System

The most recent state-of-the-art rotary kiln process for green coke calcination is designed around a countercurrent rotary kiln (primary reactor, Figure 11.1) where the green coke is subjected to heating by an open flame directly from the freeboard.

However, the system for the complete calcination process in a rotary kiln does not consist of the kiln alone but is an integration of the rotary kiln itself and several auxiliary equipment such as a rotary cooler or equivalent, where the calcined coke is cooled upon discharge from the kiln, and an afterburner also known as a pyroscrubber, thermal oxidizer, or an incinerator, where particulate matter and effluent gases from the kiln are further treated thermally before discharging into the atmosphere to meet environmental compliance (Figure 11.2). Typically in this arrangement, the pyroscrubber is complemented with a waste heat boiler to take advantage of the excess heat derived from the incinerator, which otherwise would be wasted. The inputs and outputs of each unit operation including green coke, air, and fuel entries and exit or effluent gases are as laid out in the flow sheet. Variations of such arrangement are offered by various vendors of the art, which, like any design case, would depend on space limitation. Some of the design and operational norms or characteristics of the direct-fired rotary kiln itself are presented later in the chapter.

11.3 Direct-Fired Kiln Characteristics

Historically, rotary kiln systems employed for petroleum coke calcination have been delivered by only a few companies including Alstom power energy recovery; Metso, which acquired Kennedy Van Shaun (KVS); and the Fuller Engineering Group (FFE)

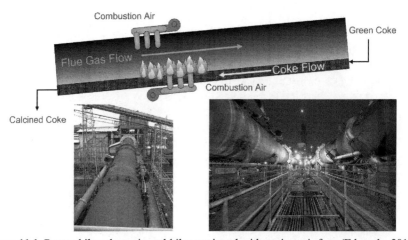

Figure 11.1 Rotary kiln schematic and kilns equipped with tertiary air fans (Edwards, 2015).

Rotary Kiln Petroleum Coke Calcination Process: Some Design Considerations

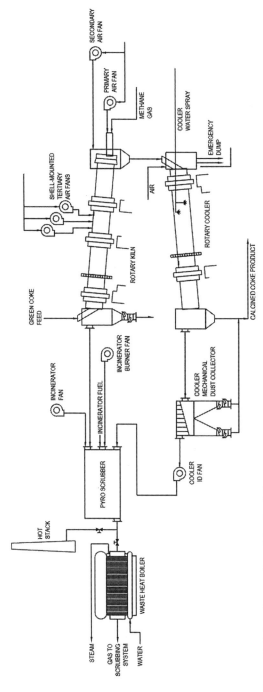

Figure 11.2 Petroleum coke calcination kiln arrangement.

Minerals (now FLSmidth (FLS)). Some of the older pet-coke calcination kilns can still be found at Rio Tinto Alcan, Canada, at its Arvida plant (Arvida kilns), Kitimat, Kingston, etc. (Sood et al., 1972). These designs were completed in the early part of the twentieth century and were based on previous designs in Brazil. However, recent designs have emerged with large capacities in the Middle East and also in Brazil that make use of modern heat integration, including PCIC in Kuwait and Coquepar in Brazil, both offered by FFE/FLS.

According to their Web site, Metso has carried out several designs and installations worldwide based on the old KVS design including Aluminum Bahrain, a 225,000 TPY (tons per year) anode-grade green coke kiln equipped with a waste heat boiler for excess steam production, and similar and multiple ones found in Romania, Saudi Arabia, Germany (Gelsenkirchen), Russia (Volgograd), Kazakhstan (Atyran), Turkmenistan, and Uzbekistan, with some exceeding 1 million TPY capacity (Saudi Arabia). Table 11.1 presents some comparative dimensions and operational characteristics of older and newer pet-coke kiln designs of the industry. Additional and generalized kiln dimensions for known operations in Brazil and elsewhere are presented in Table 11.2 for guidance. The unusually high L/D, where L denotes length and D the

Table 11.1 Characteristics of Two Kilns with Small and Large Capacities

	Smaller and Older	Larger and Newer
Capacity (MTPY)	150,000	350,000
Kiln size (ft)	9.5 × 175	14.2 × 256
L/D	18.4	18
Normal air requirement	19×10^3 cfm	119,818 Nm^3/h
Maximum air requirement	27×10^3 cfm	–
Normal oil requirement	500 lb/h bunker C	58.22 GJ/h
Maximum oil requirement	1000 lb/h	–
Particulate matter to pyroscrubber	0.7 ton/h	3423 kg/h
Maximum particulate matter	2.4 ton/h	–
Normal kiln gas velocity	41 ft/s (19.6 m/s)	12.87 m/s
Maximum kiln gas velocity	65 ft/s (21.6 m/s)	–
Normal heat input to pyroscrubber	1.529×10^8 Btu/h	58.22 GJ/h
Maximum heat input	2.773×10^8 Btu/h	–
RPM	0.5–3	0.5–3
Backend temperature	1000 °C	996 °C

Table 11.2 Some Kiln Dimensions

	L (ft)	ID (ft)	L/D	RPM	Slope
Brazil I	173	8.0	21.625	0.54–2.7	0.375–0.5
Brazil II	181	10.5	17.24	0.47–2.3	0.375–0.5
Arvida B	200	6.5	30.77	0.6–3.0	0.375–0.5
Kitimat	128	6.0	21.33	0.62–3.1	0.5
Collier	160	9.0	17.78		

ID, internal diameter.

diameter, is characteristic of pet-coke calcining because the principal objective of the kiln is to ensure complete devolatilization of the green coke followed by long exposure to specific temperature regimes, which is required for crystallization of the carbon structure (phase transformation) to improve the grain stability of CPC. A cross-sectional view of a rotary pet-coke kiln depicting axial locations of activity (Figure 11.1) shows the likelihood of an intense volatilization around midkiln, a phenomenon that could cause bed fluidization. This is an important characteristic because heat transfer rates in the region of the fluidized medium could be orders of magnitude higher than that in the regions of bulk movement or static bed media thereby increasing localized temperature excursions or gradients likely to affect product quality. However, the exact location where fluidization occurs depends on the design and must be determined to guide the location where on-board fans must be placed to burn off the volatiles in designs that use them in addition or in lieu of dedicated afterburners or incinerators.

11.4 Mass Balance

Mass quantities of materials into and out of the rotary kiln need to be accounted for and quantified as part of the design structure of any kiln, and for coke calcining kiln, these would include (1) the yearly rate of green coke to be processed (GPC); (2) the yearly product rate of lumped coke (CPC); (3) the percentage of the time the kiln will be out of operation, i.e., yearly downtime based on experience and location of the operation; (4) the input fuel (gas, liquid, or solid) to provide and sustain thermal energy for the calcination process; (5) fixed carbon recovery in percent based on the experience gained from the prototype kiln or present technology; (6) acceptable dust pickup rates of green coke or lump coke, which depends on the economics, legal limits, and interaction between kiln and auxiliary equipment in the integrated system; and lastly, (7) volatile matter burnt inside the kiln as percent of original volatiles introduced with the GPC. In rotary coke-calcining kiln designs, the volatile matter burnt in the kiln is one of the quantities that must be initially assumed based on experience and this amount can vary from as low as 30% to as high as 70%. Any design group should have hard information on downtime as the design is almost always based on the assumption of a continuous operation over the duration of a preset time interval. A reliable estimate of the shutdown time is essential and must be factored into the design; these estimates are usually based on past experience, plant location, and the skill levels of the kiln operators. Having said that, a downtime of about 13% a year may be considered as typical for many operations and is a fair estimate for operation in regions in fairly developed countries. With regard to capacity, there are two ways that a rotary pet-coke kiln can be sized, i.e., either by input or by output capacity. According to Sood et al. (1972), if the kiln is to be designed for a location at a refinery to process the output from a specific coker, then the dry green coke available per year must be specified. However, if the kiln is designed for users of CPC, then the amount of calcined coke (lump coke) per year is specified. In the latter case, the wet green coke feed per year is back-calculated. We will follow this approach and provide a calculation procedure that the designer will need to estimate the mass and heat quantities for a rotary coke calcining kiln to handle a moderate, e.g., 150,000 metric tons per year, production from which the static design parameters such as kiln L, D, slope, and rotational speed (rpm) along with air and fuel requirements may be estimated. This information is also necessary to provide input data needed to design and/or select an afterburner or pyroscrubber and a cooler as these would use the kiln exit information for their design or selection. Typically, the process flow streams we envision would be calculated following a commercial process flow design software such as ASPEN+, Pro-II, and the likes. Our presentation is only a necessary exercise to expose the interested reader and/or the student to the underlying bookkeeping and accounting of material flows through the kiln.

Like most design cases, kilns are not designed in a vacuum; it is important to base any design on an existing kiln, i.e., the prototype kiln. In Table 11.1, we presented the characteristics of two existing kilns, one old and one new, and having two very different capacities, one small and the other large. Table 11.2 also presents the static parameters of several existing or previously used pet-coke kilns from which our static design calculation example may be based.

11.5 Thermal Balance and Energy Use

Some pet-coke kilns are energy self-sufficient, i.e., use their own energy (from volatile combustion) to fuel the system. By experience, approximately 10–25 lb/t (5–12.5 kg/MT) of coke of bunker C oil is required for processing green coke in the old kiln designs. However, this is a small amount compared with other mineral processing kilns of similar designs because of the utilization of the inherent volatile combustible matter for supplemental heat. Like most kilns, the sources of heat loss are associated with the kiln shell heat loss and the discharge streams, so heat recuperation is important to improve the kiln's thermal efficiency. Coke kilns are especially vulnerable to heat generation as they are designed to process carbon, a combustible material, with exothermic heat generation, which must be dissipated to control temperature runaway.

11.6 Volatile Matter

The volatile content of petroleum coke produced in the United States ranges between 6 and 20 wt% depending on the origin. While volatile recovery depends on the heating rate, the composition of the volatile gas is not. Mean average composition (in wt%) of 19 green coke samples in the United States (Polak, 1971) was estimated as H_2, 78.6; CH_4, 18.5; CO, 1.3; CO_2, 0.5; O_2, 0.1; N_2, 0.7; C_2H_6, 0.2; and C_2H_4, <0.1. The amount of coke volatiles burnt in the kiln depends upon a great many factors including the kiln's freeboard gas velocities, the amount of air injected at the devolatilization zone, and the type of volatiles. Volatiles burnt inside the kiln as percent of original volatiles introduced into the system typically range between 30 and 70 wt% (Sood et al., 1972). As mentioned earlier, volatile combustion provides supplemental or all the heat source for the endothermic calcination process and also to immediately expose the carbon to high temperatures required to initiate structural/density changes.

11.7 Dust Output and Pickup

Dust outputs for direct-fired pet-coke kilns are never available to designers. Therefore, acceptable dust output rates must be specified when designing such kilns. A widely acceptable level of dust that does not put excessive demands on pyroscrubbers is said to be about 3 wt% of the wet green feed (Sood et al., 1972). In the design stages, one would provide the dominant particle size and density, obtain the total gas weight and acceptable amount of dust from material balance, and then compute the velocity needed to pick up the specific amount of dust. If the gas velocity is greater than the pickup velocity, then some design adjustments need to be made. It is estimated that in order to decrease dusting to 3 wt%, the feed rate has to be lowered rather than increasing the kiln diameter (Sood et al., 1972). This is an important design consideration since increased diameter for the same throughput will tend to lower the bed depth and lead to increased freeboard cross-sectional area and concomitantly reduce freeboard gas velocity. However, the lower bed depth might also result in other unintended consequences due to changes in granular flow dynamics leading to formation of patterns that can promote further dust pickup, alter heat transfer rates, etc.

11.8 Fixed Carbon Recovery and Burnout

Evidence to date indicates that in calcining wet green coke in direct-fired kilns a certain amount of fixed C will be gasified and/or burnt out. The intended goal of the pet-coke calcination process is to convert most of the carbon in the green coke into valuable product, i.e., high product yields, therefore, the amount of fixed carbon gasified constitutes to an economic loss. In the industry, 10% carbon loss is considered to be economic and experience has shown that 90% fixed carbon recovery is a fair assumption in the design of kilns, especially in the earlier designs. Of the total amount of fixed carbon gasified, some will yield CO and some will yield CO_2 with $CO:CO_2$ ratio expected to be about 1/3. Experience has shown that about 8–12 wt% of fixed C is usually burnt during direct-fired rotary kiln pet-coke calcination processes.

11.9 Kinetics and Product Quality

As mentioned earlier, the fundamental goal of coke calcination is to improve density and structural changes such as grain evolution and stability associated with product quality. This task strictly depends on the temperature—time history, i.e., it depends on the process kinetics. According to Metso, direct-fired rotary kilns offered by KVS for coke calcining can produce a quality product specification from sponge, needle, shot, fluid or tar pitch cokes. They claim that the real and vibrated bulk densities can be optimized in the KVS rotary kiln systems to achieve a true density of 2.06 g/cc for sponge calcined coke and 2.11 g/cc for needle coke from the green coke precursor with a density of 1.4 g/cc. The product quality of calcined coke as defined by the targeted properties mentioned above depends very much on the process kinetics. In other words, the apparent density, real density, porosity, and crystallinity of calcined coke are a function of the volatile content, heating rate, temperature, and holding time. Earlier studies by Rhedey (1967) and later ones by Kocaefe et al. (1995) concluded that the coke structure is affected by the heating rate and temperature. Before describing the kinetics of petroleum coke densification process and relating them to the temperature profiles of industrial kilns, some definitions of the pertinent properties could be helpful; some of these are as follows:

1. *Pore volume/porosity*: An open pore is a cavity or channel with access to an external surface. Likewise, a closed pore is a cavity or channel with access to an external surface of 32 Å or less. Pore volume is defined as the volume of open pores except otherwise stated. Particle porosity is therefore defined as the ratio of the volume of open pore to the total volume of the particle (Webb, 2001).
2. *Apparent volume*: Apparent volume of coke particle is the total volume of the particle, excluding open pores but including closed pores (pores with radius <32 Å). Hence, apparent density is the mass per unit apparent volume. Apparent density and porosity are usually determined by mercury porosimetry (Kocaefe et al., 1995).
3. *Real density*: Real (true, absolute) volume is defined as the volume excluding closed and open pores. Hence, real density is the mass of the coke particle divided by its volume excluding closed and open pores. Real density is usually measured with a pycnometer using kerosene (Kocaefe et al., 1995).
4. *Crystalline size*: The crystallinity of petroleum coke, defined by the crystal length, i.e., L_c value, is a general measure of quality affecting suitability for end use and is a function of the heat treatment, i.e., calcination. L_c is measured by X-ray diffraction following American Society for Testing and Materials (ASTM) D5187.

Figure 11.3, taken from Wallouch and Fair (1977), shows the dimensional changes of anisotropic (needle) coke (from pet coke, curve 1) and isotropic (shot) coke (from tar pitch, curve 2) particle during nonisothermal heating at 150 °C/h up to 1200 °C.

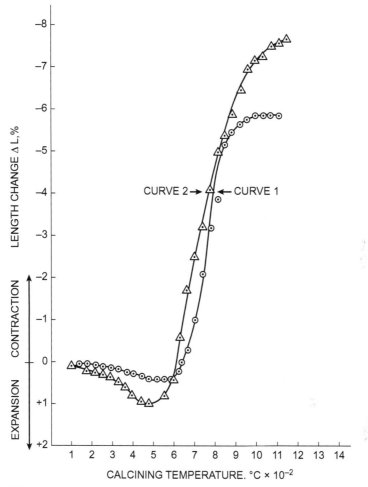

Figure 11.3 Dimensional changes of anisotropic coke (pet coke, curve 1) and isotropic coke (tar pitch, curve 2) during nonisothermal heating at 150 °C/h to 1200 °C (Wallouch and Fair, 1977).

As is evident, these curves mimic phase change curves experienced in the distribution kinetics of crystal growth or nucleation during materials processing where grain growth rate can be described by the classical Avrami equations involving the activation energy changes (Yang et al., 2004). The heat of calcination for calcined coke was given in Kocaefe et al. (1995) as:

$$\Delta H = \int_{T_1}^{T_2} C_p dt \qquad (11.1)$$

or

$$\Delta H = \frac{H^2 - H^1}{\rho_2 - \rho_1} \int_{T_1}^{T_2} \frac{d}{dt}(\rho(T)) dT \qquad (11.2)$$

where, H^1 is coke and H^2 is graphite. They established an empirical correlation between density and temperature as,

$$\rho(T) = 140.852 - 1.2519\left(\frac{T}{10}\right) + 8.1594 \times 10^{-3}\left(\frac{T^2}{100}\right)$$
$$- 1.3609 \times 10^{-5}\left(\frac{T^3}{1000}\right)$$

The heat of transformation from coke to graphite ($H^2 - H^1$), is estimated as 375 Btu/lb (208.35 kcal/kg) and the real density for green and calcined coke are, respectively, 85.488 lb/ft^3 (1369.5 kg/m^3) and 128.544 lb/ft^3 (2059.3 kg/m^3) (Sood et al., 1972). Kocaefe et al. (1995) also have provided recent correlations between calcined coke properties (density, porosity, and crystalline size) and temperature by a generalized third-degree polynomial equation by fitting their experimental data to:

$$Y = a_0 + a_1 X + a_2 X^2 + a_3 X^3 \qquad (11.3)$$

where the coefficients in Equation (11.3) are presented in Table 11.3. Following these studies, pertinent conclusions can also be drawn from the kinetics of pet-coke calcination including but not limited to the following (see Figure 11.4).

Heat rate: The apparent porosity increases slightly with increasing heating rate, whereas the real density and crystalline size decrease with heat rate. The apparent density is not affected by heating rate.

Temperature: Kocaefe et al. (1995) observed that the real and apparent densities, apparent porosity, and crystalline size increase as temperature increases.

Time: At a specific temperature and heat rate, crystalline size and real density increase with increasing holding time, whereas apparent density and porosity pass through a maximum perhaps due to pore collapse or softening.

11.10 Direct-Fired Pet-Coke Kiln Temperature Profiles

Now, how all these kinetic expressions extrapolate to direct-fired rotary kiln calcination of petroleum coke is evident in the temperature profiles or the thermal history of green coke as it journeys through the rotary kiln. Typical predicted temperature, moisture, and volatile concentration profiles for a rotary coke-calcining kiln are presented in Figures 11.5 and 11.6 (Zhang and Wang, 2010b). These are temperature profiles for coke, freeboard gas, and bed with tertiary air injection perpendicular to the coke

Table 11.3 Coefficients for Estimating the Effect of Calcined Coke Properties on Temperature and Heat Rate (1.8 wt-% moisture; 10.1 wt-% VCM; 0.3 wt-% ash), (Kocaefe et al., 1995)

Parameters[a]					
Y	X	HR	a_0	a_1	a_2	a_3
D_r	T	150	3.024	-7.548×10^{-3}	1.037×10^{-5}	-4.003×10^{-9}
D_a	T	150	3.331	-8.400×10^{-3}	1.106×10^{-5}	-4.160×10^{-9}
P_a	T	150	18.12	-5.333×10^{-2}	7.104×10^{-5}	-2.755×10^{-8}
P_f	T	43	11.52	-1.91×10^{-2}	1.863×10^{-5}	-4.375×10^{-9}
D_r	HT	—	1.995	4.060×10^{-3}	-7.87×10^{-5}	5.144×10^{-7}
D_a	HT	—	1.998	2.132×10^{-3}	-2.303×10^{-5}	—
P_a	HT	—	8.84	3.589×10^{-2}	4.111×10^{-4}	-1.506×10^{-5}
L_c	HT	—	22.02	2.601×10^{-1}	-2.177×10^{-3}	—
P_a	HR	—	7.508	9.113×10^{-3}	—	—
D_r	HR	—	2.080	-5.541×10^{-4}	—	—
L_c	HR	—	32.91	-7.715×10^{-2}	—	—

P_a – apparent porosity (%); P_f – sealed porosity (%); T – absolute temperature (K); HT – Holding time (min); HR – Heat rate (K/min); L_c – Crystalline size (Å); D_a – apparent density (g/cm³); D_r – Real density (g/cm³).
[a] $Y = a_0 + a_1 X + a_2 X^2 + a_3 X^3$.

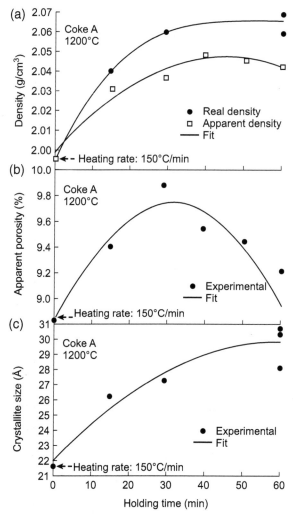

Figure 11.4 Effect of holding time on calcined coke properties and quality (Kocaefe et al., 1995).

bed. Profiles could differ due to slight variations in designs, operation, and perhaps types of green coke. For example, some profiles are based on designs that rely heavily on combustion of external heat source as energy source similar to cement kilns. Also, based on the gas temperatures, one could identify where combustion of the volatile matter could occur at the various different locations along the kiln length and where on-board fans are located. These temperature profiles can be related to the kinetics data to advise dimensional sizing of such kilns. Key points to note for the direct-fired kiln process include the following:

1. At the devolatilization zone, the bed is fluidized, thus the heat rate can be high when the volatile combustible gases are burnt in situ, thereby influencing the kinetics as described

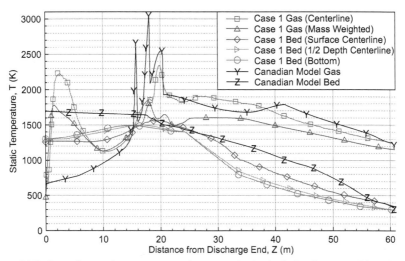

Figure 11.5 Centerline static temperatures for gas and coke beds for the case with tertiary air injection perpendicular to coke bed surface and including mass flow-weighted gas temperature (Zhang and Wang, 2010b).

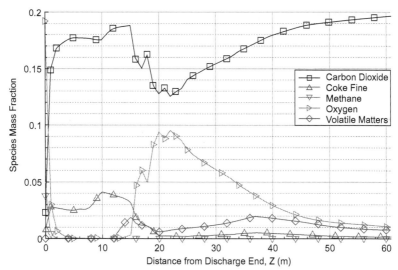

Figure 11.6 Mass-weighted species mass fraction distributions inside the kiln for the case with tertiary air injection (Zhang and Wang, 2010b), at 20 m along the kiln length.

above and the product quality such as porosity, density, and crystalline structure changes with heat rate.

2. In the calcining zone, bed temperatures can reach about 1200 °C or higher for some kilns and may be held at this temperature for almost half the kiln length; such a long exposure time is sufficient to influence the grain size and its crystallinity. All these contributes to conditioning toward product quality.

3. In some operations, cooling of the calcined coke is accomplished by direct water spray on top of the bed material in the cooler due to excessive heat of combustion liberated by oxidation reactions of carbon particles from the coke. It is well known that water has an annealing effect on grain structure in metals, but there is no evidence or study relating vapor pressure and crystalline size in direct-fired kilns, although most of the products of the combustion of the volatile matter will be water due to hydrogen-rich volatile compounds. The composition of exit gas of some modern large kilns have been estimated at about 19–21 mol% H_2O.

11.11 Design Structure

Having established the characteristics such as temperature profiles and the kinetics associated with pet-coke calcination, the next step is to use this information to develop how such kilns are sized based on the capacity to achieve the desired product quality. In this section, we present the static design process calculations that will allow the sizing of a rotary kiln for pet-coke calcination, i.e., determine the L, D, slope, and the range of rotation rates for a kiln that is capable of processing 150,000 metric tons per year of dry GPC. Although this exercise would normally be carried out following a dynamic computer model, herein we follow a static calculation procedure developed in the early years (Sood et al., 1972) so that the reader, a process engineer, or the student in the process engineering class could identify the design methodologies underlying such a computer program. The process flow calculations follow the material and energy balances that will allow one to establish the air and fuel requirements and use these to, in turn, establish the kiln exit conditions including gas flow rates, gas composition, velocity, particulate matter, kiln exit gas temperature, and hence the total energy input into the pyroscrubber. These data are needed to size and/or select the appropriate pyroscrubber. Although we will not design a pyroscrubber in this exercise, we will use a set of kiln exit gas data as input data to carry out a computational fluid dynamics (CFD) exercise that will predict the performance of an existing pyroscrubber.

Like most design cases, a rotary kiln design is usually based on a prototype with known performance data so that the newly designed kiln can be validated. After such validation, a sensitivity analysis may be carried out to inform the proposed kiln's capability in handling normal operational variations. It must be emphasized that the formulations presented herein can be captured in a model, for example, in an excel spreadsheet or any process packages such as ASPEN+, Pro-II, or even a CFD model, and therefore useful in the design of any future coke-calcining kilns or for the evaluation of existing kilns.

11.11.1 Mass Balance Scheme

Process design begins with a material balance. Given the throughput or capacity in terms of annual lumped coke the kiln is to produce and an appropriate downtime, a mass balance is established around the kiln (Figure 11.7).

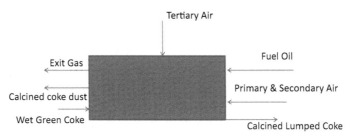

Figure 11.7 Control volume for mass balance estimation.

From the control volume above, a number of mass quantities needed for sizing the kiln and the thermal balance as well as particulate emissions are estimated. The pertinent quantities include but are not limited to the following:

Total dry green coke,...lb/h (kg/h)
Dry green coke completely oxidized,..lb/h (kg/h)
Dry green coke with a fixed carbon content
equivalent to that of the product coke,....................................lb/h (kg/h)
Mass of dry green coke whose volatile combustible matter (VCM)
content is equivalent to the amount of unburnt VCM,..................lb/h (kg/h)
Moisture associated with feed coke,...lb/h (kg/h)
Amount of gaseous N_2 in the kiln,...lb/h (kg/h)
Total VCM burnt,..lb/h (kg/h)
Total tars (heavy hydrocarbons) burnt,.....................................lb/h (kg/h)
Total H_2 burnt,...lb/h (kg/h)
Total CH_4 burnt,..lb/h (kg/h)
Total CO burnt,..lb/h (kg/h)
VCM associated with coke completely oxidized,.......................lb/h (kg/h)
Tars associated with coke completely oxidized,........................lb/h (kg/h)
Hydrogen associated with coke completely oxidized,..................lb/h (kg/h)
CH_4 associated with coke completely oxidized,.......................lb/h (kg/h)
CO as part of VCM associated with coke completely
oxidized,..lb/h (kg/h)
Amount of fixed carbon oxidized to CO,..................................lb/h (kg/h)
Amount of fixed carbon oxidized to CO_2,...............................lb/h (kg/h)
Oxygen required to burn the total (tar + H_2 + CH_4 + CO)
burnt,...lb/h (kg/h)
Oxygen required to burn the total (tar + H_2 + CH_4 + CO) burnt,
i.e., the VCM content associated with the coke
completely oxidized,..lb/h (kg/h)
Oxygen required for burning fixed carbon to CO,.....................lb/h (kg/h)
Oxygen required for burning fixed carbon to CO_2,...................lb/h (kg/h)
Total inerts in fuel oil (or main fuel used),................................lb/h (kg/h)
Moisture from fuel used,...lb/h (kg/h)
CO_2 from fuel oil,..lb/h (kg/h)

We estimate the itemized bookkeeping on mass quantities by stating the kiln capacity or yearly product rate and then calculate green coke feed rate needed based on the above estimates, after which the volumetric kiln throughput estimated may be used to establish the kiln size and operational rates such as kiln speed. With the given 150,000 TPY, we may start by assuming the yearly product rate, M, to be 100,000 TPY and then iterate to achieve the desired green coke capacity following the calculations below (Table 11.4).

We estimate that based on the inputs made available to us, our calculated capacity is 151,859 tons per year and our target is 150,000, so we are good; else we could go back and reduce the value of M from 100,000 and redo the calculations. We may use this feed rate to size the kiln dimensions. To do so we need bulk density and angle of repose and other physical properties to establish the geometry that will maintain a decent kiln loading. From the forgoing calculations, our production rate is 26,242.6 lb/h (11,903.6 kg/h). We assume a bulk density of 2 g/cc (124.94 lb/ft^3), and like the prototype kiln, we would like to operate the kiln at 0.5–3 RPM. We also need other important sizing parameters including the kiln loading in percent of the cross-section of the kiln that is filled with material, i.e., the % fill, and also choose an acceptable range of residence time that we want to design to.

By experience, we want to design to a kiln loading of 5–10% fill. Since we are using the exit conditions, 5% fill is by all means adequate. Also, a retention time of 60–180 min is economically useful. We can use these operating parameters as pleased and iterate for the ideal sizing that matches the prototype kiln (Table 11.2). From here the sizing of the kiln is a matter of geometry with the essential calculations established in Table 11.5.

11.11.2 Comments

Our prototype kiln was 9.5 × 175 ft (2.9 × 53.34 m) with an L/D of about 18. By selecting this L/D, we calculate 9.62 × 173.21 (2.93 × 52.79 m) as the design kiln dimension, which shows that we are very close. We can therefore select a standard dimension such as 10 × 180 ft (3.048 × 54.86 m) or choose the same size as the prototype 9.5 × 175 ft, whichever is available and supported by budget. With the size fixed, and also the slope fixed, we can change the bed depth and residence time by controlling the kiln speed. The next design step is to evaluate the heat duty and confirm the strength of material and mechanical/structural integrity. We will stay with the process design and focus on the heat balance. In order to be able to establish heat duty, the kiln gas flow parameters need to be estimated now that we know the kiln size and throughput. In a static design, this is done by following the itemized accounting of what is eluted and burned like we did in Table 11.5. For this the mass quantities associated with gas out and air requirements need to be computed. The total gas out is composed of moisture, CO, CO_2, and SO_2 generated due to combustion of the volatile combustion gases, fuel, etc., and the unburnt gases themselves. The total air is composed of the requirements for input fuel and the on-board fan air that we need for the combustion of volatile combustion gases evolved. Examples for accounting mass flows of moisture, CO, CO_2, etc., are presented in Table 11.6.

The exit gas at the exit temperature together with the particulate loading will set the inlet conditions for the pyroscrubber.

Table 11.4 Kiln Throughput Estimation for 150,000 TPY Product

a	Yearly product rate = M (assume 100,000 ton/year and then iterate to achieve green coke desired, i.e., 150,000)	Data: Downtime = 13% Volatile matter = 6–20% Volatile matter burnt = 30–70% $CO/CO_2 = 0.33$ mc = 8%	Calcined coke composition %: Ash = 0.35 $H_2 = 0.15$ $N_2 = 0.11$ $O_2 = 0$ $S = 1.5$	
b	Hourly product rate, P (lb/h or kg/h)	$P_r = \dfrac{M \times 2000}{365\left(1 - \frac{\%\text{DownTime}}{100}\right) \times 24}$	$\dfrac{100{,}000 \times 2000}{365 \times (1 - 0.13) \times 24}$	26,242.6
c	Amount of noncarbon in the product, nCp	$\text{nCp} = P_r \times \dfrac{\%(H_2 + S + \text{Ash} + O_2 + N_2)}{100}$	$\dfrac{26{,}242.6 \times 2.11}{100}$	553.72
d	Fixed carbon out of kiln, C_{out}	$C_{out} = P_r - \text{nCp}$	$26{,}242.6 - 553.72$	25,688.88
e	Fixed carbon in the kiln, C_{in}, where the ratio is the fixed carbon recovery, an economic value that we established as about 90% from experience based on several operations and the history of the prototype kiln	$C_{in} = C_{out} \times \left(\dfrac{C_{in}}{C_{out}}\right)$	$25{,}688.88/0.9$	28,543.2
f	Fixed carbon burnt, C_{fb}	$C_{fb} = C_{in} - C_{out}$	$28{,}543.2 - 25{,}688.88$	2854.32
g	Fixed carbon burnt to CO	$g = f_{CO} \times C_{fb}$	$2854.32 \times 0.5/3$	475.72
h	Fixed carbon burnt to CO_2	$h = (1 - f_{CO}) \times C_{fb}$	$2854.32 \times (1 - 0.167)$	2377.65
i	Ash liberated from complete oxidation of green coke may be estimated as	$i = f_{ash}(C_{in} - C_{out})$	$(0.35/100) \times 2854.32$	9.99
j	Sulfur	$j = f_S(C_{in} - C_{out})$	$(1.5/100) \times 2854.32$	42.81

Continued

Table 11.4 Kiln Throughput Estimation for 150,000 TPY Product—cont'd

k	Hydrogen	$k = f_H(C_{in} - C_{out})$	$(0.15/100) \times 2854.32$
			4.28
l	Nitrogen and oxygen liberated may be assumed negligible and therefore 0	0	0
m	Noncarbon released from the complete oxidation of coke	$m = i + j + k + 1$	$9.99 + 42.81 + 4.28 + 1$
			58.08
n	Total dry green coke feed, where n_{vf} is the nonvolatile fraction of green coke	$(m + nC_p + C_{in})/(1 - VM)$	$\dfrac{(58.08 + 553.72 + 28,543.2)}{(1 - 13/100)}$
			33,511.5
o	Dry green coke with carbon content equivalent to fixed carbon content of product coke	$P_r/(1 - VM)$	$26,242.6/0.87$
			30,163.9
p	Total wet green feed	$(P/(1-VM))/(1-\%mc)$	$30,163.9/0.92$
			32,786.85
			$30,163.9/2000 \times 24 \times 365/0.87$
			151,859.6

Table 11.5 Kiln Sizing Procedure

Production rate (lb/h)	Residence Time (min)	Bulk Density (lb/ft³)	Kiln Loading, % Fill		Tube volume to process material (in³)	Tube RPM	Slope [in/in]
26,242.6	180	124.94	5				
Volume of product required (in³/h) $V_P = P_r \rho_{bulk}$ $\frac{26,242.6}{124.94} \times 1728$ 362,943.62		Volume of material exchange within res. Time (in³/τ) $\overline{V}_P = \frac{V_P}{60 \div \tau}$ $\frac{362,943.62}{60 \div 180}$ 1,088,830.86			$\overline{V}'_P = \frac{\overline{V}_P}{(\%\text{fill} \div 100)}$ $\frac{1,088,830.86}{(5 \div 100)}$ 21,776,617.26		
L/D		D (ft)	Material speed (in/min)	L (ft)	Bed depth, h $h = r - (r^2 - 0.25c^2)^{1/2}$, where, $c = 2r \sin(\frac{1}{2}\theta)$		
		$L/(L/D)$	$= L/\tau$	$L = (V/((22/7) *(((1L/D)/2)^\wedge 2))^\wedge(1/3)$			
30		8.12	16.232	243.49	9.300	1.0	0.0317
21		9.14	12.797	191.96	10.475	1.0	0.0222
18		9.62	11.547	173.21	11.027	1.0	0.0190
15		10.23	10.226	153.39	11.718	1.0	0.0158
12		11.02	8.812	132.18	12.623	1.0	0.0127
10		11.71	7.804	117.06	13.413	1.0	0.0106

Table 11.6 Accounting for Freeboard Gas Flow

	Mass Quantities Associated with Exit Gas = Sum (a through i)	(a) Moisture, (b) CO_2, (c) CO, (d) N, (e) Tars, (f) Free H, (g) CH_4, (h) SO_2, (i) Gaseous Inert
a	*Moisture in the exit gas*	
	Moisture from burning tars $C_3H_2 + 3.5O_2 = 3CO_2 + H_2O$	$m_{VM} \times \frac{m_{tar}}{m(\text{tar} + H_2 + CH_4 + CO)} \times \frac{18}{38}$
	Moisture from burning H_2 $2H_2 + O_2 = 2H_2O$	$m_{VM} \times \frac{m_{H_2}}{m(\text{tar} + H_2 + CH_4 + CO)} \times \frac{18}{2}$
	Moisture from CH_4 combustion $CH_4 + 2O_2 = CO_2 + 2H_2O$	$m_{VM} \times \frac{m_{CH_4}}{m(\text{tar} + H_2 + CH_4 + CO)} \times \frac{18 \times 2}{16}$
	Moisture associated with oxidation of firmly bound H liberated from completely burnt coke, moisture from green coke, oxidation of H component of fuel, etc.	$mH_{\text{burnt-coke}} \times \frac{18}{2}$ $\%mc \times p$ (Table A.1) $m_{\text{fuel}} \times \left(\frac{\%H_2}{100}\right) \times \frac{18}{2}$ $m_{\text{fuel}} \times \left(\frac{\%\text{inert}}{100}\right)$
b	*Total CO_2 in the exit gas*	
	CO_2 from fixed carbon oxidation	$m_{fC} \times (1 - \%fC_{\to CO_2}) \times \frac{44}{12}$
	From tar $C_3H_2 + 3.5O_2 = 3CO_2 + H_2O$	$m_{VM} \times \frac{m_{tar}}{m(\text{tar} + H_2 + CH_4 + CO)} \times 3 \times \frac{44}{38}$
	From CH_4 combustion $CH_4 + 2O_2 = CO_2 + 2H_2O$	$m_{VM} \times \frac{m_{CH_4}}{m(\text{tar} + H_2 + CH_4 + CO)} \times \frac{44}{16}$
	From CO oxidation	$m_{VM} \times \frac{m_{CO}}{m(\text{tar} + H_2 + CH_4 + CO)} \times \frac{44}{28}$
	CO_2 introduced as volatiles (use n from Table A.1 as m_p)	$m_p \times f_{CO_2}$
	From fuel combustion	$\dot{m}_{\text{fuel}} \times \left(\frac{\%C}{100}\right) \times \frac{44}{12}$
	Sum of b	
c	*Total CO in the exit gas*	
	From oxidation of fixed C	$\dot{m}_{fC} \times fC_{\to CO}\frac{28}{12}$
	CO contained in the unburnt VCM	$m_{VM_{\text{un-burnt}}} \times \frac{m_{CO}}{m(\text{tar} + H_2 + CH_4 + CO)}$
	Sum of c	
d	*Nitrogen in exit gas*	
	N_2 introduced with air inlet (assuming not oxidized)	$\dot{m}_{\text{air}} \times f_N$
	N_2 librated from burning of coke completely oxidized	
	Sum of d	

Table 11.6 Accounting for Freeboard Gas Flow—cont'd

	Mass Quantities Associated with Exit Gas = Sum (a through i)	(a) Moisture, (b) CO_2, (c) CO, (d) N, (e) Tars, (f) Free H, (g) CH_4, (h) SO_2, (i) Gaseous Inert
e	Tar material out	$m_{VM_{un-burnt}} \times \frac{m_{tar}}{m(tar + H_2 + CH_4 + CO)}$
f	Total free H_2 exited	$m_{VM_{un-burnt}} \times \frac{m_{H_2}}{m(tar + H_2 + CH_4 + CO)}$
g	Total methane out	$m_{VM_{un-burnt}} \times \frac{m_{CH_4}}{m(tar + H_2 + CH_4 + CO)}$
h	SO_2 in exit gas	
	From fuel	$\dot{m}_{fuel} \times \left(\frac{\%S}{100}\right) \times \frac{64}{32}$
	From oxidation of S liberated from completely burnt coke	$\dot{m}_S \times \frac{64}{32}$
	Sum h	
	Total exit gas	Sum (a, b, c, d, e, f, g, h)
	Conversion factor from mass to volume @ STP	$CF = 359 \times \frac{T_{ExitGas}}{(460 + 32)}$

11.12 Computational Fluid Dynamic Modeling (Pyroscrubber)

In the past, the rotary kiln was considered as a black box and coke-calcining kilns are not an exception. These days, CFD modeling is commonly used to predict what is happening in this black box especially what temperature profiles one might expect in a kiln. Owing to the complex chemical interactions between volatile evolution and temperature and the subsequent gas–solid reactions both in the bed and in the freeboard of the coke-calcining kiln, a reactive flow CFD modeling appears to be an obvious choice not only for designing a pet-coke kiln but also for evaluating an existing kiln and/or a newly designed kiln, and then optimizing the design features or the operation. For some recent CFD simulations of the rotary calcining kiln, the reader is referred to works by the Wang group at the University of New Orleans (Zhang and Wang, 2010a,b) but herein we will provide an example whereby a CFD modeling is used in optimizing the design features of how a rotary calcining kiln interfaces with a pyroscrubber and a cooler. We do so by evaluating the turbulent diffusional mixing and combustion performance of the pyroscrubber (or incinerator) "as designed" and subsequent design modifications to optimize design performance.

Figure 11.8 presents the section simulated and it includes the kiln exit leading to the afterburner or pyroscrubber. In this arrangement, some or all of the volatile matter, mainly hydrogen, are burnt by on-board fans to provide supplemental heat or in some cases the entire heat requirement. The stream of unburnt gases devolatilized from the calcination of pet coke in the primary rotary kiln processor (coke kiln gas)

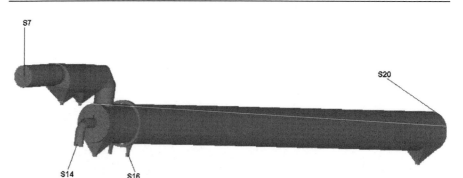

Figure 11.8 Petroleum coke-calcining kiln incinerator (pyroscrubber). Modeling by Alpha Thermal Process, LLC (2006).

plus particulates elutriated thereof, typically enter, in this particular design case, a twin-cyclone settling chamber (S7) where some of the particles are dropped out. The particle-laden settling chamber exit gas stream enters the pyroscrubber (or incinerator, thermal oxidizer) vertically through a right-angled duct (down-comer) where it mixes with a stream of air−water−solids mixture from the product (calcined coke, CPC) cooler (S14) to a pyroscrubber. Downstream of the mixing point in the pyroscrubber, an air shroud introduces a secondary air stream (S16) to complete the combustion of the devolatilized gas and particulates prior to exiting the pyroscrubber (S20). CFD is used to evaluate the turbulent diffusional mixing and combustion performance of the incinerator "as designed" and subsequent design modifications to optimize design performance. The cooler air and kiln exit gases at the incinerator entrance are mixed with tertiary air from the air shrouds for firing to ensure that kiln off gasses are completely oxidized before discharge into the atmosphere. The simulation was carried out by Alpha Thermal Process, LLC, a kiln process consultant, using CFD software offered by FLUENT.

In such a case, the modeling assignment may be divided into two phases: (1) base-case modeling, which we model "as is" or "as designed" followed by (2) optimization models to correct areas needing design and/or operational changes for improvement. The complicated physical and chemical phenomena of the kiln's exhaust gas composition as well as combustion could be simulated by a set of generalized conservation equations similar to that described in Chapter 6. The conservative equations include the conservation of mass, momentum, and energy in both the gas and discrete phases. Physical models are needed for turbulence, green coke devolatilization, volatile combustion, heterogeneous char reaction, particle dispersion, radiation, and pollutant emission. FLUENT treats the gas phase of the two-phase reactive flows as an Eulerian system, while the particulate phase is treated in the Lagrangian system.

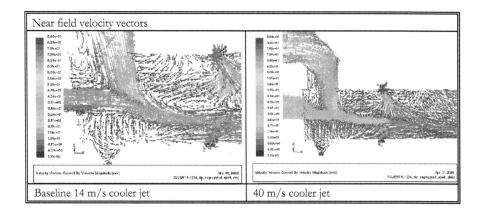
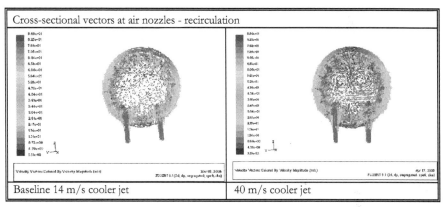

Figure 11.9 Aerodynamics: baseline versus optimum. The self-similar turbulent eddies seen on the right is an indication of increased period of recirculation frequency of particle vortex rotation and therefore increased particle retention time in the frame zone.

In cases like this, there are two combustion traits of interest needing optimization, i.e., aerodynamic mixing of the incoming streams that dictate the successful and complete combustion of the effluent gases from the kiln and the combustion reactions. Pertinent results from the aerodynamic modeling of the "as-design" case indicated the following. (1) About 47% of particles dropped out of the settling chamber and most of the particles dropped out in the second chamber. (2) The calcining kiln exit gas approached the incinerator at a velocity of 12.87 m/s (2533 fpm) meeting the cooler air/water stream having a discharge velocity of 14.07 m/s (2770 fpm). These velocities were considered low, resulting in too low a momentum to induce any significant mixing of the remaining streams. (3) Still and animated results showed that the two streams were unmixed with the gas stream from the kiln, forcing down the cooler air and resulting in little or weak turbulent mixing. Although the air nozzle jet velocity was high (80 m/s), it occurred downstream of the initial mixing zone and played little role

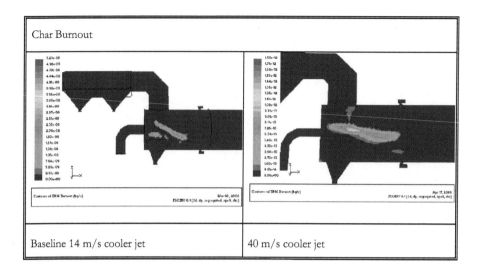

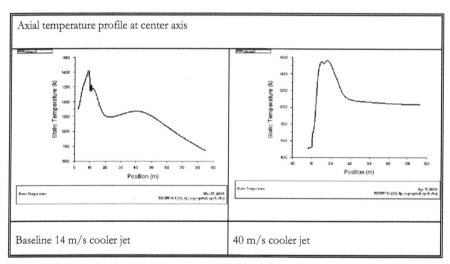

Figure 11.10 Combustion: baseline versus optimum. Top, char burnout; bottom, flame temperature profile. Bifurcation of flame front for coke particles has collapsed into one as a result of higher diffusion time and higher Damkohler II number (diffusion time/reaction time).

without the S14 jet penetration there. (4) The concomitant effect is settling of the gas-borne particles into the lower portion of the incinerator with no chance of entering into combustion (Figure 11.9). Based on the combustion modeling, it was apparent that the homogeneous calcining coke kiln gas was completely burnt in the incinerator but the fate of the particulate matter entering the incinerator after the settling chamber with respect to combustion was unknown due to poor mixing. The peak temperature was predicted to be 1533 K (2300 F) occurring just upstream of the air shroud.

From this, the calculated coke combustion efficiency based on the amount entering the settling chamber (i.e., domain boundary condition) was 48% indicating about 89.5% burnout. Although this is a relatively good combustion performance, the burnout was not uniform showing bifurcations that lead to double flame fronts (Figure 11.10). The oxygen exiting from the incinerator after combustion was calculated as 7.8%. This was deemed to be a result of either poor mixing or particle capture in the settling chamber, thereby resulting in lean combustion overall. After establishing the base case and its shortcomings, the model was used to improve the design and operation. In order to improve the mixing of the entrance streams, the cooler air velocity needed to be increased to 40 m/s. This resulted in a dramatically improved mixing of the settling chamber and cooling air streams. The higher tip velocity induced a strong turbulent diffusion mixing at the incinerator centerline with symmetrical exterior recirculation zones, ensuring better flame penetration and propagation (Figure 11.10). The burnout collapsed into a single flame front ending just before the air shroud. Although the combustion efficiencies were about the same as the baseline case, oxygen at the exit had reduced to 6.9%. The peak temperature was about 1683 K, about 150° higher than baseline. More importantly, the centerline temperature profile was more uniform. The backend temperature reduced with increased tip velocity. The residence time increased by about 80%. It could be concluded that the physical design configuration was adequate for the intended incinerator operation. However, there was poor aerodynamic mixing when the system was operated at the design velocities. CFD was able to recommended that the cooler air injection velocity be adjusted to maintain sufficient momentum to sustain particle suspension and burnout and that the tip velocity be 40 m/s if the cooling air fan could deliver sufficient pressure.

References

Alpha Thermal Process, LLC, 2006. Personal Communications.
Edwards, L., 2015. The history and future challenges of calcined petroleum coke production and use in aluminum smelting. JOM 67, 308−321.
Edwards, L., Backhouse, N., Darmstadt, H., Dion, M.-J., 2012. Evolution of anode grade coke quality light metals 2012. In: Suarez, C.E. (Ed.), TMS (The Minerals, Metals & Materials Society), pp. 1207−1212.
Kocaefe, D., Charette, A., Castonguay, L., 1995. Green coke pyrolysis: investigation of simultaneous changes in gas and solid phases. Fuel 74, 791−799.
Polak, S.L., 1971. Volume and Composition of Volatiles Evolved during Calcination. Report No. A-RR-1475-71-08. Alcan Research and Development Ltd, Arvida, Canada.
Rhedey, P., 1967. Structural changes in petroleum coke during calcination. Trans. Met. Soc. AIME 239, 1084.
Sood, R.R., Stokes, D.M., Clark, R., 1972. Static Design of Coke Calcination Kilns with Special Reference to the Proposed Brazilian Kiln. GE Report No. AWT-72-B1-1.
Wallouch, R.W., Fair, F.V., 1977. Kinetics of the coke shrinkage process during calcination. Carbon 18, 147−153.
Webb, P.A., 2001. Volume and Density Determinations for Particle Technologists. Micromeritics Instrument Corp, USA.

Yang, J., McCoy, B.J., Madras, G., 2004. Distribution kinetics of polymer crystallization and the Avrami equation. http://eprints.iisc.ernet.in/3184/1/Distribution_kinetics.pdf.

Zhang, Z., Wang, T., 2010a. Simulation of combustion and thermal-flow inside a petroleum coke rotary calcining kiln—Part I: process review and modeling. J. Therm. Sci. Eng. Appl. 2/021006/1−8.

Zhang, Z., Wang, T., 2010b. Simulation of combustion and thermal-flow inside a petroleum coke rotary calcining kiln—Part II: analysis of effects of tertiary airflow and rotation. J. Therm. Sci. Eng. Appl. 2/021007/1−7.

Rotary Kiln Environmental Applications

12

This chapter introduces the reader to the technical basics of the US environmental regulations that are applicable when an industrial rotary kiln is permitted to burn hazardous waste for energy recovery. The first rotary kiln to burn hazardous waste in the United States was a cement kiln operated by Keystone Cement Company[1] in 1976. In addition, the economics of power generation through waste heat recovery are covered with consideration to economic benefits. Lastly, emerging technology regarding carbon capture and control or reduction is reviewed.

12.1 Basic Regulatory Framework for Waste Burning Kilns

Until 1991, the burning of hazardous waste ("waste") was exempt from the Environmental Protection Agency (EPA) regulations provided that the waste was being burned for energy recovery rather than being burned just for destruction (i.e., incineration) purposes. On February 21, 1991, the EPA published in the federal register the Boiler and Industrial Furnace (BIF) rule (40 CFR §266, Subpart H). With this statue, kilns that were burning waste for energy recovery had to demonstrate compliance with the BIF rule in August 1991. For regulatory purposes, the EPA considers all energy recovery kilns to be "industrial furnaces." The EPA specifically distinguishes industrial furnaces that produce a product from incinerators that do not (EPA530-R-92-011, March 1992). The BIF rule placed regulations on the energy recovery kilns, which can be enumerated in four general categories:

- Emission Standards (Particulate Matter (PM), 10 metals, HCl, Cl_2, CO, total hydrocarbon (THC), and dioxin and furan (D/F)), and a performance requirement of 99.99% destruction and removal efficiency (DRE) of the waste burned. The 10 metals are identified by the EPA as antimony, arsenic, barium, beryllium, cadmium, chromium, lead, mercury, silver, and thallium.
- Operating Requirements (analysis of fuels; feed rate monitoring for all kiln feeds, i.e., both fuel and raw materials; feed rate limits; continuous emissions monitors (CEMs); automatic waste feed cutoff controls; and recordkeeping).
- Nontechnical Requirements including personnel training, Preparedness, Prevention and Contingency plans, closure requirements, or exit costs.
- Other Air Emission Requirements, e.g., tank and piping leak monitoring and prevention requirements, which the EPA refers to as Subparts AA, BB, and CC.

---

[1] Source; http://www.keystone-cement.com/History.html.

It should be pointed out that the EPA began to regulate the storage of hazardous waste at kiln operations in the early 1980s and that regulatory framework still exists (40 CFR §§264 and 265). Accordingly, in 1991, the EPA expanded its regulations of hazardous waste to include energy recovery in boilers and industrial furnaces.

On September 30, 1999, the EPA began to regulate the emissions of waste burning kilns under the Clean Air Act Amendments (CAAA) of 1990 referred to as the Maximum Achievable Control Technology (MACT) standards versus the BIF rule. Under the CAAA, the emissions limitations shifted from the BIF rule (§266, Subpart H) to the CAAA MACT; these requirements are found at 40 CFR Part 63, Subpart EEE (referred to as "triple E"). The EPA accomplished this transition by moving certain requirements out of BIF rule and into MACT EEE. Generally, the CAAA standards are divided into two categories, "new" and "existing" sources, which are based on the date of first publication by the EPA of a notice of a change in the regulations. A summary of the MACT EEE emissions limits (as of the date of publication of this book) is presented below.

The results of D/F data are expressed in terms of the total toxicity equivalent quotient (TEQ), which provides a calculation of the toxicity of a sample. Using the TEQ approach, each individual 2-, 3-, 7-, and 8-substituted D/F (there are 17 of these) is assigned a toxicity equivalency factor (TEF). The TEF correlates the toxicity of each of the 2-, 3-, 7-, and 8-substituted D/F to that of 2-, 3-, 7-, 8-Tetrachlorodibenzodioxin (TCDD), which is considered to be the most toxic of all D/F. Standard conditions for these EPA emissions limits are 68 °F (293 K). For existing cement plants, the limits to comply with for D/Fs have the following two options:

1. Emissions in excess of 0.20 ng TEQ per dry standard cubic meter (ng TEQ/dscm) corrected to 7% oxygen or
2. Emissions in excess of 0.40 ng TEQ/dscm corrected to 7% oxygen provided the maximum control device inlet (e.g., baghouse, electrostatic precipitator (ESP)) dry particulate matter control device is not greater than 400 °F. Note that these are based on the average of each test runs average temperature.

Oxygen corrections (EPA530-R-92-011, March 1992) for all standards are performed with the following formula, using CO as an example:

$$CO_c = CO_m \times \frac{(21 - 7)}{(21 - Y)}$$

where:

CO_c = corrected CO level
CO_m = measured CO level
Y = measured O_2 concentration in the gas on a dry gas basis

If O_2-enriched air is used for combustion, the enriched O_2 percent level (E) is substituted for the constant "21" in the above formula. The 7% correction in the enumerator is based on "21−7" (i.e., 14); in Europe, kiln emission limits are typically corrected to 10% oxygen or "21−10."

$$CO_c = CO_m \times \frac{(21-7)}{(E-Y)}$$

For mercury (Hg), there are the following two emission requirements:

1. An average as-fired concentration of mercury in all hazardous waste feed streams in excess of 3.0 ppm by weight and
2. Either:
   a. Emissions in excess of 120 µg/dscm, corrected to 7% oxygen, or
   b. A hazardous waste feed maximum theoretical emission concentration (MTEC) in excess of 120 µg/dscm.

The MTEC calculation is performed by dividing the measured concentration of metals or HCl/Cl_2, expressed in µg/dscm, by the feed rate and the stack gas flow rate.

For example, let us suppose that a kiln test was conducted, which resulted in the following data:

1. Waste fuel maximum Hg concentration = 0.08 ppm
2. Waste fuel firing rate = 7.0 TPH (tons per hour)
3. Stack minimum air flow (Q_{stack}) = 245,000 dscf/min (dry standard cubic feet per minute)
4. Stack oxygen (O_2) = 12.0%

Calculations: Kiln waste fuel Hg input will be as follows:

$$\frac{Hg}{h} = 0.08/(1E^6) \times 7.0 \left[\frac{ton}{h}\right] \times 2000 \left[\frac{lb}{ton}\right] = 0.00112 \text{ lb/h}$$

$$Hg\left[\frac{\mu g}{h}\right] = 0.00112 \text{ lb}\left[\frac{Hg}{h}\right] \times 453.6\left[\frac{g}{lb}\right] \times \left[\frac{1E^6}{\mu g}\right] = 508,032 \text{ µgHg/h}$$

$$Q_{STACK}\left[\frac{dscm}{h}\right]_{@7\%O_2} = \left(245,000\left[\frac{dscf}{min}\right] \div 35.315\right) \times \left(\frac{21-12}{21-7}\right) \times 60$$

$$= 267,592 \text{ dscm/h Hg [MTEC]}$$

$$= 508,032\left[\frac{\mu g}{h}\right] \div 267,592\left[\frac{dscm}{h}\right] = 1.89 \text{ µg/dscm}$$

∴ 1.89 < 120 limitation or 1.58% of standard

The kiln restrictions would be minimum airflow of 245,000 dscf/min and the maximum Hg concentration in waste fuel would be 0.08 ppm.

For semivolatile metals like cadmium (Cd) and lead (Pb), emissions are combined and the combination has two requirements as follows:

1. Emissions in excess of 7.6×10^{-4} lb, combined emissions of cadmium and lead, attributable to the hazardous waste per million Btu heat input from the hazardous waste; and
2. Emissions in excess of 330 µgm/dscm, combined emissions, corrected to 7% oxygen.

For metals with low volatility (arsenic (As), beryllium (Be), and chromium (Cr)), the emissions are also combined and the combination also has two requirements as follows:

1. Emissions in excess of 2.1×10^{-5} lb, combined emissions of arsenic, beryllium, and chromium attributable to the hazardous waste per million Btu heat input from the hazardous waste
2. Emissions in excess of 56 μgm/dscm, combined emissions, corrected to 7% oxygen.

Carbon monoxide and hydrocarbon emissions: Carbon monoxide and hydrocarbons are also regulated; depending upon the kiln design, there are many options to choose from and these include, for example, bypass monitoring (split stream) and main stack monitoring. In order to accomplish split stream monitoring, a facility would use two THC monitors and spilt the limitation in half, for example, 10 ppm THC (7% oxygen corrected) at each location. One location would be following the calciner and the other location would be at the bypass duct leading to the bypass baghouse. The latter option may be preferred when high organic matter is present in the raw ores from the quarry where naturally occurring organics are driven off the raw feed and migrate through the main baghouse to the main stack. These naturally occurring organics are not indicative of good or poor combustion, therefore the split stream monitoring may be preferred to a single monitor in the main stack.

Generally, carbon monoxide (CO), which is required to be monitored with a CEM in the bypass duct is limited to 100 parts per million by volume (ppmv), over an hourly rolling average, dry basis, and corrected to 7% oxygen.

As an alternative to CO monitoring, kilns can elect to comply with hydrocarbons (THC) in the bypass duct at a limit of 10 ppmvd with CEMs and 7% O_2 and reported as propane (C_3H_8). For kiln systems not equipped with a bypass duct, their limits are either THC 20 ppmv or CO 100 ppmvd in the main stack.

Hydrochloric acid and chlorine emissions: Hydrochloric acid and chlorine gas are limited to 120 ppmvd, corrected to 7% oxygen.

Particulate matter emissions: For particulate matter, both concentration and opacity limits need to be complied with, i.e.,

1. Emissions in excess of 0.028 g/dscf corrected to 7% oxygen and
2. Opacity greater than 20%, unless the kiln is using a bag leak detection system under §63.1206(c)(8) or a particulate matter detection system (PMDS) under §63.1206(c)(9). A PMDS is a system capable of continuously recording and monitoring particulate matter emissions at concentrations of 1.0 mg per actual cubic meter (mg/acm). These systems have set points that are established during particulate matter tests and are used to demonstrate real-time compliance. The set points are established using EPA Method 5, which is the EPA's source testing method for particulate matter tests. It should be noted that particulate matter is used as a surrogate for what the EPA terms the "nonenumerated metals." The nonenumerated metals consist of toxic metals that the EPA did not establish a specific limit for during the transition from BIF to MACT EEE, but instead substituted a low PM limit as a basis to provide controls of these metal emissions. Examples are thallium, antimony, barium, beryllium, and silver (Table 12.1).

Table 12.1 Summary of Waste Burning Cement Kiln EPA Emission Standards

Regulated Pollutant	Existing Kilns	New Kilns
PM	0.028 g/dscf and 20% opacity	0.0069 g/dscf and 20% opacity
Pb and Cd[a]	7.6E-4 lb/MMBtu and 330 µg/dscmss	6.2E-5 lb/MMBtu and 180 µg/dscm
As, Be, Cr[a]	2.1E-5 lb/MMBtu and 56 µg/dscm	1.5E-5 lb/MMBtu and 54 µg/dscm
Hg	Waste limit 3.0 ppmw and 120 µg/dscm or 120 µg/dscm	Waste limit 1.9 ppmw and 120 µg/dscm or 120 µg/dscm
HCl	120 ppmv	86 ppmv
CO	See note below	See note below
THC	See note below	See note below
D/F	0.2 or 0.4 and $T < 400\,°F$ at APCD inlet	0.2 or 0.4 and $T < 400\,°F$ at APCD inlet
DRE	99.99% each POHC	99.99% each POHC

APCD, Air Pollution Control Device.
[a]Standards are expressed as mass of pollutant contributed by hazardous waste per million Btu contributed by the hazardous waste.

12.1.1 Destruction and Removal Efficiency

DRE in energy recovery kilns is scientifically based on the combustion principle of the three Ts and excess air, where the three Ts are temperature, turbulence, and time (EPA/625/6-89/019, January 1989). A result of 99.99% DRE is required in order to pass this performance requirement. In order to allow for organic (both the semivolatile and volatile fraction of the wastes) variations in wastes fed to the kiln, the operator selects a principal organic hazardous constituent (POHC). In cement kilns, POHCs are usually two for each burning location (e.g., main burner and calciner). Accordingly, for kilns that burn in the kilns' main burner and another location (e.g., calciner), two POHCs are selected for each test condition at each burning location or four total. When an in-line raw mill is part of the cement-making process, typically two other possible testing conditions exist involving the scenarios of raw mill being off and raw mill being on. The table below (Table 12.2) summarizes the typical testing objectives.

POHC selection (EPA/625/6-89/019, January, 1989) criteria are typically based on the following:

- POHC destruction difficulty and
- Analysis of the organics typically found in the waste.

From the analysis of the waste or the permitted chemicals, the facility may accept that in the waste the POHCs must be at least as difficult to destroy (99.99%) as any compound that is permitted to be in the waste fuel. The most common POHC ranking

Table 12.2 Summary of Typical Testing Objectives for a Waste Burning Kiln

Activity	Scenario A (Raw Mill Off)	Scenario B (Raw Mill On)
Tests		
• Particulate matter	1	1
• Metals	1	1
• D/Fs	1	1
• Chlorine/hydrochloric acid	1	1
• Calciner DRE	1	1
• Rotary kiln DRE	1	1
• THC (monitor in first stage exit and bypass) or main stack	2	2
Operating Parameter Limits		
• Minimum combustion zone temperature in rotary kiln	3	3 or 4
• Minimum combustion zone temperature in calciner	5	5
• Maximum WF feed rate rotary kiln (pumpable and total)	5	5
• Maximum WF feed rate to calciner (pumpable and total)	5	5
• Maximum flue gas flow rate	5	5
• Maximum feed rate of semivolatile and low-volatility metals	5	5
• Maximum feed rate of total chlorine/chloride	5	5
• Maximum feed rate of total mercury (for non-MTEC approach only)	5	5
• Maximum APCD inlet temperature (main and bypass)	5	5

WF, waste fuel.
1, Emissions testing will be performed as noted. Operating conditions not being established during the respective scenario will be minimized or maximized as close as practical to demonstrated values.
2, Monitoring THC in bypass and first stage exit with 10 ppm limit at each location or 20 ppm main stack.
3, A facility may conduct DRE testing for the calciner combustion zone only during a comprehensive performance test (CPT) with agency approval. Therefore, a new minimum combustion zone temperature for the rotary kiln will not be established during this CPT.
4, The minimum combustion zone temperatures established during the raw mill off scenario may serve as the raw mill on operating parameter limits (OPLs) with agency approval.
5, Operating limitations will be demonstrated where designated.

system is the thermal stability index. This index was developed by the University of Dayton Research Institute under an EPA contract. The index is abbreviated as "TSLoO2." TSLoO2 or the thermal stability index at low oxygen is a ranking of POHCs by the degree of difficulty in destroying them. See the example TSLoO2 table below (Table 12.3) (Taylor, 1991).

From Table 12.3, benzene, ranked 3, is a higher order ranking POHC than trichlorobenzene (TCB) (ranking 28−29) and therefore it is more difficult to destroy (e.g., higher temperatures are required). The middle column in Table 12.3 explains the technical research basis for the ranking, for example, benzene has a 99% DRE with only a 2-s residence time at a T of 1150 °C. In practice, when rotary kilns are employed, the residence time (T_r) may be derived using the following formula:

$$T_r = \frac{L \times A}{Q}$$

where L and A are the kiln length and cross-sectional area, respectively.

Since kiln freeboard gas forms a pressure gradient along the kiln's long axis, the airflow measurement (Q) may be considered constant. Accordingly, the length times the area ($L \times A$) of the kiln is the effective or working volume (V_w) (i.e., the inside diameter of the refractory) of the kiln. Typically, more accurate calculations are desired to account for kiln feed volume displacement, some of which have already been discussed in the earlier chapters. Accordingly, since the kiln is a production unit, the total available V_a must be calculated first but the depth of the kiln feed must be taken into account and the volume subtracted so that the feed displacement volume is properly accounted for when calculating T_r.

The practical constraints in selecting POHCs for proper DRE testing involve avoiding the selection of a POHC that is a product of incomplete combustion of the fuels or the other POHCs. This is a necessary technical consideration because it could negatively impact the test results. In addition, the POHC should not be present in the waste stream being burned during the actual testing, otherwise it might skew the data or test results. Other sources of POHCs should be screened and tested prior to, and during, the DRE testing, e.g., quench or evaporative cooling of water. In a waste burning kiln, two POHCs are typically selected and injected into the system directly from containers (i.e., 55-gallon drums). This approach is considered the most technically sound practice because it avoids potential stratification concerns, which could occur if the chemical were fed directly into waste fuel storage tanks.

Table 12.3 **Selected POHC Thermal Stability Index Example**

POHC	T 99 (2) at (°C)	Rank
Benzene	1150	3
Dichlorobenzene	970	24−25
Trichlorobenzene	955	28−29
Trichloroethene	865	44−45

The most common method involves a slipstream approach when the POHCs are injected after waste fuel flow measurements and prior to the burner nozzle in order to obtain proper atomization and droplet sizes in the flame zone (burning zone) of the kiln. The chemical drums of POHCs are either placed on a certified scale or data are collected from a mass flow measurement, which only typically measures the mass of POHCs being injected into the system. The last key consideration in POHC DRE test injection is being cognizant of the detection limit of the stack test data in order to ensure that enough POHC is injected into the system to actually prove a result to the significance of 99.99% or four nines. The rule of thumb is to inject POHCs at a rate such that five nines (99.999 DRE) can be theoretically proven. The following formula provides the basis to check the POHC rates just prior to the test. This POHC feed rate is necessary to confirm the acceptability of method detection limits with the planned POHC feed rates.

For example, a comprehensive performance test (CPT) test with a planned feed rate of TCB, rank 28–29, at 250 lb/h (1890 g/min) would be proven as detectable prior to starting the CPT as follows.

$$\text{Stack output}\left(E_{(99.99)}\right) = (1890 \times (1 - 0.9999))$$

$$E_{(99.99)} = 0.1890 \text{ g/min}$$

$$\text{Stack output}\left(E_{(99.9999)}\right) = (1890 \times (1 - 0.999999))$$

$$E_{(99.9999)} = 0.001890 \text{ g/min}$$

The POHC expected in the stack gas would be as follows with an expected flow (Q) in dry standard cubic feet per minute (DSCFM):

$$Q = 80{,}000 \text{ DSCFM } (2265 \text{ dry standard cubic meters per minute (DSCMM)})$$

$$\text{Concentration of TCB at } E_{(99.99)} = (0.1890/2265)$$

$$E_{(99.99)} = 8.3433\text{e}^{-5} \text{ g/dscm}$$

$$E_{(99.9999)} = 8.3433\text{e}^{-7} \text{ g/dscm}$$

The POHC catch weights (CWs) from the stack sampling would calculate to the following with 110 dscf of sample withdrawn from the stack.

Formula:

$$CW = \text{g/dscm} \times 10\text{e}^{6} \mu\text{g/g} \times \text{m}^3/35.314 \text{ ft}^3 \times 110 \text{ dscf}$$

$$CW = 259.9 \text{ μg} \left(\text{at } E_{(99.99)}\right)$$

$$CW = 25.99 \, \mu g \left(\text{at } E_{(99.999)} \right)$$

$$CW = 2.6 \, \mu g \left(\text{at } E_{(99.99999)} \right)$$

Typical laboratory detection limits for TCB are 20 μg using EPA Method 8270. Since the predicted CW at 99.999% (five nines) is greater than the laboratory detection limit, a POHC federate of 250 lb/h would be acceptable for proving a DRE of (four nines) 99.99%.

One further point needed to be made is that, unlike metals, POHCs are typically chlorinated organics and their equilibrium in the systems' exit gases are rather quick and pre-POHC spiking of 15 min is considered sufficient to ensure a valid test result.

12.1.2 Flame Turbulence and DRE

As we know by now from Chapter 3, the rotary kiln flame is an enclosed turbulent jet diffusion flame, i.e., a confined flame and hence possesses recirculation eddies. The flame is physically surrounded by the kiln, which is a high-temperature combustion chamber with a high velocity of freeboard gas. Reynolds numbers (Re) in the kiln exceeds $1E^6$ (Re = 10^6). Inside the kiln, a pressure gradient (dp/dx) exists but turbulence remains dominant thereby aiding in achieving the 99.99% DRE required by the EPA Subpart EEE regulations.

Based on the TSLoO2 index, it is generally accepted that the first 99% of the DRE occurs in the combustion zone of the rotary kiln where there is intense recirculation, with the remaining 0.99% occurring postcombustion in the kiln due to additional residence time, temperature, and some turbulence albeit minor.

As the freeboard post flame gas (fluid) moves through the kiln, in theory and in practice, it eventually becomes more steady and laminar as depicted in Figure 12.1 below.

In Figure 12.1, primary air is drawn into the system by the induced draft fan. Secondary and atomizing air streams are forced through the kiln main burner. There typically exists a short distance Lo where the atomized fuel/air mixture is separated from the burner tip or the precombustion zone. The combustion zone follows and the postcombustion high-turbulence zone, as depicted in Figure 12.1.

12.2 Hazardous Waste Incineration

An incinerator is a combustion device that uses a closed flame technology to incinerate or destroy waste materials. These devices are not classified as industrial furnaces because they are not necessarily permitted to make any useful product. Incinerator regulations date back to 1980 and are found in 40 CFR Part 264/265, Subpart O. The types of standards for incinerators are similar to those for industrial furnaces; however, incinerators are not necessarily engaged in "energy recovery" but rather waste

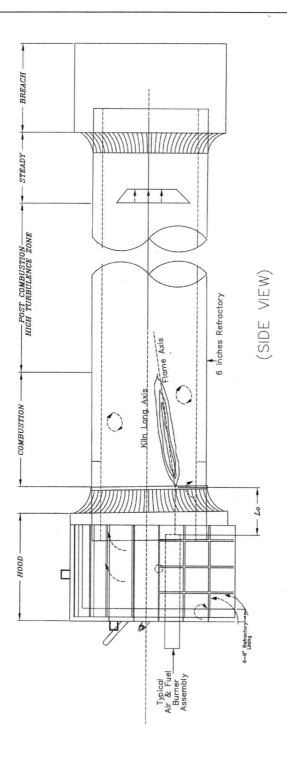

Figure 12.1 Flame front aerodynamics showing recirculation zones.

destruction. Because of this unique aspect of incineration, many of the more toxic wastes, for example, D/F wastes, are only burned in incinerators. The EPA recognized the human health risk involved in burning these wastes and similar to industrial furnaces, incinerators typically have to meet the DRE of four nines (99.99%) for POHCs. In addition, many incinerators burn and destroy D/F waste and because of this they have to meet the most stringent standard of a DRE of six nines (99.9999%) when burning these materials.

12.2.1 Risk Assessment

Both industrial furnaces such as rotary kilns and incinerators are required under the omnibus authority of the Resource Recovery and Conservation Act (RCRA) to perform site-specific risk assessments (SSRAs). The omnibus provision of RCRA is found in Section 3005 of the act (§3005(c)(3)). This provision requires that all RCRA permits include conditions necessary to protect human health and the environment. This provision was originally part of the EPA's combustion strategy in 1994. The risk assessment process continues to be an essential part of RCRA permitting for kilns. The typical drivers for permit conditions resulting from SSRAs are mercury limitations and/or more stringent limitations on the regulatory prescribed limits found in MACT EEE. If the SSRA data (typically stack emissions data) were not gathered during the CPT, then under the omnibus authority, a facility will be required to conduct a "risk burn." This is one of the reasons why it is technically and economically desirable to coordinate risk burn testing with "MACT performance testing" (i.e., CPT). Figure 12.2 provides an overview of the comprehensive risks that needed to be evaluated in a risk assessment.

12.3 Dual Use—Combustion Systems

Depending on the regional alternative fuel market conditions and power costs, many waste burning kilns consider the manufacturing cost reduction benefits of converting waste heat to power. This is in addition to any revenue enhancement provided by burning the wastes in place of fossil fuels, i.e., coal.

12.3.1 Municipal Solid Waste and Power Generation

Power generation from MSW typically involves utilization of excess waste heat from various parts of the process. For example, in cement manufacturing process, the use of a steam Rankine cycle to generate electrical power is considered the lowest capital choice. In municipal solid waste systems (MSWs), the heat released from the burning of the solid waste is used to convert water into steam. The steam is then routed to a turbine generator to produce electricity. MSWs will typically produce 14 MW of electricity. The volume reduction (V_r) of the waste to ash is expected to be 90%. The estimates below provide an example of the energy produced from a ton of waste materials.

Risk Evaluated Direct Emissions and Indirect Exposure

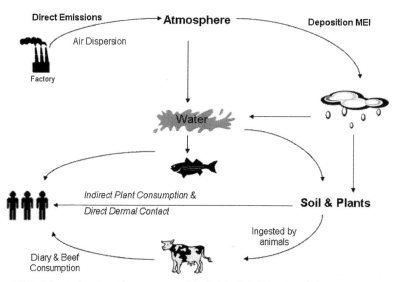

Figure 12.2 Schematic of maximum exposed individual (MEI). Note: MEI is typically a maximum modeled predicted air dispersion deposition based on "worst-case" emissions (i.e., taken from the CPT) and coupled with the worst-case local meteorology.

Given: MSW plant produces 470–930 kWh of electricity per ton of waste material burned.

Landfill fill prices vary by region and country but in most of North America they vary from $20 to $100 per ton (2012 dollars), for burial, which includes transportation and disposal.

Assume: The electricity benefit is $0.11/kWh for each kWh sent back to the grid.

Then: revenue produced (R_p) plus cost avoidance (C_a) (example $30/ton landfill price) calculates the following gross economic benefit (G_e):

$$R_p = 550 \text{ kWh} \times 0.11 = \$60.50/\text{ton waste}$$

$$V_r = 0.9 \times 1 \text{ ton} = 0.1 \text{ ton ash for disposal}$$

$$C_a = \$30/\text{ton} - (0.1 \text{ ton} \times \$30/\text{ton}) = \$27/\text{ton waste}$$

$$G_e = \$87.50/\text{ton waste converted to energy}$$

12.3.1.1 Rotary Kiln Dual Use

A modern rotary kiln manufacturing facility (e.g., cement, lime, lightweight aggregate) has two ideal sources of waste heat that could be effectively recuperated. The first source is the calciner/preheater tower preceding the main kiln where raw materials are initially preprocessed in the manufacturing process. The typical flue gas exit temperature at the top of a preheater tower is 1100 °F (866.5 K) and in a typical lightweight kiln, the exit temperatures are 900 °F (755.4 K). These exhaust gases from the calciner, tower, or postkiln are typically cooled to 400 °F (477.6 K) using evaporative cooling in a water spray quench chamber or by tempering air coupled with long ductworks to dissipate radiant heat. A waste heat recovery system may typically be installed in the preheater to serve the dual purposes of cooling the exhaust gases and of generating electricity to offset electrical demand. A simple calculation can predict such power availability as follows.

Assuming in this example a volumetric flow rate (V) of 175,000 cfm of air with a density of $\rho = 0.050$ lb/ft^3 and a heat capacity, C_p of 0.25 Btu/lb-°F with a corresponding change in temperature of 700 °F, the available energy, Q, may be estimated as

$$Q = V \times \rho \times 60 \times C_p \times \Delta T \times n \text{ or}$$

$$Q = 175,000 \times 0.050 \times 60 \times 0.25 \times 700 \times 0.80 = 73.5 \text{ MBtu/hr}$$

where $n = 80\%$ is the conversion efficiency. This amount converts to 21.5 MW of energy; however, taking into account the typically encountered efficiency of 20% to convert thermal to electrical energy, this yields 4.3 MW gross annual plant-wide energy savings, exclusive of operating, maintenance, and capital amortization costs.

Therefore, for a typical specific power consumption of 1 million ton per year, a rotary kiln clinker-producing plant would generate between 100 and 120 kWh/mton of cement (110–130 kWh/ton) (Alsop, 2005). With 90% uptime and 95% availability, the annual hours of operation are 7490 hr. Therefore, the total annual megawatts would calculate to 16 MW. The gross economic benefit of reducing power consumption from 16 to 11.7 MW would yield an annual savings of approximately $3.6 million annually, estimated as follows:

Baseline costs	C_b = (120 kWh/ton) (1e6 ton) ($0.11/kWh)
	C_b = $13.2 million
% Reduction	(1 − (4.3 MW/16 MW)) = 0.73 (73%)
New costs	C_r = $9.6 million or $3.6 million avoided costs

The second stage in the cement manufacturing process that may be considered a source of waste heat is the clinker cooler dust collector, where exhaust temperatures of up to 600 °F (598.7 K) are typically available.

Based on an air volumetric flow rate of 75,000 cfm, this location can add a second potential design area for waste heat recovery of up to 9.5 GJ/h (9 MMBtu/h), which can also be used to offset electrical demand.

12.4 Carbon Emissions, Reduction, and Capture

As we showed in Chapter 10 earlier, the rotary kiln cement industry produces about 95% of its CO_2 emissions through the calcination of its primary raw material calcium carbonate ($CaCO_3$):

$$CaCO_3 \rightarrow CaO + CO_2$$

The CaO is typically combined with silica, iron, and alumina to produce the gray powder referred to as cement. Only 1% of the CaO remains uncombined as "free lime." The CO_2 is released to the atmosphere after air pollution control contributing to greenhouse gas (GHG) sources. CO_2 capture is therefore an important proposition in the kiln industry. One approach is the increased use of waste lignocellulosic materials as a carbon source to avoid or reduce CO_2 emissions. In October 2008, the EPA reported that cement manufacturers that use alternative fuels and raw materials can achieve reduced energy costs and reduce GHG emissions (EPA, 2008).

12.4.1 Oxycombustion Cement Plant

Carbon capture and storage (CCS) technologies involve two process steps, i.e., separation and storage (United States Global CCS Institute, 2012). Oxyfuel combustion processes have been studied for use in the power industry. This oxycombustion process involves the removal of bulk nitrogen from the air precombustion. After combustion, the by-products may have CO_2 levels at 90%. After combustion, a CO_2 purification unit would compress the CO_2 and then send it to storage. There is ongoing research on the possibility of adapting oxycombustion CO_2 capture in rotary kiln cement operations. Although the economic feasibility is yet to be demonstrated, it is still considered a technology that could prove to be adaptable in the cement industry. In modern cement plants, fuels are burned in two locations, i.e., in the precalciner and in the rotary kiln proper. Oxyfuel combustion technology requires that the kiln operates with a mixture of recycled CO_2 and O_2 in place of ambient air. In doing so, the exhaust gas would gradually result in a pure CO_2 exhaust stream.

Since cement operations are not typically conducted in a relatively well-sealed vessel or system, experience has shown that a significant amount of "false air" or in-leakage air gets infiltrated into the space, which complicates the use of oxycombustion as a successful CCS alternative (Barker et al., 2009).

Many technical issues will have to be overcome in order for the CCS technology to be feasible in cement plant. Some of the major technical challenges are as follows:

- False air minimization and/or elimination. Sources of false air will dilute the high-concentration CO_2 flue gas.
- Process complications for clinker. Research has yet to prove that traditional clinker formation in a cement kiln will be achieved in different atmospheres.
- Oxycombustion requires the use of an air separation unit (ASU). This unit will increase the plant's electric demand requiring typically 200 kWh/tO_2 (Barker et al., 2009).

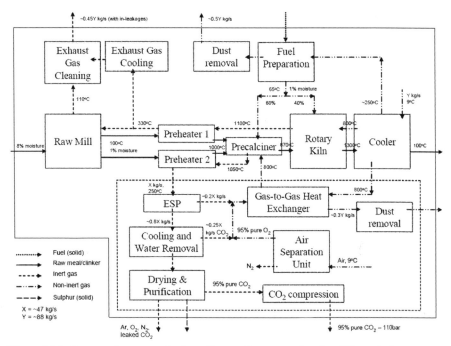

Figure 12.3 Cement plant example flow diagram for oxycombustion CO_2 emissions reduction.
Barker et al. (2009).

Figure 12.3 from Barker et al. (2009) shows a flow diagram of a cement plant with oxycombustion CO_2 capture. Three key features that can be distinguished from a conventional cement manufacturer include:

1. The precalciner combustion air is segregated prior to the precalciner by using an ASU. Research by Barker et al. reports the molar concentration to the precalciner following the ASU to be 95% O_2, 2% N_2, and 3% Ar.
2. The precalciner exit gases are recirculated back to the precalciners' in-line burners at 50%.
3. The remaining precalciner exit gases are cooled and compressed to 110 bar (1595 psig).

Barker et al. (2009) report that oxycombustion in the precalciner alone avoids approximately 61% of a typical cement plant's CO_2. However, the increased power demands for the system reduce the CO_2 reduction to 52%. The cost per ton of CO_2 avoided was reported as €40/ton in 2009, which amounts to $44/ton. It is worth mentioning that oxycombustion is capital intensive but at 50% reduction being observed by researchers, its development is expected to continue.

12.4.2 Cement Plant CO_2 Limitations in the United States

In the United States, the EPA requires CO_2 CEM monitoring and reporting by many industries including cement. The regulatory framework for setting CO_2 limits is the

Table 12.4 **Summary of Air Permit Limitations for GHGs in the U.S. Cement Sector**

Date	Description	CO_2 Limitation
December 20, 2011	5-stage PH/PC cement	0.95 ton/ton clinker 12-month rolling average
January 27, 2010	Multistage PH/PC cement	1860 lb/ton clinker 12-month rolling average

EPA's GHG "Tailoring Rule." This regulation was first issued in May 2010 to establish a regulatory permitting framework for GHG emissions under the Clean Air Act of the United States. Table 12.4 provides a summary of recent regulatory decisions regarding GHG emissions in the cement industry (where PH is "Preheater" and PC is "Precalciner").

As of the time of the writing of this book, the EPA tends to recognize that no feasible CO_2 technology exists for the rotary kiln industry. Therefore, in addition to the limits summarized in Table 12.4, the EPA places a pollution prevention add-on control as follows:

Add-on description:

1. Continued use of the modern cement process design including the preheater/precalciner kiln system
2. Continued use of high energy-efficiency equipment systems
3. Continued implementation of a sustainability program to reduce overall GHG emissions with the use of new additives, raw materials, and fuels consistent with availability and cost while maintaining the quality of the cement product manufactured.

Clearly, CO_2 emission reductions for the rotary kiln industry is being researched and regulated. The trends would predict that once the technology is feasible, the regulated community will begin to see add-on control permit requirements. These could be expensive as they will most likely involve high capital costs.

References

Alsop, P.A., January 2005. International Cement Review. The Cement Plant Operations Handbook, fourth ed.

Barker, D.J., Turner, S.A., Napier-Moore, P.A., Clark, M., Davison, J.E., 2009. CO_2 capture in the cement industry. Energy Procedia 1, 87–94. Elsevier Ltd.

Global CCS Institute, January 2012. CO_2 Capture Technologies | OXY Combustion with CO_2 Capture, (Derived for the Report Sponsored by the Global Carbon Capture and Storage Institute, Canberra, Australia. CO_2 Capture Technologies July 2011).

Taylor, P., The University of Dayton Letter to S. Garg, US-EPA, Thermal Stability-Based Incinerability Ranking. Letter dated, April 30, 1991.

United States Environmental Protection Agency, March 1992. Technical Implementation Document for EPA's Boiler and Industrial Furnace Regulations. EPA530-R-92–011. Office of Solid Waste and Emergency Response.

United States Environmental Protection Agency, January 1989. Guidance on Setting Permit Conditions and Reporting Trial Burn Results. Volume II of the Hazardous Waste Incineration Guidance Series. EPA/625/6—89/019. Office of Solid Waste and Emergency Response.

United States Environmental Protection Agency, October 2008. Cement Sector Trends in Beneficial Use of Alternative Fuels and Raw Materials. Table ES-1, Page 9.

Appendix

Nomenclature

a	Velocity of sound (m/s)
A	Interfacial area, or bed cross-sectional area, m^2 (ft^2)
A_p, A_E, A_w, A_N, A_S	Coefficients in finite difference formulations
b	Body force (N)
b'	Coefficients resulting from velocity profile
c	Particle local velocity vector (m/s)
C	Velocity fluctuation vector (m/s)
C_F	Flotsam concentration (−)
C_f'	Skin friction coefficient
C_i	Bagnold's constant for grain inertia
C_J	Jetsam concentration (−)
C_o	Normalized bed surface velocity (−)
C_p	Specific heat capacity at constant pressure (kJ/kg)
C_{sp}	Characteristic particle spacing (1/m)
C_v	Specific heat of water vapor (kJ/kg)
C'	Apparent viscosity defined
C_1, C_2	Constants in Plank's equation
CFD	Computational fluid dynamics
CRZ	Central recirculation zone
d_p	Particle diameter (m)
D	Kiln diameter, m (ft), or diffusion coefficient (m^2/s)
D_I, D_{II}	Damkohler number
DPM	Dry particle mass burnout rate (kg/s)
e	Emissivity, also void fraction where appropriate
e_p	Coefficient of restitution of particles (−)
E	Emissive power (W/m^2)
$\overline{E}$	Mean void diameter ratio (−)
E_h	Thermal energy, sensible heat (W)
E_k	Kinetic energy associated with local average velocity
E_m	Void diameter ratio that results in spontaneous percolation
E_{PT}	Pseudothermal energy associated with velocity fluctuations
Eu	Euler number (−)
ERZ	External recirculation zone
F	Feed rate (kg-dry/h or lb-dry/h), also view factor
f_c, % fill	Percent fill
f_p	Frequency of precession

Fr	Rotational Froude number, $\omega^2 R/g$ $(-)$
g	Acceleration due to gravity (m/s^2)
g_0	Pair distribution function in collisional theory
$g_1(v), g_2(v), \ldots$	Terms defined in granular flow constitutive equations
G	Mass flow rate of gas (kg/s) or freeboard gas velocity (lb/h/ft^2), irradiation (W/m^2), conductance, Gibbs free energy, and also turbulence
$\dot{G}_0$	Linear momentum
h	Heat transfer coefficient (W/m^2) or specific enthalpy (kJ/kg)
H	Bed depth (m) or molar enthalpy (kJ/mol)
HHV	High heating value (kJ/kg)
ΔH_0	Molar enthalpy of reaction ($H_{\text{prod}} - H_{\text{react}}$)
H_i, H_j	Species enthalpy (kJ/kg)
J	Radiosity (W/m^2)
k	Thermal conductivity (W/mK)
K	Dissociation constant (atm units), also a factor in Craya–Curtet number determination
$\bar{k}$	Segregation flux (m^{2}s^2)
k_{av}	Ratio of mean voids projected area and mean projected total area
K_c, K_v, k_s	Rate constant
L, L_c	Distance from apex of bed cross section to midchord, chord length
Le'	Lewis number $(-)$
LHV	Low heating value (kJ/kg)
$\dot{L}_0$	Angular momentum
$\dot{m}$	Mass flow rate (kg/s)
m_f	Mass fraction component
M	Craya–Curtet number $(-)$, also molar mass (kJ/kmol) or yearly production rate
Ma	Mach number $(-)$
M/N	Ratio of number of voids to number of particles in a layer
n	Number of revolutions (l/s)
$\dot{n}$	Species fraction
N_c	Number of cascades per kiln revolution
Nu	Nusselt number $(-)$
P	Total stress tensor
p	Absolute pressure (bar), also hourly production rate (lb/h or kg/h)
Pe	Peclet number $(-)$
Pr	Prandtl number, $c_p\mu/k$ $(-)$
q	Energy flux (W/m^2), or bed's axial volumetric flow rate, m^3/s (ft^3/h)
q''	Heat flux (W/m^2)
$\dot{q}$	Heat source term (W/m^2)
q_d	Dimensionless axial volumetric flow rate $(-)$
q_h, q, Q	Flux of sensible energy
q_{PT}	Flux of pseudothermal energy
Q	Solids mass flow rate (kg/s) or volume flow rate (m^3/s), or net heat transfer (W), also a factor in Craya–Curtet number determination
r	Radius of particle path in bed, m (ft)
r_0	Radius at bed surface, m (ft)
$\bar{r}$	Diffusion flux (1/s)
R	Cylinder radius or kiln internal radius (m, ft), or specific gas constant (kJ/kg K), resistance to heat flow, also Rosin–Rammler relation

Re	Reynolds number (−)
R′, R₀	Universal gas constant (kJ/kmol K)
s	Kiln slope (degree, radians, ft/ft)
S	Deviatoric stress
S	Molar entropy (kJ/kmol K)
St$_p$	Strouhal number
t	Temperature (°C) or time (s)
T	Temperature (K)
TKE	Turbulent kinetic energy
u	Velocity parallel to bed surface (m/s), or specific internal energy (kJ/kg)
U	Velocity (m/s) or molar internal energy (kJ/kmol)
u_{ax}	Averaged axial bulk velocity, m/s (ft/s)
$\overline{U}$	Tangential velocity of rotary drum (m/s)
V_s	Solids velocity in axial direction (m/s)
v_p	Percolation velocity (m/s)
VM	Volatile matter
W	Mass of water in material (free moisture) (kg)
w_s	Reaction rate (1/s)
y	Distance variable, arbitrary distance in active layer (m) or bed depth, m (ft)
Y	Mass fraction
z	Axial distance (m)
z_0	Particle axial advance per cascade, m (ft)

Greek

α	Thermal diffusivity (m²/s), also dimension in active layer (m)
α_g	Absorbtivity
β	Kiln slope (radians), or ratio relating hydrodynamic force to weight of particle, or coefficient of expansion
ε	Emissivity
γ	Energy dissipation due to inelastic collisions, also specific heat ratio
Γ	Effective mass flux (kg/ms)
δ, δ_x	Active layer depth at distance, x, from apex (m)
Δ	Active layer depth at midchord (m)
η	Expression for coefficient of restitution of particle (−), also ceramic or aggregate expansion factor
$\overline{\eta}$	Particle number ratio
θ	Angle subtended by bed material at cylinder center (radians) or half bed angle (degree, radians)
θ'	Variance on residence time
κ	Ratio of surface velocity to plug flow velocity near yield line (−)
$\overline{\lambda}$	Dilation factor
λ	Conductivity (granular of thermal), also, latent heat of water vapor and mean free path
μ	Dynamic viscosity (kg/m·s)
ν	Solids volume concentration (solids fraction) (−), or kinematic viscosity (m²/s)

ρ		Bulk density (kg/m³)
ρ_p		Particle density (kg/m³)
σ		Stefan–Boltzmann's constant
σ_{xx}, σ_{yy}		Normal stress components
$\bar{\sigma}$		Particle size ratio (–)
σ_{ox}		Oxidizer mass fraction
τ		Shear stress, also time
$\bar{\tau}$		Residence time
$\bar{\tau}_d$		Dimensionless residence time
$\tilde{T}$		Granular temperature (grain temperature) (m²/s²)
ϕ		Static angle of repose of material or kiln slope in angular measure (degree, radians)
Φ		Equivalent ratio
ω		Angular velocity (1/s)
ψ		Angle between surface of bed material and kiln axis (degree, radian)
ξ		Dynamic angle of repose

Subscript

ax	Axial active layer
b	Bed
c	Collisional
cb	Covered bed
cw	Covered wall
eb	Exposed bed
ew	Exposed wall
eff	Effective
f	Frictional
g	Freeboard gas
k	Kinetic
L	Large
m	Mean
PF	Plug flow
s	Small
w	Wall
Shell	Outer wall (shell)

SI to British Conversion Factors

Mass: 1 kg = 1/0.45359237 lb = 2.205 lb
Length: 1 m = 1/0.3048 ft = 3.281 ft
Volume: 1 m³ = 10^3 dm³ (l) = 35.31 ft³ = 220.0 gal (UK) = 264 gal (US)
Time: 1 s = 1/60 min = 1/3600 h
Temperature unit: 1 K = 1.8 R

Force: 1 N (or kg/m/s^2) = 10^5 dyn = 1/9.80665 kg$_f$ = 7.233 pdl = 7.233/32.174 or 0.2248 lb$_f$

Pressure p: 1 bar = 10^5 N/m^2 (or Pa) = 14.50 lb$_f$/in^2 = 750 mmHg = 10.20 mH$_2$O

Specific volume v: 1 m^3/kg = 16.02 ft^3/lb

Density ρ: 1 kg/m^3 = 0.06,243 lb/ft^3

Energy: 1 kJ = 10^3 Nm = 1/4.1868 kcal = 0.9478 Btu = 737.6 ft·lb$_f$

Power: 1 kW = 1 kJ/s = 10^3/9.80,665 kg$_f$ m/s = 10^3/(9.80,667 × 75) metric hp = 737.6 ft·lb$_f$/s = 737.6/550 or 1/0.7457 British hp = 3412 Btu/h

Specific energy (u, h): 1 kJ/kg = 1/2.326 Btu/lb = 0.4299 Btu/lb

Specific heat (c, R, s): 1 kJ/kg K = 1/4.1868 Btu/lb R = 0.2388 Btu/lb R

Thermal conductivity k: 1 kW/mK = 577.8 Btu/ft h R

Heat transfer coefficient: 1 kW/m^{2}K = 176.1 Btu/ft^2 h R

Dynamic viscosity μ: 1 kg/m/s = 1 N s/m^2 = 1 Pa·s = 10 dyn s/cm^2 (or poise) = 2419 lb/ft·h = 18.67 × 10^{-5} pdl·h/ft^2

Kinematic viscosity v: 1 m^2/s = 10^4 cm^2/s (or stokes) = 38,750 ft^2/h

General Information

Standard acceleration: g_o = 9.80665 m/s^2 = 32.1740 ft/s^2

Standard atmospheric pressure: 1 Atm = 1.01,325 bar = 760 mmHg = 10.33 mH$_2$O = 1.0332 kg$_f$/cm^2 = 29.92 in/Hg = 33.90 ft/H$_2$O = 14.696 lb$_f$/in^2

Molar (universal) gas constant: R_o = 8.3144 kJ/kmol K[1] = 1.986 Btu/lb·mol R = 1545 ft·lb$_f$/lb·mol R

1 kmol occupies 22.41 m^3 at 1 Atm and 0 °C

1 lb·mol occupies 359.0 ft^3 at 1 Atm and 32 °F

Composition of Air

The Stefan–Boltzmann constant

$\sigma = 56.7 \times 10^{-12}$ kW/m^2 K^4 = 0.171 × 10^{-8} Btu/ft^2 h R^4

Mass

1 metric ton = 1000 kg
1 short ton (sht) = 2000 lb = 907.2 kg
1 long ton (ton) = 2240 lb = 1015.9 kg

Velocity

1 rpm = 0.1047 radian/s
1 mph = 0.4470 m/s = 1.609 km/h
1 knot = 0.5144 m/s

---

[1] The kilomole (kmol) is the amount of substance of a system that contains as many elementary entities as there are atoms in 12 kg of carbon 12. The elementary entities must be specified, but for problems involving mixtures of gases and combustion they will be molecules or atoms.

Temperature

$$C = \frac{5}{9} \times (F - 32)$$

$$C = \frac{9}{5} \times (C + 32)$$

$K = C + 273.15$
$R = F + 459.67$

Heat Content

1 cal/g = 1.80 Btu/lb = 4187 J/kg
1 cal/cm^3 = 112.4 Btu/ft^3
1 kcal/m^3 = 0.1124 Btu/ft^3 = 4187 J/m^3
1 cal/g °C = 1 Btu/lb °F = 4187 J/kg °K
1 Btu/lb = 0.5556 cal/g = 2326 J/kg
1 Btu/US gal = 0.666 kcal/l
1 hp = 33,000 ft·lb/min = 745.7 J/s = 641.4 kcal/h

Pressure

Technical Atmosphere

1 bar = 1 Atm = 1 kg/cm^2
10 m/H$_2$O = 100 kPa

Normal Atmosphere

$$\begin{aligned}
1 \text{ Atm} &= 101.3 \text{ kPa} = 101,325 \text{ N/m}^2 \\
&= 10,330 \text{ mm/H}_2\text{O} = 407.3 \text{ in/H}_2\text{O} \\
&= 760.0 \text{ mmHg} = 29.92 \text{ in/Hg} \\
&= 235.1 \text{ oz/in}^2 = 14.70 \text{ lb/in}^2 \\
&= 1.033 \text{ kg/cm}^2 \\
&= 1.013 \text{ bar}
\end{aligned}$$

Thermodynamic Tables—Gases

Reference: Rogers, C.F.C., Mayhew, Y.R., 1982. Thermodynamic and Transport Properties of Fluids. Basil Blackwell Publishers, Oxford.
T = temperature in Kelvin
c_p = molar heat in cal/mol
G = free enthalpy in kcal/mol
H = enthalpy in kcal/mol
B = B-function $B(T) = -10^3 G(T)/4.575T$

The specific heats of atomic H, N, and O are given with adequate accuracy by $c_p = 2.5 R_o/M$, where M is the molar mass of the atomic species.

H, U, S, G for Some Gases and Vapors

By definition $U = H - R_o T$ and $G = H - TS$. H and U are virtually independent of pressure. S and G are tabulated for states at 1 Atm and are denoted by S^a and G^a (note that superscript "0" has been used in some parts of the text). At any other pressure p, S and G at a given temperature T can be found from

$$S - S^a = R_o \ln \left[\frac{p}{\text{Atm}}\right]$$

$$G - G^a = (H - H^a) - T(S - S^a) = +R_o T \ln \left[\frac{p}{\text{Atm}}\right]$$

For individual gases and vapors, changes in S and G between states (p_1, T_1) and (p_2, T_2) are given by

$$S_2 - S_1 = \left(S_2 - S_2^a\right) + \left(S_2^a - S_1^a\right) + \left(S_1^a - S_1\right)$$

$$= \left(S_2^a - S_1^a\right) - R_o \ln \frac{p_2}{p_1}$$

$$G_2 - G_1 = \left(G_2 - G_2^a\right) + \left(G_2^a - G_1^a\right) + \left(G_1^a - G_1\right)$$

$$= \left(G_2^a - G_1^a\right) + R_o T_2 \ln \frac{p_2}{[\text{Atm}]} - R_o T_1 \ln \frac{p_1}{[\text{Atm}]}$$

where p_1 and p_2 are partial pressures of a constituent mixture.
Density at sea level $\rho_o = 1.225$ kg/m^3.

Thermodynamic Tables—Inorganic Materials

Reference: Knacke, O., Kubaschewski, O., Hesselmann, K. (Eds.), 1991. Thermochemical Properties of Inorganic Substances, Volumes I and II. Springer-Verlag, New York.

T = temperature in Kelvin
c_p = molar heat in cal/mol
G = free enthalpy in kcal/mol
H = enthalpy in kcal/mol
B = B-function $B(T) = -10^3 G(T)/4.575T$

Table A.1 Composition of Air

	Volume Basis	Mass Basis
Nitrogen (N_2 = 28.013 kg/kmol)	0.7809	0.7553
Oxygen (O_2 = 31.999 kg/kmol)	0.2095	0.2314
Argon (Ar = 39.948 kg/kmol)	0.0093	0.0128
Carbon dioxide (CO_2 = 44.010 kg/kmol)	0.0003	0.0005
Molar mass M = 28.96 kg/kmol		
Specific gas constant		
R = 0.2871 kJ/kg		
K = 0.06856 Btu/lb		
R = 53.35 ft·lb_f/lb R		

Table A.2 Useful Conversion Factors

Physical Quantity	Symbol	SI to English Conversion	English to SI Conversion
Length	L	1 m = 3.2808 ft	1 ft = 0.3048 m
Area	A	1 m^2 = 10.7639 ft^2	1 ft^2 = 0.092903 m^2
Volume	V	1 m^3 = 35.3134 ft^3	1 ft^3 = 0.028317 m^3
Velocity	V	1 m/s = 3.2808 ft/s	1 ft/s = 0.3048 m/s
Density	P	1 kg/m^3 = 0.06243 lb$_m$/ft^3	1 lb$_m$/ft^3 = 16.018 kg/m^3
Force	F	1 N = 0.2248 lb$_f$	1 lb$_f$ = 4.4482 N
Mass	M	1 kg = 2.20462 lb$_m$	1 lb$_m$ = 0.45359237 kg
Pressure	P	1 N/m^2 = 1.45038 × 10^{-4} lb$_f$/in^2	1 lb$_f$/in^2 = 6894.76 N/m^2
Energy, heat	Q	1 kJ = 0.94783 Btu	1 Btu = 1.05504 kJ
Heat flow	Q	1 W = 3.4121 Btu/h	1 Btu/h = 0.29307 W
Heat flux per unit area	q/A	1 W/m^2 = 0.317 Btu/h·ft^2	1 Btu/h·ft^2 = 3.154 W/m^2
Heat flux per unit length	q/L	1 W/m = 1.0403 Btu/h·ft	1 Btu/h·ft = 0.9613 W/m
Heat generation per unit volume	$\dot{q}$	1 W/m^3 = 0.096623 Btu/h·ft^3	1 Btu/h·ft^3 = 10.35 W/m^3
Energy per unit mass	q/m	1 kJ/kg = 0.4299 Btu/lb$_m$	1 Btu/lb$_m$ = 2.326 kJ/kg
Specific heat	C	1 kJ/kg·°C = 0.23884 Btu/lb$_m$·°F	1 Btu/lb$_m$·°F = 4.1869 kJ/kg·°C
Thermal conductivity	K	1 W/m·°C = 0.5778 Btu/h·ft·°F	1 Btu/h·ft·°F = 1.7307 W/m·°C
Convection heat transfer coefficient	H	1 W/m^2·°C = 0.1761 Btu/h·ft^2·°F	1 Btu/h·ft^2·°F = 5.6782 W/m^2·°C
Dynamic viscosity	M	1 kg/M·s = 0.672 lb$_m$/ft·s = 2419.2 lb$_m$/ft·h	1 lb$_m$/ft·s = 1.4881 kg/m·s
Kinematic viscosity and thermal diffusivity		1 m^2/s = 10.7639 ft^2/s	1 ft^2/s = 0.092903 m^2/s

Table A.3 Dry Air at Low Pressure

$T/(K)$	c_p (kJ/kg K)	c_v	γ	$\mu\ 10^{-5}$ (kg/ms)	$k\ 10^{-5}$ (kW/mK)	Pr	ρ (kg/m³) At 1 Atm	$\nu\ 10^{-5}$ (m²/s) At 1 Atm
175	1.0023	0.7152	1.401	1.182	1.593	0.744	2.017	0.586
200	1.0025	0.7154	1.401	1.329	1.809	0.736	1.765	0.753
225	1.0027	0.7156	1.401	1.467	2.020	0.728	1.569	0.935
250	1.0031	0.7160	1.401	1.599	2.227	0.720	1.412	1.132
275	1.0038	0.7167	1.401	1.725	2.428	0.713	1.284	1.343
300	1.0049	0.7178	1.400	1.846	2.624	0.707	1.k 77	1.568
325	1.0063	0.7192	1.400	1.962	2.816	0.701	1.086	1.807
350	1.0082	0.7211	1.398	2.075	3.003	0.697	1.009	2.056
375	1.0106	0.7235	1.397	2.181	3.186	0.692	0.9413	2.317
400	1.0135	0.7264	1.395	2.286	3.365	0.688	0.8824	2.591
450	1.0206	0.7335	1.391	2.485	3.710	0.684	0.7844	3.168
500	1.0295	0.7424	1.387	2.670	4.041	0.680	0.7060	3.782
550	1.0398	0.7527	1.381	2.849	4.357	0.680	0.6418	4.439
600	1.0511	0.7640	1.376	3.017	4.661	0.680	0.5883	5.128
650	1.0629	0.7758	1.370	3.178	4.954	0.682	0.5430	5.853
700	1.0750	0.7879	1.364	3.332	5.236	0.684	0.5043	6.607

750	1.0870	0.7999	1.359	3.482	5.509	0.687	0.4706	7.399
800	1.0987	0.8116	1.354	3.624	5.774	0.690	0.4412	8.214
850	1.1101	0.8230	1.349	3.763	6.030	0.693	0.4153	9.061
900	1.1209	0.8338	1.344	3.897	6.276	0.696	0.3922	9.936
950	1.1313	0.8442	1.340	4.026	6.520	0.699	0.3716	10.83
1000	1.1411	0.8540	1.336	4.153	6.754	0.702	0.3530	11.76
1050	1.1502	0.8631	1.333	4.276	6.985	0.704	0.3362	12.72
1100	1.1589	0.8718	1.329	4.396	7.209	0.707	0.3209	13.70
1150	1.1670	0.8799	1.326	4.511	7.427	0.709	0.3069	14.70
1200	1.1746	0.8875	1.323	4.626	7.640	0.711	0.2941	15.73
1250	1.1817	0.8946	1.321	4.736	7.849	0.713	0.2824	16.77
1300	1.1884	0.9013	1.319	4.846	8.054	0.715	0.2715	17.85
1350	1.1946	0.9075	1.316	4.952	8.253	0.717	0.2615	18.94
1400	1.2005	0.9134	1.314	5.057	8.450	0.719	0.2521	20.06
1500	1.2112	0.9241	1.311	5.264	8.831	0.722	0.2353	22.36
1600	1.2207	0.9336	1.308	5.457	9.199	0.724	0.2206	24.74
1700	1.2293	0.9422	1.305	5.646	9.554	0.726	0.2076	27.20
1800	1.2370	0.9499	1.302	5.829	9.899	0.728	0.1961	29.72
1900	1.2440	0.9569	1.300	6.008	10.233	0.730	0.1858	32.34

Continued

Table A.3 Dry Air at Low Pressure—cont'd

T/(K)	c_p (kJ/kg K)	c_v	γ	$\mu\, 10^{-5}$ (kg/ms)	$k\, 10^{-5}$ (kW/mK)	Pr	ρ (kg/m³)	$v\, 10^{-5}$ (m²/s)
							At 1 Atm	
2000	1.2505	0.9634	1.298	—	—		0.1765	—
2100	1.2564	0.9693	1.296	—	—		0.1681	—
2200	1.2619	0.9748	1.295	—	—		0.1604	—
2300	1.2669	0.9798	1.293	—	—		0.1535	—
2400	1.2717	0.9846	1.292	—	—		0.1471	—
2500	1.2762	0.9891	1.290	—	—		0.1412	—
2600	1.2803	0.9932	1.289	—	—		0.1358	—
2700	1.2843	0.9972	1.288	—	—		0.1307	—
2800	1.2881	1.0010	1.287	—	—		0.1261	—
2900	1.2916	1.0045	1.286	—	—		0.1217	—
3000	1.2949	1.0078	1.285	—	—		0.1177	—

Table A.4 Specific Heat c_p for Some Gases and Vapors

$T/(K)$	CO_2	CO	H_2	N_2	O_2	H_2O	CH_4	C_2H_4	C_2H_6
175	0.709	1.039	13.12	1.039	0.910	1.850	2.083	1.241	–
200	0.735	1.039	13.53	1.039	0.910	1.851	2.087	1.260	–
225	0.763	1.039	13.83	1.039	0.911	1.852	2.121	1.316	–
250	0.791	1.039	14.05	1.039	0.913	1.855	2.156	1.380	1.535
275	0.819	1.040	14.20	1.039	0.915	1.859	2.191	1.453	1.651
300	0.846	1.040	14.31	1.040	0.918	1.864	2.226	1.535	1.766
325	0.871	1.041	14.38	1.040	0.923	1.871	2.293	1.621	1.878
350	0.895	1.043	14.43	1.041	0.928	1.880	2.365	1.709	1.987
375	0.918	1.045	14.46	1.042	0.934	1.890	2.442	1.799	2.095
400	0.939	1.048	14.48	1.044	0.941	1.901	2.525	1.891	2.199
450	0.978	1.054	14.50	1.049	0.956	1.926	2.703	2.063	2.402
500	1.014	1.064	14.51	1.056	0.972	1.954	2.889	2.227	2.596
550	1.046	1.075	14.53	1.065	0.988	1.984	3.074	2.378	2.782
600	1.075	1.087	14.55	1.075	1.003	2.015	3.256	2.519	2.958
650	1.102	1.100	14.57	1.086	1.017	2.047	3.432	2.649	3.126
700	1.126	1.113	14.60	1.098	1.031	2.080	3.602	2.770	3.286
750	1.148	1.126	14.65	1.110	1.043	2.113	3.766	2.883	3.438
800	1.168	1.139	14.71	1.122	1.054	2.147	3.923	2.989	3.581
850	1.187	1.151	14.77	1.134	1.065	2.182	4.072	3.088	3.717
900	1.204	1.163	14.83	1.146	1.074	2.217	4.214	3.180	3.846
950	1.220	1.174	14.90	1.157	1.082	2.252	4.348	3.266	–
1000	1.234	1.185	14.98	1.167	1.090	2.288	4.475	3.347	–
1050	1.247	1.194	15.06	1.177	1.097	2.323	4.595	3.423	–
1100	1.259	1.203	15.15	1.187	1.103	2.358	4.708	3.494	–
1150	1.270	1.212	15.25	1.196	1.109	2.392	4.814	3.561	–
1200	1.280	1.220	15.34	1.204	1.115	2.425			–

Continued

Table A.4 **Specific Heat c_p for Some Gases and Vapors—cont'd**

T/(K)	CO_2	CO	H_2	N_2	O_2	H_2O	CH_4	C_2H_4	C_2H_6
1250	1.290	1.227	15.44	1.212	1.120	2.458	T/(K)	C_6H_6	C_8H_{18}
1300	1.298	1.234	15.54	1.219	1.125	2.490			
1350	1.306	1.240	15.65	1.226	1.130	2.521	250	0.850	1.308
1400	1.313	1.246	15.77	1.232	1.134	2.552	275	0.957	1.484
1500	1.326	1.257	16.02	1.244	1.143	2.609	300	1.060	1.656
1600	1.338	1.267	16.23	1.254	1.151	2.662	325	1.160	1.825
1700	1.348	1.275	16.44	1.263	1.158	2.711	350	1.255	1.979
1800	1.356	1.282	16.64	1.271	1.166	2.756	375	1.347	2.109
1900	1.364	1.288	16.83	1.278	1.173	2.798	400	1.435	2.218
2000	1.371	1.294	17.D1	1.284	1.181	2.836	450	1.600	2.403
2100	1.377	1.299	17.18	1.290	1.188	2.872	500	1.752	2.608
2200	1.383	1.304	17.35	1.295	1.195	2.904	550	1.891	2.774
2300	1.388	1.308	17.50	1.300	1.202	2.934	600	2.018	2.924
2400	1.393	1.311	17.65	1.304	1.209	2.962	650	2.134	3.121
2500	1.397	1.315	17.80	1.307	1.216	2.987	700	2.239	3.232
2600	1.401	1.318	17.93	1.311	1.223	3.011	750	2.335	3.349
2700	1.404	1.321	18.06	1.314	1.230	3.033	800	2.422	3.465
2800	1.408	1.324	18.17	1.317	1.236	3.053	850	2.500	3.582
2900	1.411	1.326	18.28	1.320	1.243	3.072	900	2.571	3.673
3000	1.414	1.329	18.39	1.323	1.249	3.090	—	—	—
3500	1.427	1.339	18.91	1.333	1.276	3.163	—	—	—
4000	1.437	1.346	19.39	1.342	1.299	3.217	—	—	—
4500	1.446	1.353	19.83	1.349	1.316	3.258	—	—	—
5000	1.455	1.359	20.23	1.355	1.328	3.292	—	—	—
5500	1.465	1.365	20.61	1.362	1.337	3.322	—	—	—
6000	1.476	1.370	20.96	1.369	1.344	3.350	—	—	—

Table A.5 H, U, S, G for Some Gases and Vapors

	Carbon Dioxide (CO_2)				T/(K)	Water Vapor (H_2O)			
H (kJ/kmol)	U (kJ/kmol)	S^a (kJ/kmol K)	G^a (kJ/kmol)			H (kJ/kmol)	U (kJ/kmol)	S^a (kJ/kmol K)	G^a (kJ/kmol)
−9364	−9364	0	−9364		0	−9904	−9904	0	−9904
−6456	−7287	178.90	−24,346		100	−6615	−7446	152.28	−21,843
−3414	−5077	199.87	−43,387		200	−3280	−4943	175.38	−38,356
0	−2479	213.69	−63,710		298.15	0	−2479	188.72	−56,268
67	−2427	213.92	−64,108		300	63	−2432	188.93	−56,616
4008	683	225.22	−86,082		400	3452	126	198.67	−76,017
12,916	7927	243.20	−133,000		600	10,498	5509	212.93	−117,260
22,815	16,164	257.41	−183,110		800	17,991	11,340	223.69	−160,960
33,405	25,091	269.22	−235,810		1,000	25,978	17,664	232.60	−206,620
44,484	34,507	279.31	−290,680		1,200	34,476	24,499	240.33	−253,920
55,907	44,266	288.11	−347,440		1,400	43,447	31,806	247.24	−302,690
67,580	54,277	295.90	−405,860		1,600	52,844	39,541	253.51	−352,780
79,442	64,476	302.88	−465,750		1,800	62,609	47,643	259.26	−404,060
91,450	74,821	309.21	−526,970		2,000	72,689	56,060	264.57	−456,450
103,570	85,283	314.99	−589,400		2,200	83,036	64,744	269.50	−509,860

Continued

Table A.5 H, U, S, G for Some Gases and Vapors—cont'd

	Carbon Dioxide (CO_2)				T/(K)	Water Vapor (H_2O)			
H (kJ/kmol)	U (kJ/kmol)	S^a (kJ/kmol K)	G^a (kJ/kmol)			H (kJ/kmol)	U (kJ/kmol)	S^a (kJ/kmol K)	G^a (kJ/kmol)
115,790	95,833	320.30	−652,940		2,400	93,604	73,650	274.10	−564,230
128,080	106,470	325.22	−717,490		2,600	104,370	82,752	278.41	−619,490
140,440	117,160	329.80	−782,990		2,800	115,290	92,014	282.45	−675,580
152,860	127,920	334.08	−849,390		3,000	126,360	10,1420	286.27	−732,460
165,330	138,720	338.11	−916,620		3,200	137,550	110,950	289.88	−790,080
177,850	149,580	341.90	−984,620		3,400	148,850	120,590	293.31	−848,390
190,410	160,470	345.49	−1,053,360		3,600	160,250	130,320	296.57	−907,390
203,000	171,400	348.90	−1,122,800		3,800	171,720	140,130	299.67	−967,010
215,630	182,370	352.13	−1,192,900		4,000	183,280	150,020	302.63	−102,725

Table A.6 H, U, S, G for Some Gases and Vapors

Hydrogen (H$_2$)				T/(K)	Carbon Monoxide (CO)			
H (kJ/kmol)	U (kJ/kmol)	S^a (kJ/kmol K)	G^a (kJ/kmol)		H (kJ/kmol)	U (kJ/kmol)	S^a (kJ/kmol K)	G^a (kJ/kmol)
−8468	8468	0	−8468	0	−8699	−8669	0	−8669
−5293	−6124	102.04	−15,496	100	−5770	−6601	165.74	−22,344
−2770	−4433	119.33	−26,635	200	−2858	−4521	185.92	−40,041
0	−2479	130.57	−38,931	298.15	0	−2479	197.54	−58,898
54	−2440	130.75	−39,172	300	54	−2440	197.72	−59,263
2958	−368	139.11	−52,684	400	2975	−351	206.12	−79,475
8812	3823	150.97	−81,769	600	10,196	5208	218.20	−12,0730
14,703	8051	159.44	−112,850	800	15,175	8524	227.16	−166,550
20,686	12,371	166.11	−145,430	1,000	21,686	13,371	234.42	−212,740
26,794	16,817	171.68	−179,220	1,200	28,426	18,449	240.56	−260,250
33,062	21,422	176.51	−214,050	1,400	35,338	23,698	245.89	−308,910
39,522	26,219	180.82	−249,790	1,600	42,384	29,081	250.59	−358,560
46,150	31,184	184.72	−286,350	1,800	49,522	34,556	254.80	−409,110
52,932	36,303	188.30	−323,660	2,000	56,739	40,110	258.60	−460,460
59,860	41,569	191.60	−361,650	2,200	64,019	45,728	262.06	−512,520

Continued

Table A.6 H, U, S, G for Some Gases and Vapors—cont'd

Hydrogen (H_2)					Carbon Monoxide (CO)			
H (kJ/kmol)	U (kJ/kmol)	S^a (kJ/kmol K)	G^a (kJ/kmol)	T/(K)	H (kJ/kmol)	U (kJ/kmol)	S^a (kJ/kmol K)	G^a (kJ/kmol)
66,915	46,960	194.67	−400,290	2,400	71,346	51,391	265.25	−565,260
74,090	52,473	197.54	−439,510	2,600	78,714	57,096	268.20	−618,610
81,370	58,090	200.23	−479,280	2,800	86,115	62,835	270.94	−672,530
88,743	63,799	202.78	−519,590	3,000	93,542	68,598	273.51	−726,980
96,199	69,592	205.18	−560,390	3,200	101,000	74,391	275.91	−781,930
103,740	75,469	207.47	−601,650	3,400	108,480	80,210	278.18	−837,340
111,360	81,430	209.65	−643,370	3,600	115,980	86,044	280.32	−893,190
119,060	87,469	211.73	−685,510	3,800	123,490	91,900	282.36	−949,460
126,850	93,589	213.73	−728,060	4,000	131,030	97,769	284.29	−1,006,120

Table A.7 H, U, S, G for Some Gases and Vapors

Oxygen (O_2)				T/(K)	Nitrogen (N_2)			
H (kJ/kmol)	U (kJ/kmol)	S^a (kJ/kmol K)	G^a (kJ/kmol)		H (kJ/kmol)	U (kJ/kmol)	S^a (kJ/kmol K)	G^a (kJ/kmol)
−8682	−8682	0	−8682	0	−8669	−8669	0	−8669
−5778	−6610	173.20	−23,098	100	−5770	−6601	159.70	−21,740
−2866	−4529	193.38	−41,541	200	−2858	−4521	179.88	−38,833
0	−2479	205.03	−61,131	298.15	0	−2479	191.50	−57,096
54	−2440	205.21	−61,509	300	54	−2440	191.68	−57,450
3029	−297	213.76	−82,477	400	2971	−355	200.07	−77,058
9247	4258	226.35	−126,560	600	8891	3902	212.07	−118,350
15,841	9189	235.81	−172,810	800	15,046	8394	220.91	−161,680
22,707	14,392	243.48	−220,770	1,000	21,460	13,145	228.06	−206,600
29,765	19,788	249.91	−270,120	1,200	28,108	18,131	234.12	−252,830
36,966	25,325	255.45	−320,670	1,400	34,936	23,296	239.38	−300,190
44,279	30,976	260.34	−372,260	1,600	41,903	28,600	244.03	−348,540
51,689	36,723	264.70	−424,770	1,800	48,982	34,016	248.19	−397,770

Continued

Table A.7 H, U, S, G for Some Gases and Vapors—cont'd

	Oxygen (O_2)					Nitrogen (N_2)		
H (kJ/kmol)	U (kJ/kmol)	S^a (kJ/kmol K)	G^a (kJ/kmol)	T/(K)	H (kJ/kmol)	U (kJ/kmol)	S^a (kJ/kmol K)	G^a (kJ/kmol)
59,199	42,571	268.65	−478,110	2,000	56,141	39,512	251.97	−447,800
66,802	48,510	272.28	−532,210	2,200	63,371	45,079	255.41	−498,540
74,492	54,537	275.63	−587,010	2,400	70,651	50,696	258.58	−549,940
82,274	60,657	278.74	−642,440	2,600	77,981	56,364	261.51	−601,950
90,144	66,864	281.65	−698,490	2,800	85,345	62,065	264.24	−654,530
98,098	73,155	284.40	−755,100	3,000	92,738	67,795	266.79	−707,640
106,130	79,521	286.99	−812,240	3,200	100,160	73,555	269.19	−761,230
114,230	85,963	289.44	−869,880	3,400	107,610	79,339	271.45	−815,310
122,400	92,467	291.78	−928,010	3,600	115,080	85,149	273.58	−869,800
130,630	99,034	294.01	−986,590	3,800	122,570	90,976	275.60	−924,730
138,910	105,660	296.13	−1,045,590	4,000	130,080	96,819	277.53	−980,040

Table A.8 H, U, S, G for Some Gases and Vapors

Hydroxyl (OH)				T/(K)	Nitric Oxide (NO)			
H (kJ/kmol)	U (kJ/kmol)	S^a (kJ/kmol K)	G^a (kJ/kmol)		H (kJ/kmol)	U (kJ/kmol)	S^a (kJ/kmol K)	G^a (kJ/kmol)
−9171	−9171	0	−9171	0	−9192	−9192	0	−9192
−6138	−6969	149.48	−21,086	100	−6071	−6902	176.92	−23,763
−2975	−4638	171.48	−37,271	200	−2950	−4613	198.64	−42,678
0	−2479	183.60	−54,740	298.15	0	−2479	210.65	−62,806
54	−2440	183.78	−55,080	300	54	−2440	210.84	−63,198
3033	−292	192.36	−73,909	400	3042	284	219.43	−84,729
8941	3953	204.33	−113,660	600	9146	4158	231.78	−129,920
14,878	8227	212.87	−115,420	800	15,548	8896	240.98	−177,240
20,933	12,618	219.62	−198,690	1,000	22,230	13,915	248.43	−226,200
27,158	17,181	225.30	−243,200	1,200	29,121	19,143	254.71	−276,540
33,568	21,928	230.23	−288,760	1,400	36,166	24,526	260.14	−328,030
40,150	26,847	234.63	−335,250	1,600	43,321	30,018	264.92	−380,550
46,890	31,924	238.59	−382,580	1,800	50,559	35,594	269.18	−433,960
53,760	37,131	242.22	−430,670	2,000	57,861	41,232	273.03	−488,190

Continued

Table A.8 H, U, S, G for Some Gases and Vapors—cont'd

Hydroxyl (OH)				T/(K)	Nitric Oxide (NO)			
H (kJ/kmol)	U (kJ/kmol)	S^a (kJ/kmol K)	G^a (kJ/kmol)		H (kJ/kmol)	U (kJ/kmol)	S^a (kJ/kmol K)	G^a (kJ/kmol)
60,752	42,460	245.55	−479,450	2,200	65,216	46,924	276.53	−543,150
67,839	47,885	248.63	−528,870	2,400	72,609	52,655	279.75	−598,780
75,015	53,397	251.50	−578,890	2,600	80,036	58,418	282.72	−655,030
82,266	58,985	254.19	−629,460	2,800	87,492	64,211	285.48	−711,860
89,584	64,640	256.71	−680,540	3,000	94,977	70,034	288.06	−769,220
96,960	70,354	254.09	−732,130	3,200	102,480	75,873	290.48	−827,070
10,4390	76,118	261.34	−784,170	3,400	110,000	81,733	292.77	−885,410
111,860	81,927	263.48	−836,670	3,600	117,550	87,613	294.92	−944,170
119,380	87,783	265.51	−889,550	3,800	125,100	93,507	296.96	−1,003,360
126,940	93,680	267.45	−942,860	4,000	132,670	99,417	298.90	−1,062,950

Table A.9 H, U, S, G for Some Gases and Vapors

Methane Vapor (CH$_4$)					Ethylene Vapor (C$_2$H$_4$)			
H (kJ/kmol)	U (kJ/kmol)	S^a (kJ/kmol K)	G^a (kJ/kmol)	T/(K)	H (kJ/kmol)	U (kJ/kmol)	S^a (kJ/kmol K)	G^a (kJ/kmol)
−10,025	−10,025	0	−10,025	0	−10,519	−10,519	0	−10,519
−6699	−7530	149.39	−21,638	100	−7192	−8024	180.44	−25,236
−3368	−5031	172.47	−37,863	200	−3803	−5466	203.85	−44,573
0	−2479	186.15	−55,499	298.15	0	−2479	291.22	−65,362
67	−2427	186.37	−55,843	300	79	−2415	291.49	−65,767
3862	536	197.25	−75,038	400	4883	−1557	233.24	−88,412
13,129	8141	215.88	−116,400	600	17,334	12,346	258.24	−137,610
24,673	18,022	232.41	−161,260	800	32,849	26,197	280.47	−191,520
38,179	29,865	247.45	−209,270	1,000	50,664	42,350	300.30	−249,640
53,271	43,293	261.18	−260,150	1,200	70,254	60,276	318.13	−311,510
69,609	57,969	273.76	−313,660	1,400	91,199	79,558	334.27	−376,780
86,910	73,607	285.31	−369,590	1,600	113,180	99,878	348.94	−445,120
104,960	89,994	295.93	−427,720	1,800	135,970	121,010	362.36	−516,270
123,600	106,970	305.75	−487,900	2,000	159,390	142,760	374.69	−589,990

Table A.10 International Standard Atmosphere

Z (m)	p (bar)	T/(K)	ρ/ρ_0	$v\,10^{-5}$ (m^2/s)	$k\,10^{-5}$ (kW/mK)	a (m/s)	$\lambda\,10^{-8}$ (m)
−2500	1.3521	304.4	1.2631	1.207	2.661	349.8	5.251
−2000	1.2778	301.2	1.2067	1.253	2.636	347.9	5.497
−1500	1.2070	297.9	1.1522	1.301	2.611	346.0	5.757
−1000	1.1393	294.7	1.0996	1.352	2.585	344.1	6.032
−500	1.0748	291.4	1.0489	1.405	2.560	342.2	6.324
0	1.01325	288.15	1.0000	1.461	2.534	340.3	6.633
500	0.9546	284.9	0.9529	1.520	2.509	338.4	6.961
1000	0.8988	281.7	0.9075	1.581	2.483	336.4	7.309
1500	0.8456	278.4	0.8638	1.646	2.457	334.5	7.679
2000	0.7950	275.2	0.8217	1.715	2.431	332.5	8.072
2500	0.7469	271.9	0.7812	1.787	2.405	330.6	8.491
3000	0.7012	268.7	0.7423	1.863	2.379	328.6	8.936
3500	0.6578	265.4	0.7048	1.943	2.353	326.6	9.411
4000	0.6166	262.2	0.6689	2.028	2.327	324.6	9.917
4500	0.5775	258.9	0.6343	2.117	2.301	322.6	10.46
5000	0.5405	255.7	0.6012	2.211	2.275	320.5	11.03
5500	0.5054	252.4	0.5694	2.311	2.248	318.5	11.65
6000	0.4722	249.2	0.5389	2.416	2.222	316.5	12.31
6500	0.4408	245.9	0.5096	2.528	2.195	314.4	13.02
7000	0.4111	242.7	0.4817	2.646	2.169	312.3	13.77
7500	0.3830	239.5	0.4549	2.771	2.142	310.2	14.58
8000	0.3565	236.2	0.4292	2.904	2.115	308.1	15.45
8500	0.3315	233.0	0.4047	3.046	2.088	306.0	16.39
9000	0.3080	229.7	0.3813	3.196	2.061	303.8	17.40
9500	0.2858	226.5	0.3589	3.355	2.034	301.7	18.48
10,000	0.2650	223.3	0.3376	3.525	2.007	299.5	19.65
10,500	0.2454	220.0	0.3172	3.706	1.980	297.4	20.91

Table A.10 **International Standard Atmosphere—cont'd**

Z (m)	p (bar)	T/(K)	ρ/ρ₀	$v\, 10^{-5}$ (m²/s)	$k\, 10^{-5}$ (kW/mK)	a (m/s)	$\lambda\, 10^{-8}$ (m)
11,000	0.2270	216.8	0.2978	3.899	1.953	295.2	22.27
11,500	0.2098	216.7	0.2755	4.213	1.952	295.1	24.08
12,000	0.1940	216.7	0.2546	4.557	1.952	295.1	26.05
12,500	0.1793	216.7	0.2354	4.930	1.952	295.1	28.18
13,000	0.1658	216.7	0.2176	5.333	1.952	295.1	30.48
13,500	0.1533	216.7	0.2012	5.768	1.952	295.1	32.97
14,000	0.1417	216.7	0.1860	6.239	1.952	295.1	35.66
14,500	0.1310	216.7	0.1720	6.749	1.952	295.1	38.57
15,000	0.1211	216.7	0.1590	7.300	1.952	295.1	41.72
15,500	0.1120	216.7	0.1470	7.895	1.952	295.1	45.13
16,000	0.1035	216.7	0.1359	8.540	1.952	295.1	48.81
16,500	0.09572	216.7	0.1256	9.237	1.952	295.1	52.79
17,000	0.08850	216.7	0.1162	9.990	1.952	295.1	57.10
17,500	0.08182	216.7	0.1074	10.805	1.952	295.1	61.76
18,000	0.07565	216.7	0.09930	11.686	1.952	295.1	66.79
18,500	0.06995	216.7	0.09182	12.639	1.952	295.1	72.24
19,000	0.06467	216.7	0.08489	13.670	1.952	295.1	78.13
19,500	0.05980	216.7	0.07850	14.784	1.952	295.1	84.50
20,000	0.05529	216.7	0.07258	15.989	1.952	295.1	91.39
22,000	0.04047	218.6	0.05266	22.201	1.968	296.4	126.0
24,000	0.02972	220.6	0.03832	30.743	1.985	297.7	173.1
26,000	0.02188	222.5	0.02797	42.439	2.001	299.1	237.2
28,000	0.01616	224.5	0.02047	58.405	2.018	300.4	324.0
30,000	0.01197	226.5	0.01503	80.134	2.034	301.7	441.3
32,000	0.00889	228.5	0.01107	109.62	2.051	303.0	599.4

Table A.11 **Alpha-Aluminum Oxide, Al_2O_3**

Phase	T	c_p	H	S	G	B
Solid						
	298	18.871	−400.4	12.174	−404.030	295.202
	300	18.979	−400.365	12.291	−404.052	294.391
	400	22.987	−398.243	18.369	−405.590	221.634
	500	25.179	−395.826	23.754	−407.702	178.231
	600	26.656	−393.230	28.482	−410.319	149.479
	700	27.797	−390.505	32.680	−413.381	129.081
	800	28.757	−387.677	36.456	−416.841	113.891
	900	29.354	−384.770	39.878	−420.661	102.164
	1000	29.845	−381.809	42.997	−424.807	92.854
	1100	30.265	−378.803	45.862	−429.252	85.296
	1200	30.638	−375.758	48.512	−433.972	79.048
	1300	30.976	−372.677	50.978	−438.948	73.804
	1400	31.290	−369.563	53.285	−444.162	69.346
	1500	31.586	−366.420	55.454	−449.600	65.516
	1600	31.868	−363.247	57.501	−455.249	62.192
	1700	32.139	−360.046	59.441	−461.097	59.286
	1800	32.402	−356.819	61.286	−467.134	56.725
	1900	32.658	−353.566	63.045	−473.351	54.455
	2000	32.909	−350.288	64.726	−479.740	52.431
	2100	33.156	−346.985	66.338	−486.294	50.616
	2200	33.399	−343.657	67.886	−493.006	48.982
	2300	33.639	−340.306	69.376	−499.869	47.505
	2327	33.703	−339.396	69.769	−501.748	47.130
Liquid	2327	34.623	−311.096	81.930	−501.748	47.130
	2400	34.623	−308.568	83.000	−507.768	46.245
	2500	34.623	−305.106	84.413	−516.139	45.127
	2500	34.623	−301.644	85.771	−524.649	44.107
	2700	34.623	−298.181	87.078	−533.292	43.173
	2800	34.623	−294.719	88.337	−542.063	42.316
	2900	34.623	−291.257	89.552	−550.958	41.527

Table A.11 Alpha-Aluminum Oxide, Al_2O_3—cont'd

Phase	T	c_p	H	S	G	B
	3000	34.623	−287.795	90.726	−559.972	40.799
	3100	34.623	−284.332	91.861	−569.101	40.127
	3200	34.623	−280.870	92.960	−578.343	39.504
	3300	34.623	−277.408	94.025	−587.692	38.926
	3400	34.623	−273.945	95.059	−597.147	38.389
	3500	34.623	−270.483	96.063	−606.703	37.889

Table A.12 Aluminum Silicate (Andalusite), $Al_2O_3 \cdot SiO_2$

Phase	T	c_p	H	S	G	B
Solid						
	400	29.177	−619.52	22.28	−626.163	459.051
	300	29.363	−619.466	22.461	−626.204	456.251
	400	36.117	−616.148	31.960	−628.932	343.679
	500	39.556	−612.348	40.424	−632.561	276.529
	600	41.692	−508.279	47.838	−536.982	232.052
	700	43.212	−504.030	54.385	−542.099	200.499
	800	44.404	−599.547	60.235	−647.835	177.004
	900	45.405	−595.156	65.525	−554.128	158.865
	1000	46.287	−590.570	70.355	−650.925	144.455
	1100	47.092	−585.901	74.805	−558.185	132.774
	1200	47.844	−581.154	78.935	−675.875	123.110
	1300	48.559	−576.333	82.793	−683.964	115.000
	1400	49.247	−571.443	85.417	−692.425	108.107
	1500	49.915	−566.484	89.838	−701.241	102.184
	1600	50.568	−551.450	93.080	−710.388	97.048
	1700	51.210	−556.371	95.155	−719.851	92.556
	1800	51.842	−551.218	99.110	−729.616	88.599
	1900	52.457	−546.003	101.930	−739.669	85.093
	2000	53.085	−540.725	104.637	−749.998	81.957

Table A.13 Aluminum Disilicate Dihydrate, $Al_2O_3 \cdot 2SiO_2 \cdot 2H_2O$

Phase	T	c_p	H	S	G	B
Solid						
	298	59.141	−964.94	40.5	−977.015	716.268
	300	59.316	−964.830	40.866	−977.090	711.906
	400	66.671	−958.504	59.017	−982.110	536.672
	500	71.972	−951.562	74.485	−988.804	432.264
	600	76.464	−944.136	88.012	−996.943	363.185
	700	80.574	−936.281	100.111	−1006.359	314.242
	800	84.480	−928.028	111.127	−1016.929	277.850
	900	88.268	−919.389	121.297	−1028.557	249.801

Table A.14 Aluminum Disilicate, $Al_2O_3 \cdot 2SiO_2$

Phase	T	c_p	H	S	G	B
Solid						
	298	53.559	−767.5	32.612	−777.223	569.797
	300	53.623	−767.401	32.944	−777.284	566.327
	400	56.195	−761.898	48.757	−781.401	426.995
	500	57.858	−756.191	61.485	−786.933	344.015
	600	59.163	−750.338	72.153	−793.629	289.118
	700	60.300	−744.364	81.360	−801.315	250.216
	800	61.346	−138.281	89.481	−809.865	221.275
	900	62.340	−732.096	96.764	−819.184	198.952
	1000	63.302	−725.814	103.382	−829.196	181.245
	1100	64.242	−719.436	109.460	−839.842	166.884
	1200	65.168	−712.966	115.089	−851.073	155.022
	1300	66.084	−706.403	120.342	−862.847	145.077
	1400	66.992	−699.749	125.272	−875.131	136.632
	1500	67.895	−693.005	129.925	−887.893	129.383
	1600	68.794	−686.170	134.336	−901.108	123.102
	1700	69.690	−679.246	138.533	−914.753	117.615
	1800	70.583	−672.233	142.542	−928.808	112.788

Table A.15 **Trialumino Disilicate (Mullite), $3Al_2O_3 \cdot 2SiO_2$**

Phase	T	c_p	H	S	G	B
Solid						
	298	77.559	−1519.25	55.7	−1538.838	1201.053
	300	78.054	−1519.105	55.181	−1638.950	1194.142
	400	95.503	−1510.250	91.503	−1545.851	899.924
	500	105.071	−1500.089	114.151	−1557.169	724.445
	500	112.079	−1589.152	131.055	−1659.501	508.234
	700	116.348	−1577.730	151.580	−1583.906	525.810
	800	119.638	−1555.924	157.439	−1699.876	454.447
	900	122.311	−1553.823	181.590	−1717.344	417.084
	1000	124.559	−1541.476	194.695	−1735.172	379.491
	1100	125.491	−1528.921	205.551	−1756.248	348.981
	1200	128.172	−1515.186	217.741	−1777.475	323.755
	1300	129.443	−1503.294	228.059	−1799.771	302.510
	1400	130.931	−1490.254	237.715	−1823.055	284.632
	1500	132.054	−1477.113	245.788	−1847.295	259.187
	1500	133.025	−1453.858	255.342	−1872.405	255.793
	1700	133.855	−1450.513	253.432	−1898.348	244.082
	1800	134.548	−1437.091	271.103	−1925.078	233.758
	1900	135.110	−1423.608	278.394	−1952.555	224.525
	2000	135.544	−1410.074	285.335	−1980.745	216.475

Table A.16 **Kaolinite, $Al_2O_3 \cdot 2SiO_2 \cdot 2H_2O$**

Phase	T	c_p	H	S	G	B
Solid						
	298	58.618	−979.47	48.5	−993.930	728.669
	300	58.885	−979.361	48.863	−994.020	724.241
	400	69.445	−972.893	67.399	−999.853	546.368
	500	76.115	−965.596	83.654	−1007.423	440.404
	600	81.253	−957.720	98.000	−1016.520	370.317
	700	85.668	−949.369	110.864	−1026.974	320.679

Table A.17 Calcium Carbonate, $CaCO_3$

Phase	T	c_p	H	S	G	B
Solid Aragonite						
	298	19.568	−288.4	21.03	−294.670	216.028
	300	19.663	−288.364	21.151	−294.709	214.724
	323	20.730	−287.899	22.644	−295.213	199.775
Solid Calcite						
	323	20.730	−287.854	22.783	−295.213	199.775
	400	23.201	−286.154	27.494	−297.152	162.378
	500	25.120	−283.730	32.895	−300.177	131.225
	600	26.402	−281.151	37.594	−303.707	110.640
	700	27.383	−278.460	41.740	−307.678	96.074
	800	28.203	−275.579	45.452	−312.041	85.257
	900	28.931	−272.822	48.816	−316.757	76.929
	1000	29.600	−269.895	51.899	−321.795	70.338
	1100	30.232	−265.903	54.750	−327.129	55.003
	1200	30.837	−253.850	57.407	−332.738	50.508

Table A.18 $CaO \cdot MoO_3$

Phase	T	c_p	H	S	G	B
Solid						
	298	117.04	−1546.783	122.591	−1583.333	277.386
	300	117.41	−1546.566	123.316	−1583.561	275.715
	400	131.19	−1534.058	159.203	−1597.740	208.638
	500	139.13	−1520.514	189.393	−1615.211	168.736
	600	144.79	−1506.306	215.282	−1635.475	142.377
	700	149.35	−1491.593	237.954	−1658.160	123.730
	800	153.34	−1476.454	258.162	−1682.984	109.885
	900	157.00	−1460.935	276.437	−1709.729	99.228
	1000	160.44	−1445.062	293.158	−1738.220	90.793
	1100	163.75	−1428.852	308.605	−1768.318	83.968

Table A.18 $CaO \cdot MoO_3$—cont'd

Phase	T	c_p	H	S	G	B
	1200	166.96	−1412.316	322.992	−1799.906	78.346
	1300	170.11	−1395.462	336.481	−1832.886	73.644
	1400	173.22	−1378.295	349.201	−1867.176	69.663
	1500	176.28	−1360.820	361.257	−1902.704	66.256
	1600	179.32	−1343.039	372.731	−1939.408	63.314
	1700	182.34	−1324.955	383.693	−1977.233	60.751
	1718	182.89	−1321.668	385.616	−1984.157	60.325

Table A.19 $CaO \cdot WO_3$

Phase	T	c_p	H	S	G	B
Solid						
	298	113.23	−1624.019	126.357	−1661.692	291.114
	300	113.61	−1623.809	127.058	−1661.927	289.360
	400	127.55	−1611.666	161.896	−1676.425	218.913
	500	135.12	−1598.502	191.240	−1694.122	176.979
	600	140.17	−1584.724	216.346	−1714.532	149.260
	700	144.04	−1570.507	238.255	−1737.285	129.635
	800	147.27	−1555.937	257.705	−1762.101	115.051
	900	150.14	−1541.064	275.220	−1788.762	103.814
	1000	152.78	−1525.916	291.177	−1817.093	94.913
	1100	155.27	−1510.512	305.857	−1846.955	87.702
	1200	157.66	−1494.865	319.470	−1878.229	81.755
	1300	159.98	−1478.982	332.182	−1910.819	76.776
	1400	162.25	−1462.870	344.121	−1944.640	72.554
	1500	164.47	−1446.534	355.391	−1979.621	68.935
	1600	166.67	−1429.976	366.077	−2015.699	65.804
	1700	168.85	−1413.200	376.246	−2052.819	63.074
	1800	171.01	−1396.207	385.959	−2090.933	60.676
	1853	172.15	−1387.113	390.938	−2111.521	59.521

Table A.20 $CaO \cdot Cr_2O_3$

Phase	T	c_p	H	S	G	B
Solid						
	298	146.65	−1,829,663	125.227	−1,866,999	327.082
	300	147.00	−1,829,369	126.209	−1,867,232	325.106
	400	159.98	−1,813,937	170.517	−1,882,136	245.776
	500	166.70	−1,797,572	207.005	−1,901,060	198.597
	600	170.96	−1,780,676	237.799	−1,923,333	167.437
	700	174.05	−1,763,419	264.395	−1,948,466	145.393
	800	176.53	−1,745,886	287.803	−1,976,092	129.022
	900	178.64	−1,728,125	308.720	−2,005,929	116.418
	1000	180.53	−1,710,166	327.641	−2,037,754	106.439
	1100	182.27	−1,692,025	344.929	−2,071,388	98.360
	1200	183.91	−1,673,715	360.860	−2,106,680	91.699
	1300	185.48	−1,655,245	375.643	−2,143,507	86.125
	1400	187.01	−1,636,620	389.445	−2,181,762	81.400
	1500	188.49	−1,617,845	402.398	−2,221,353	77.352
	1600	189.95	−1,598,923	414.609	−2,262,202	73.851
	1700	191.38	−1,579,856	426.168	−2,304,238	70.799
	1800	192.80	−1,560,647	437.147	−2,347,401	68.118
	1900	194.20	−1,541,298	447.609	−2,391,636	65.749
	1918	194.45	−1,537,800	449.441	−2,399,708	65.352

Table A.21 $3CaO \cdot V_2O_5$

Phase	T	c_p	H	S	G	B
Solid						
	298	257.03	−3777.608	274.889	−3859.566	676.163
	300	257.22	−3777.132	276.479	−3860.076	672.082
	400	267.35	−3750.904	351.863	−3891.649	508.184
	500	277.48	−3723.662	412.609	−3929.967	410.550
	600	287.62	−3695.407	464.096	−3973.865	345.947

Table A.21 $3CaO \cdot V_2O_5$—cont'd

Phase	T	c_p	H	S	G	B
	700	297.75	−3666.139	509.193	−4022.574	300.161
	800	307.88	−3635.857	549.614	−4075.548	266.099
	900	318.02	−3604.562	586.463	−4132.378	239.831
	1000	328.15	−3572.254	620.494	−4192.747	219.001
	1100	338.28	−3538.932	652.245	−4256.401	202.115
	1200	348.42	−3504.597	682.114	−4323.133	188.176
	1300	358.55	−3469.248	710.402	−4392.771	176.499
	1400	368.69	−3432.886	737.345	−4465.169	166.593
	1500	378.82	−3395.511	763.127	−4540.202	158.100
	1600	388.95	−3357.122	787.899	−4617.761	150.751
	1653	394.32	−3336.366	800.661	−4659.859	147.247

Table A.22 $2CaO \cdot V_2O_5$

Phase	T	c_p	H	S	G	B
Solid						
	298	213.90	−3082.683	220.497	−3148.424	551.517
	300	214.12	−3082.287	221.821	−3148.833	548.241
	400	226.22	−3060.270	285.016	−3114.300	414.511
	500	238.32	−3031.043	336.856	−3205.411	334.865
	600	25Q.42	−3012.606	381.316	−3241.432	282.185
	100	262.52	−2986.959	420.888	−3281.580	244.868
	800	214.62	−2960.102	456.132	−3325.481	211.121
	900	286.12	−2932.034	489.111	−3312.833	195.149
	1000	298.82	−2902.151	520.612	−3423.369	118.814
	1100	310.92	−2812.270	549.660	−3416.896	165.100
	1200	323.02	−2840.513	511.233	−3533.252	153.195
	1288	333.61	−2811.619	600.465	−3585.011	145.389

Table A.23 $CaO \cdot V_2O_5$

Phase	T	c_p	H	S	G	B
Solid						
	298	170.75	−2328.772	179.075	−2382.163	417.335
	300	170.97	−2328.456	180.132	−2382.496	414.819
	400	182.89	−2310.763	230.951	−2403.143	313.810
	500	194.81	−2291.878	273.042	−2428.399	253.687
	600	206.72	−2271.801	309.612	−2457.569	213.945
	700	218.64	−2250.533	342.374	−2490.195	185.816
	800	230.56	−2228.073	372.347	−2525.951	164.924
	900	242.47	−2204.422	400.190	−2564.593	148.841
	1000	254.39	−2179.579	426.354	−2605.933	136.117
	1051	260.46	−2166.451	439.157	−2628.005	130.608

Table A.24 $3CaO \cdot 2TiO_2$

Phase	T	c_p	H	S	G	B
Solid						
	298	239.59	−4003.962	234.722	−4073.945	713.720
	300	240.41	−4003.518	236.207	−4074.380	709.395
	400	269.83	−3977.807	309.971	−4101.796	535.626
	500	284.29	−3950.030	371.895	−4135.978	432.072
	600	292.88	−3921.139	424.545	−4175.866	363.532
	700	298.69	−3891.545	470.154	−4220.653	314.941
	800	303.02	−3861.450	510.333	−4269.717	278.777
	900	306.48	−3830.970	546.230	−4322.577	250.870
	1000	309.42	−3800.172	578.676	−4378.848	228.722
	1100	312.00	−3769.099	608.290	−4438.218	210.748
	1200	314.34	−3737.780	635.540	−4500.427	195.894
	1300	316.52	−3706.235	660.787	−4565.259	183.430
	1400	318.58	−3674.479	684.320	−4632.528	172.837
	1500	320.54	−3642.523	706.367	−4702.074	163.737

Table A.24 $3CaO \cdot 2TiO_2$—cont'd

Phase	T	c_p	H	S	G	B
	1600	322.44	−3610.373	727.116	−4773.758	155.843
	1700	324.29	−3578.036	746.719	−4847.459	148.941
	1800	326.09	−3545.517	765.306	−4923.068	142.860
	1900	327.86	−3512.819	782.985	−5000.490	137.470
	2000	329.61	−3479.945	799.846	−5079.638	132.663
	2013	329.83	−3475.659	801.982	−5090.050	132.077

Table A.25 $4CaO \cdot 3TiO_2$

Phase	T	c_p	H	S	G	B
Solid						
	298	337.76	−5671.663	328.444	−5769.589	1010.782
	300	338.94	−5671.037	330.537	−5770.198	1004.655
	400	381.17	−5634.745	434.653	−5808.607	758.507
	500	401.87	−5595.490	522.163	−5856.572	611.816
	600	414.11	−5554.645	596.598	−5912.604	514.725
	700	422.34	−5512.800	661.087	−5975.561	445.891
	800	428.44	−5470.247	717.899	−6044.567	394.660
	900	433.30	−5427.152	768.652	−6118.949	355.125
	1000	437.40	−5383.612	814.522	−6198.135	323.749
	1100	440.98	−5339.690	856.382	−6281.710	298.286
	1200	444.23	−5295.427	894.894	−6369.300	277.241
	1300	447.24	−5250.851	930.571	−6460.594	259.583
	1400	450.07	−5205.985	963.820	−6555.333	244.576
	1500	452.77	−5160.842	994.964	−6653.288	231.682
	1600	455.37	−5115.434	1024.269	−6754.264	220.499
	1700	457.90	−5069.770	1051.951	−6858.088	210.718
	1800	460.37	−5023.857	1078.194	−6964.607	202.103
	1900	462.79	−4977.699	1103.150	−7073.684	194.464
	2000	465.17	−4931.301	1126.948	−7185.198	187.653
	2078	467.00	−4894.946	1144.780	−7273.800	182.837

Table A.26 Calcium Oxide, CaO

Phase	T	c_p	H	S	G	B
Solid						
	298	10.315	−151.6	9.5	−154.432	113.217
	300	10.340	−151.581	9.564	−154.450	112.532
	400	11.254	−150.495	12.680	−155.568	85.010
	500	11.736	−149.344	15.248	−155.968	68.620
	600	12.047	−148.154	17.417	−158.604	57.779
	700	12.277	−146.937	19.292	−160.441	50.099
	800	12.465	−145.700	20.944	−162.455	44.387
	900	12.627	−144.445	22.422	−164.624	39.982
	1000	12.774	−143.175	23.760	−166.935	36.488
	1100	12.911	−141.891	24.984	−169.373	33.656
	1200	13.041	−140.593	26.113	−171.928	31.317
	1300	13.166	−139.283	27.162	−174.593	29.356
	1400	13.287	−137.960	28.142	−177.358	27.691
	1500	13.406	−136.625	29.062	−180.219	26.261
	1600	13.523	−135.279	29.931	−183.169	25.023
	1700	13.639	−133.921	30.755	−186.204	23.941
	1800	13.753	−132.551	31.538	−189.319	22.990
	1900	13.866	−131.170	32.284	−192.510	22.147
	2000	13.978	−129.778	32.998	−195.774	21.396
	2100	14.090	−128.374	33.683	−199.109	20.724
	2200	14.202	−126.960	34.341	−202.510	20.120
	2300	14.313	−125.534	34.975	−205.976	19.575
	2400	14.423	−124.097	35.585	−209.504	19.081
	2500	14.533	−122.649	35.177	−213.093	18.631
	2500	14.643	−121.191	36.749	−216.739	18.221
	2700	14.753	−119.721	37.304	−220.442	17.846
	2800	14.853	−118.240	37.843	−224.199	17.502
	2888	14.959	−116.928	38.304	−227.550	17.222

Table A.26 Calcium Oxide, CaO—cont'd

Phase	T	c_p	H	S	G	B
Liquid						
	2888	15.000	−97.928	44.883	−227.550	17.222
	2900	15.000	−97.748	44.945	−228.089	17.192
	3000	15.000	−96.248	45.454	−232.509	16.948
	3100	15.000	−94.748	45.945	−237.179	16.723
	3200	15.000	−93.248	45.422	−241.798	16.516
	3300	15.000	−91.748	45.883	−246.463	16.325
	3400	15.000	−90.248	47.331	−251.174	16.147
	3500	15.000	−88.748	47.766	−255.929	15.983

Table A.27 Magnesium Oxide, MgO

Phase	T	c_p	H	S	G	B
Solid						
	298	8.855	−143.7	6.44	−145.620	106.757
	300	8.895	−143.684	6.495	−145.632	106.107
	400	10.299	−142.714	9.273	−146.424	80.013
	500	10.989	−141.647	11.653	−147.473	64.469
	600	11.398	−140.526	13.696	−148.743	54.187
	700	11.675	−139.371	15.475	−150.204	46.902
	800	11.881	−138.193	17.048	−151.831	41.484
	900	12.045	−136.997	18.457	−153.608	37.306
	1000	12.185	−135.785	19.733	−155.518	33.993
	1100	12.307	−134.560	20.900	−157.551	31.307
	1200	12.418	−133.324	21.976	−159.695	29.088
	1300	12.522	−132.077	22.974	−161.943	27.229
	1400	12.619	−130.820	23.906	−164.288	25.650
	1500	12.712	−129.553	24.780	−156.73	24.295
	1000	12.802	−128.277	25.603	−169.242	23.121
	1700	12.889	−126.993	26.382	−171.842	22.095

Continued

Table A.27 Magnesium Oxide, MgO—cont'd

Phase	T	c_p	H	S	G	B
	1800	12.974	−125.700	27.121	−174.517	21.192
	1900	13.058	−124.398	27.825	−177.265	20.393
	2000	13.141	−123.088	28.496	−180.081	19.681
	2100	13.222	−121.770	29.140	−182.963	19.044
	2200	13.303	−120.444	29.757	−185.908	18.471
	2300	13.383	−119.109	30.350	−188.914	17.953
	2400	13.462	−117.767	30.921	−191.977	17.484
	2500	13.541	−116.417	31.472	−195.095	17.058
	2600	13.619	−115.059	32.005	−198.271	16.668
	2700	13.697	−113.693	32.520	−201.497	16.312
	2800	13.775	−112.320	33.020	−204.775	15.986
	2900	13.852	−110.938	33.504	−208.101	15.685
	3000	13.930	−109.549	33.975	−211.47S	15.408
	3098	14.005	−108.180	34.424	−214.827	15.157
Liquid	3098	14.500	−89.680	40.396	−214.827	15.157
	3100	14.500	−89.651	40.405	−214.907	15.153
	3200	14.500	−88.201	40.866	−218.971	14.957
	3300	14.500	−86.751	41.312	−223.080	14.776
	3400	14.500	−85.301	41.745	−227.233	14.608
	3500	14.500	−83.851	42.165	−231.429	14.453
	3533	14.500	−83.373	42.301	−232.822	14.404

Table A.28 Calcium Magnesium Dioxide, CaO·MgO

Phase	T	c_p	H	S	G	B
Solid						
	298	19.021	−297.1	15.85	−301.826	221.274
	300	19.085	−297.065	15.968	−301.855	219.931
	400	21.387	−295.026	21.817	−303.753	165.985

Table A.28 Calcium Magnesium Dioxide, CaO·MgO—cont'd

Phase	T	c_p	H	S	G	B
	500	22.551	−292.824	26.727	−306.187	133.852
	600	23.267	−290.530	30.906	−309.074	112.595
	700	23.771	−288.177	34.532	−312.350	97.533
	800	24.163	−285.780	37.733	−315.966	86.330
	900	24.489	−283.347	40.598	−319.885	77.689
	1000	24.774	−280.883	43.194	−324.077	70.836
	1100	25.033	−278.393	45.567	−328.517	65.279
	1200	25.273	−275.878	47.756	−333.184	60.689
	1300	25.501	−273.339	49.788	−338.063	56.841
	1400	25.720	−270.778	51.685	−343.137	53.573
	1500	25.931	−268.195	53.467	−348.396	50.768
	1600	26.138	−265.592	55.147	−353.827	48.337
	1700	26.340	−262.968	56.738	−359.422	46.213
	1800	26.539	−260.324	58.249	−365.172	44.344

Table A.29 2-Calcium 3-Silicate 5/2-Hydrate, $2CaO·3SiO_2·5/2H_2O$

Phase	T	c_p	H	S	G	B
Solid						
	298	70.550	−1175.7	64.9	−1185.050	876.113
	300	70.860	−1175.569	65.337	−1195.170	870.798
	400	83.021	−1167.814	87.564	−1202.840	657.239
	500	90.600	−1159.111	106.953	−1212.888	530.093
	600	96.375	−1149.753	123.999	−1224.152	445.957
	700	101.298	−1139.864	139.233	−1237.327	386.363
	800	105.768	−1129.508	153.055	−1251.952	342.063
	900	109.973	−1118.719	165.757	−1267.901	307.930
	1000	114.015	−1107.519	177.555	−1285.073	280.890

Table A.30 Magnesium Carbonate, $MgCO_3$

Phase	T	c_p	H	S	G	B
Solid						
	298	18.055	−262.0	15.7	−266.681	195.509
	300	18.138	−261.967	15.812	−266.710	194.324
	400	21.540	−259.968	21.537	−268.583	146.767
	500	23.856	−257.693	26.604	−270.995	118.468
	600	25.744	−255.211	31.125	−273.886	99.776
	700	27.431	−252.551	35.222	−277.206	86.559
	800	29.010	−249.728	38.989	−280.919	16.754

Table A.31 Calcium Magnesium Carbonite (Dolomite), $CaCO_3 \cdot MgCO_3$

Phase	T	c_p	H	S	G	B
Solid						
	298	37.262	−556.	28.2	−564.408	413.778
	300	37.369	−555.931	28.431	−564.460	411.264
	400	41.801	−551.955	39.840	−567.890	310.323
	500	44.886	−547.614	49.513	−572.370	250.217
	600	47.441	−542.995	57.928	−577.751	210.474
	700	49.745	−538.134	65.416	−583.925	182.334
	800	51.916	−533.050	72.202	−590.811	161.424
	900	54.009	−527.753	78.438	−598.347	145.318
	1000	56.054	−522.250	84.234	−606.484	132.565
	1100	58.068	−516.543	89.672	−615.182	122.242
	1200	60.060	−510.637	94.810	−624.409	113.736

Table A.32 Iron Oxide (Wutsite), FeO

Phase	T	c_p	H	S	G	B
Solid						
	298	48.04	−265.955	59.409	−283.668	49.696
	300	48.10	−265.867	59.706	−283.778	49.409
	400	50.34	−260.935	73.879	−290.486	37.933

Table A.32 Iron Oxide (Wutsite), FeO—cont'd

Phase	T	c_p	H	S	G	B
	500	51.83	−255.823	85.279	−298.462	31.179
	600	53.02	−250.579	94.836	−307.481	26.768
	700	54.07	−245.224	103.088	−317.386	23.683
	800	55.04	−239.769	110.372	−328.066	21.420
	843	55.45	−237.393	113.264	−332.875	20.625
	843	55.45	−237.393	113.264	−332.875	20.625
	900	55.97	−234.218	116.909	−339.436	19.700
	1000	56.88	−228.575	122.853	−351.429	18.356
	1100	57.76	−222.843	128.316	−363.991	17.284
	1200	58.64	−217.023	133.380	−377.079	16.413
	1300	59.51	−211.115	138.108	−390.655	15.696
	1400	60.37	−205.121	142.549	−404.691	15.099
	1500	61.22	−199.042	146.743	−419.157	14.596
	1600	62.08	−192.877	150.722	−434.032	14.169
	1645	62.46	−190.075	152.449	−440.854	13.998

Table A.33 Iron Oxide (Magnetite), Fe_3O_4

Phase	T	c_p	H	S	G	B
Solid						
	300	152.15	−1115.198	147.171	−1159.349	201.856
	400	172.35	−1098.973	193.708	−1176.456	153.626
	500	192.54	−1080.729	234.335	−1197.896	125.140
	600	212.74	−1060.464	271.225	−1223.200	106.486
	700	232.94	−1038.181	305.536	−1252.056	93.427
	800	253.13	−1013.877	337.959	−1284.244	83.850
	900	273.33	−987.554	368.940	−1319.600	76.586
	900	200.83	−987.554	368.940	−1319.600	76.586
	1000	200.83	−967.470	390.100	−1357.570	70.910
	1100	200.83	−947.387	409.241	−1397.553	66.363

Continued

Table A.33 Iron Oxide (Magnetite), Fe₃O₄—cont'd

Phase	T	c_p	H	S	G	B
	1200	200.83	−927.304	426.716	−1439.363	62.652
	1300	200.83	−907.221	442.791	−1482.849	59.580
	1400	200.83	−887.138	457.674	−1527.882	57.005
	1500	200.83	−867.054	471.530	−1574.350	54.822
	1600	200.83	−846.971	484.492	−1622.158	52.957
	1700	200.83	−826.888	496.667	−1671.222	51.349
	1800	200.83	−806.805	508.146	−1721.468	49.954
	1870	200.83	−792.747	515.808	−1757.308	49.086

Table A.34 Silicon Dioxide (Quartz), SiO₂

Phase	T	c_p	H	S	G	B
Solid A						
	298	10.660	−217.7	9.91	−220.655	161.766
	300	10.709	−217.680	9.976	−220.673	160.782
	400	12.761	−216.499	13.361	−221.843	121.226
	500	14.209	−215.147	16.371	−223.333	97.632
	600	15.420	−213.665	19.071	−225.107	82.006
	700	16.518	−212.067	21.531	−227.139	70.925
	800	17.556	−210.363	23.805	−229.407	62.680
	847	18.031	−209.527	24.821	−230.550	59.496
Solid B						
	847	16.113	−209.353	25.026	−230.550	59.496
	900	16.240	−208.495	26.008	−231.902	56.321
	1000	16.480	−206.859	27.731	−234.591	51.277
	1100	16.720	−205.199	29.313	−237.444	47.182
	1200	16.900	−203.515	30.779	−240.450	43.798
	1300	17.200	−201.807	32.146	−243.597	40.958
	1400	17.440	−200.075	33.429	−246.876	38.544
	1500	17.680	−198.319	34.640	−250.280	36.471
	1600	17.920	−196.539	35.789	−253.802	34.672
	1696	18.150	−194.808	36.840	−257.288	33.159

Table A.35 Silicon Dioxide (Cristobalite), SiO$_2$

Phase	T	c_p	H	S	G	B
Solid A						
	298	10.740	−217.1	10.372	−220.192	161.427
	300	10.788	−217.080	10.439	−220.212	160.446
	400	12.713	−215.897	13.830	−221.429	120.999
	500	14.009	−214.558	16.813	−222.964	97.471
	543	14.479	−213.945	17.988	−223.712	90.053
Solid B						
	543	14.198	−213.624	18.579	−223.712	90.053
	600	14.797	−212.797	20.027	−224.813	81.899
	700	15.530	−211.278	22.367	−226.935	70.862
	800	16.022	−209.699	24.475	−229.279	62.644
	900	16.373	−208.078	26.383	−231.823	56.302
	1000	16.637	−206.427	28.123	−234.550	51.268
	1100	16.845	−204.752	29.719	−237.443	47.182
	1200	17.013	−203.059	31.192	−240.489	43.805
	1300	17.154	−201.351	32.559	−243.678	40.971
	1400	17.276	−199.629	33.835	−246.998	38.563
	1500	17.382	−197.896	35.031	−250.442	36.494
	1600	17.478	−196.153	36.156	−254.002	34.700
	1700	17.564	−194.401	37.218	−257.671	33.130
	1800	17.644	−192.640	38.224	−261.444	31.748
	1900	17.719	−190.872	39.180	−265.314	30.522
	1996	17.787	−189.168	40.055	−269.118	29.471
Liquid						
	1996	20.500	−186.878	41.202	−269.118	29.471
	2000	20.500	−186.796	41.244	−269.283	29.430
	2100	20.500	−184.746	42.244	−273.458	28.463
	2200	20.500	−182.696	43.197	−277.730	27.594
	2300	20.500	−180.646	44.109	−282.096	26.809
	2400	20.500	−178.596	44.981	−286.550	20.097
	2500	20.500	−176.546	45.818	−291.091	25.451

Continued

Table A.35 **Silicon Dioxide (Cristobalite), SiO_2—cont'd**

Phase	T	c_p	H	S	G	B
	2600	20.500	−174.496	46.622	−295.713	24.860
	2700	20.500	−172.446	47.396	−300.414	24.320
	2800	20.500	−170.396	48.141	−305.191	23.824
	2900	20.500	−168.346	48.861	−310.041	23.368
	3000	20.500	−166.296	49.556	−314.962	22.948

Table A.36 **Magnesium Metasilicate, $MgO \cdot SiO_2$**

Phase	T	c_p	H	S	G	B
Solid 1						
	298	19.586	−370.20	16.2	−375.030	274.941
	300	19.659	−370.164	16.321	−375.060	273.268
	400	22.523	−368.040	22.412	−377.005	206.013
	500	24.271	−365.695	27.538	−379.514	155.908
	600	25.580	−353.200	32.183	−382.509	139.348
	700	26.682	−360.585	36.211	−385.933	120.510
	800	27.672	−357.867	39.839	−389.738	106.486
	900	28.599	−355.053	43.152	−393.890	95.662
	903	28.626	−354.967	43.248	−394.020	95.376
Solid 2						
	903	28.752	−354.807	43.425	−394.020	S5.376
	1000	28.762	−352.017	46.360	−398.377	87.077
	1100	28.762	−349.141	49.101	−403.152	80.110
	1200	28.762	−346.265	51.503	−408.189	74.351
	1258	28.762	−344.597	52.961	−411.222	71.450
Solid 3						
	1258	29.262	−344.207	53.271	−411.222	71.450
	1300	29.262	−342.978	54.232	−413.479	69.522
	1400	29.262	−340.051	56.401	−419.012	55.420
	1500	29.262	−337.125	58.420	−424.755	51.895

Table A.36 Magnesium Metasilicate, $MgO \cdot SiO_2$—cont'd

Phase	T	c_p	H	S	G	B
	1600	29.262	−334.199	60.308	−430.692	58.838
	1700	29.262	−331.273	62.082	−436.812	56.164
	1800	29.262	−328.347	53.755	−443.105	53.808
	1850	29.262	−326.884	54.556	−445.313	52.732
Liquid						
	1850	35.000	−308.884	74.286	−446.313	52.732
	1900	35.000	−307.134	75.219	−450.051	51.775
	2000	35.000	−303.634	77.015	−457.663	50.018
	2100	35.000	−300.134	78.722	−465.451	48.447
	2200	35.000	−296.634	80.351	−473.405	47.035
	2300	35.000	−293.134	81.906	−481.518	45.761
	2400	35.000	−289.634	83.396	−489.764	44.507
	2500	35.000	−286.134	84.825	−498.195	43.558
	2600	35.000	−282.634	86.197	−506.747	42.602
	2700	35.000	−279.134	87.518	−515.433	41.727
	2800	35.000	−275.634	88.791	−524.249	40.925

Table A.37 Magnesium Orthosilicate, $2MgO \cdot SiO_2$

Phase	T	c_p	H	S	G	B
Solid						
	298	28.126	−520.3	22.75	−527.083	386.414
	300	28.263	−520.248	22.924	−527.125	384.062
	400	33.302	−517.138	31.838	−529.873	289.548
	500	35.938	−513.664	39.578	−533.453	233.203
	600	37.628	−509.981	46.289	−537.154	195.903
	700	38.872	−506.153	52.186	−542.684	169.456
	800	39.818	−502.214	57.445	−548.110	149.773
	900	40.745	−498.182	62.193	−554.156	134.585
	1000	41.527	−494.068	66.527	−560.595	122.534

Continued

Table A.37 **Magnesium Orthosilicate, $2MgO \cdot SiO_2$—cont'd**

Phase	T	c_p	H	S	G	B
	1100	42.252	−489.878	70.519	−567.450	112.757
	1200	42.939	−485.619	74.225	−574.689	104.679
	1300	43.599	−481.292	77.688	−582.286	97.904
	1400	44.240	−476.899	80.943	−590.220	92.150
	1500	44.866	−472.444	84.017	−598.469	81.209
	1600	45.481	−467.927	86.932	−607.018	82.926
	1700	46.088	−463.348	89.707	−615.851	79.184
	1800	46.688	−458.709	92.359	−624.955	75.890
	1900	47.282	−454.011	94.899	−634.319	12.973
	2000	47.873	−449.253	97.339	−643.932	70.375
	2100	48.460	−444.436	99.689	−653.784	68.049
	2171	48.875	−440.981	101.307	−660.919	66.542
Liquid						
	2171	49.000	−423.981	109.138	−660.919	66.542
	2200	49.000	−422.560	109.788	−664.094	65.981
	2300	49.000	−417.660	111.966	−675.182	64.166
	2400	49.000	−412.760	114.052	−686.484	62.521
	2500	49.000	−407.860	116.052	−697.990	61.026
	2600	49.000	−402.960	117.914	−709.692	59.663
	2700	49.000	−398.060	119.823	−721.582	58.416
	2800	49.000	−393.160	121.605	−733.654	57.272
	2900	49.000	−388.260	123.325	−145.901	56.220
	3000	49.000	−383.360	124.986	−158.317	55.251

Table A.38 **Serpentine, $3MgO \cdot 2SiO_2 \cdot 2H_2O$**

Phase	T	c_p	H	S	G	B
Solid						
	298	65.465	−1043.04	53.1	−1058.872	776.278
	300	65.767	−1042.919	53.506	−1058.970	771.563
	400	77.472	−1035.696	74.205	−1015.378	582.174
	500	84.588	−1027.571	92.306	−1073.724	469.387

Table A.38 Serpentine, $3MgO \cdot 2SiO_2 \cdot 2H_2O$—cont'd

Phase	T	c_p	H	S	G	B
	600	89.897	−1018.837	108.215	−1083.766	394.814
	700	94.352	−1009.619	122.415	−1095.310	342.017
	800	98.353	−999.981	135.279	−1108.204	302.788
	900	102.090	−989.957	147.081	−1122.330	272.576
	1000	105.662	−979.569	158.023	−1137.592	248.654

Table A.39 Anthophyllite, $7MgO \cdot 8SiO_2 \cdot H_2O$

Phase	T	c_p	H	S	G	B
Solid						
	298	150.332	−2888.72	133.6	−2928.553	2146.976
	300	151.119	−2888.441	134.532	−2928.801	2133.917
	400	179.952	−2871.706	182.479	−2944.698	1609.124
	500	195.130	−2852.886	224.411	−2965.092	1296.215
	600	204.932	−2832.854	260.907	−2989.399	1089.034
	700	212.197	−2811.983	293.066	−3017.129	942.117
	800	218.108	−2790.459	321.798	−3047.898	832.759
	900	223.233	−2768.387	347.769	−3081.398	748.366
	1000	227.870	−2745.829	371.552	−3117.381	681.395
	1100	232.188	−2722.824	393.475	−3158.847	627.054

Table A.40 Nickel Carbide, Ni_3C

Phase	T	c_p	H	S	G	B
Solid						
	298	25.491	9.00	25.4	1.427	−1.046
	300	25.500	9.047	25.558	1.380	−1.005
	400	26.000	11.622	32.962	−1.563	0.854
	500	26.500	14.241	38.818	−5.162	2.256
	600	27.000	16.922	43.693	−9.294	3.386
	700	27.500	19.647	47.893	−13.878	4.333
	800	28.000	22.422	51.598	−18.856	5.152

Table A.41 Nickel Carbonate, NiCO$_3$

Phase	T	c_p	H	S	G	B
Solid						
	298	21.454	−162.5	20.4	−168.582	123.591
	300	21.512	−162.460	20.533	−168.620	122.856
	400	23.876	−160.181	27.075	−171.011	93.448
	500	25.470	−157.710	32.582	−174.001	76.066
	600	26.761	−155.096	37.343	−177.502	64.664
	700	27.908	−152.362	41.556	−181.451	56.659

Table A.42 Nickel Monoxide, NiO

Phase	T	c_p	H	S	G	B
Solid A						
	298	10.591	−57.5	9.1	−60.213	44.143
	300	10.606	−57.480	9.166	−60.230	43.883
	400	12.473	−56.340	12.434	−61.313	33.505
	500	15.356	−54.953	15.516	−62.711	27.415
	525	16.151	−54.560	16.284	−63.109	26.275
Solid B						
	525	13.880	−54.560	16.284	−63.109	26.275
	565	13.880	−54.004	17.303	−63.781	24.675
Solid C						
	565	12.321	−54.004	17.303	−63.781	24.675
	600	12.392	−53.572	18.046	−64.399	23.461
	700	12.594	−52.323	19.971	−66.302	20.703
	800	12.796	−51.053	21.666	−68.386	18.685
	900	12.998	−49.763	23.185	−70.630	17.154
	1000	13.200	−48.453	24.565	−73.018	15.960
	1100	13.402	−47.123	25.832	−75.539	15.010
	1200	13.604	−45.773	27.007	−78.182	14.241
	1300	13.806	−44.403	28.104	−80.938	13.609

Table A.42 Nickel Monoxide, NiO—cont'd

Phase	T	c_p	H	S	G	B
	1400	14.008	−43.012	29.135	−83.800	13.084
	1500	14.210	−41.601	30.108	−86.763	12.643
	1600	14.412	−40.170	31.032	−89.820	12.271
	1700	14.614	−38.719	31.911	−92.968	11.953
	1800	14.816	−37.247	32.752	−96.201	11.682
	1900	15.018	−35.755	33.559	−99.517	11.449
	2000	15.220	−34.243	34.334	−102.912	11.247
	2100	15.422	−32.711	35.082	−106.383	11.073
	2200	15.624	−31.159	35.80 ft	−109.927	10.922
	2257	15.739	−30.265	36.205	−111.980	10.845

Table A.43 Zinc Carbonate, ZnCO₃

Phase	T	c_p	H	S	G	B
Solid						
	298	19.139	−194.2	19.7	−200.074	146.578
	300	19.200	−194.165	19.819	−200.110	145.800
	400	22.500	−192.080	25.794	−202.397	110.500
	500	25.800	−189.655	31.169	−205.249	89.726

Table A.44 Zinc Oxide, ZnO

Phase	T	c_p	H	S	G	B
Solid						
	298	9.621	−83.2	10.4	−86.301	63.269
	300	9.654	−83.182	10.460	−86.320	62.893
	400	10.835	−82.150	13.421	−87.518	47.824
	500	11.448	−81.033	15.910	−88.988	38.902
	600	11.836	−79.868	18.034	−90.688	33.038
	700	12.119	−78.669	19.881	−92.586	28.911
	800	12.345	−77.446	21.514	−94.657	25.863

Continued

Table A.44 Zinc Oxide, ZnO—cont'd

Phase	T	c_p	H	S	G	B
	900	12.539	−76.201	22.980	−96.883	23.530
	1000	12.712	−74.939	24.310	−99.249	21.694
	1100	12.872	−73.659	25.529	−101.742	20.217
	1200	13.023	−72.365	26.656	−104.352	19.008
	1300	13.167	−71.055	27.704	−107.070	18.003
	1400	13.307	−69.731	28.685	−109.890	17.157
	1500	13.443	−68.394	29.607	−112.805	16.438
	1600	13.577	−67.043	30.479	−115.810	15.821
	1700	13.709	−65.679	31.306	−118.899	15.288
	1800	13.839	−64.301	32.094	−122.070	14.823
	1900	13.968	−62.911	32.845	−125.317	14.417
	2000	14.095	−61.508	33.565	−128.638	14.059
	2100	14.223	−60.092	34.256	−132.029	13.742
	2200	14.349	−58.663	34.920	−135.488	13.461
	2243	14.403	−58.045	35.199	−136.996	13.350

Index

Note: Page numbers followed by "f" and "t" indicate figures and tables respectively.

A

acid-alkali modeling, 128–129
acid rain, 117
active layer thickness, 55–57, 78f
active layer velocity, 69, 74–76, 77f, 78
aggregate, lightweight (LWA), 2, 6t, 7–8, 118, 119f, 231, 247–255
  feedstock mineralogy, 250–252
  raw material characterization, 250
  thermal history, 252–255
aggregate kilns, 35, 35t
air
  composition, 208, 316t
  infiltration, 27–28, 122, 258
  at low pressure (dry), 318t–320t
  water percentage calculations, 213, 219t
air-fuel ratio, 113–114, 113f, 128
air requirements for combustion
  acid-alkali modeling, 128–129
  determining, 112
  pulverized coal or coke, 118–121, 142–144
  sources of, 110–111
air velocities, typical values of, 35, 35t
alpha-aluminum oxide (Al_2O_3), 334t–335t
Alsop, P.A., 303
aluminum disilicate ($Al_2O_3 \cdot 2SiO_2$), 336t
aluminum disilicate dihydrate ($Al_2O_3 \cdot 2SiO_2 \cdot 2H_2O$), 336t
aluminum silicate ($Al_2O_3 \cdot SiO_2$), 335t
annular shift kiln, 3, 4f
anode-grade coke, 265
anthophyllite ($7MgO \cdot 8SiO_2 \cdot H_2O$), 355t
Arrhenius equation, 33, 127
Ash Grove Cement Plant, 41–42
Aspdin, Joseph, 1
atmosphere
  international standard measurements, 332t–333t
  units of conversion
    normal, 314
    technical, 314
atmospheric discharge, 27–28
atomization, 123–126
axial motion, 21–24
axial segregation, 86

B

Bac, N., 164–165
Bagnold, R. A., 43–45, 52, 67–68
banding, 86
Barker, D. J., 305
Barr, P. V., 54–56, 165, 167–168
Becher process, 231, 243–247, 246f
Becker, H. A., 38
bed, *See also specific types*
  depth, determining, 24–25
  heat transfer
    flame to bed, 164
    introduction, 145
    paths, 14–15, 14f
    rotating bed mode, 181–182
    wall to bed, 166–167, 174–177
  segregation model, 91–94
  thermal model, 184–187
bed behavior diagram, 19, 20f
bed motion
  modes of, 17f
  transverse, 16–19
bed phenomenon, 13–15
$2CaO \cdot 3SiO_2 \cdot 5/2H_2O$, 347t
$2CaO \cdot V_2O_5$, 341t
Blasius, H., 31–32
Blasius problem, 65–66
bloating, 248–249, 252–253
  agents, 251
  temperature, 196–198
Boateng, A. A., 54–57, 165

Boiler and Industrial Furnace (BIF), 107–108, 291
boiler and industrial furnace act (EPA), 117
Boudouard reaction, 244–245
Boynton, R. S., 236–237
Brimacombe, J. K., 245

C

calcination, 209, 209f
  heat consumed by, 218, 226t
  limestone dissociation, 232–236
calcination calculation, 209, 217t
calcined petroleum coke (CPC), 265
calciners, fluidized-bed, 3–4, 4f
calcining
  coke calcining, 269–270, 272
  pet-coke calcining, 265, 268–269
  rotary calcining, 266
calcium carbonate ($CaCO_3$), 338t
calcium magnesium carbonite ($CaCO_3 \cdot MgCO_3$), 348t
calcium oxide (CaO), 344t–345t
  Cr_2O_3, 340t
  MgO, 346t–347t
  MoO_3, 338t–339t
  WO_3, 339t
  V_2O_5, 342t
Campbell, C. S., 57
carbon black, 118
carbon capture and storage (CCS) technologies, 304
carbon dioxide (CO_2), 110, 114–115, 190
carbon emissions, reduction, and capture, 304–306
  oxycombustion cement plant, 304–305
  cement plant CO_2 limitations, in the United States, 305–306
carbon monoxide (CO), 110
  modeling, 135
carbothermic reduction, 243–244
  roasting of titaniferous Materials
  rotary kiln SL/RN process, 244–245
Carnot cycle, 204
cascading bed, 16–17, 17f, 19, 20f
cataracting bed, 16–17, 17f
cement, history of, 1–2
cement kilns
  air infiltration, 27–28, 38
  air velocities, 35, 35t

Craya-Curtet parameter, 38, 129
  energy usage, 242–243
  fuel types used, 117
  history of, 1–2
  schematic, 238, 240f
cement process chemistry, 238
  decomposition zone, 240
  sintering zone, 241–242
  transition zone, 240–241
cenosphere, 123–124
centrifuging bed, 16–17, 17f, 19, 20f
CFD, *See* computational fluid dynamics (CFD)
char combustion, 122–124, 130–132
clay, 247–253
  metal oxide composition of, 251t
  densityetemperature curves for, 253f
coal
  combustion chemistry, 110–111
  combustion requirements, 142–144
  conveying air, 210t, 221t–222t
  delivery and firing systems, 120–121
  firing in kilns, 118–119, 129
  heat transfer, 165–166
  sensible heat calculations, 221t–222t
  tires combined with, 117
  types of, 115–116
Cohen, E. S., 165
coke, 116–117
  firing in kilns, 118–119
combustion, *See also specific types, e.g.* char combustion
  air requirements for
    acid-alkali modeling, 128–129
    determining, 112
    pulverized coal or coke, 118–121, 142–144
    sources of, 110–111
  CFD modeling, 129–131
  heat release calculations, 225t
  introduction, 107–108
  energy balance module, 213, 221t–222t
  mole and mass fractions of intended fule, 108–110
  practical stoichiometry, 112
  radiative effect of combustion gases, 164
combustion chemistry, 110–111
combustion efficiency, 27, 39–40, 136–140, 286–289

Index   361

combustion modeling
   acid-alkali, 128—129
   Hawthrone method, 128
   overview, 126—128
combustion systems, dual use, 301—303
   municipal solid waste and power
      generation, 301—303
comprehensive performance test (CPT), 298
computational fluid dynamics (CFD)
   modeling rotary kiln processes, 129—131
      gas-phase conservation equations used
         in, 131—132
   modeling, 285—289
      aerodynamics, 286—289, 287f
      afterburner, 285—286, 286f
      combustion, 286—289, 288f
   pulverized fuel burner evaluation, 135—140
concentration transport equation, 33, 127
concrete, 2, 247—249
   replacement with LWA structural concrete,
      249f
concrete ships, 247—248
confined flame theory, 38
confined jet-free jet relation, 38
confined jets, 27—28, 37—39, 129
continuity equation, 33, 64, 74, 127,
   131, 190
conversion factors, SI to British, 312—313,
   317t
coolers, 6—9, 9f
Couette flow, 28—30, 29f
countercurrent flow rotary kiln, 2—3, 3f
cracking, 41—42, 116
Craya, A., 38
Craya-Curtet parameter, 38—39, 122,
   127—129, 135—140, 136t, 238
cristobalite, 351t—352t
Curtet, R., 38

D

Damkohler I and II, 34
Davies, T. R. H., 50
destruction and removal efficiency (DRE),
   295—299
   flame turbulence and, 299
direct-fired kiln
   characteristics, 266—269
   fixed carbon recovery and burnout,
      271—272

key points, 276—278
kiln dimensions, 269t
direct-fired pet-coke kiln temperature
   profiles, 274—278
dirty fans, 46
dolomite ($CaCO_3 \cdot MgCO_3$), 211t, 231, 233,
   235—236, 238, 241—242, 348t
dryers, 6—9
dry kilns, 6—7
   long dry kilns, 7—8
   short dry kilns, 8
dumbell-type rotary kiln, 5—6, 6f
dust, site survey, 209, 212t
dust constituents, 218, 224t
dust entrainment, 43—46, *See also*
   entrainment; saltation
dust output and pickup, 271

E

Edison, Thomas A., 1—2
efficiency, thermal, 206—207, 209
Ehlers, E. G., 251
emissions modeling
   modeling of nitric oxide, 133—135
      fuel NO_x, 133—134
      prompt NO, 135
      thermal NO, 134—135
   modeling of carbon monoxide, 135
emissivity, 159—160, 164, 179
   gas emissivity, 167
   monochromatic emissivity, 160—161
   surface emmisivity, 161
energy cost savings, 118, 119f
energy savings, maximizing, 112
enthalpy transport equation, 33, 127
entrainment, dust, 43—46, *See also* dust
   entrainment
   jet properties and
      confined jets, 37—39
      precessing jets, 40—42
      particle-laden jet, 42—43
      swirling jets, 39—40
      turbulent jet, development of, 35—37
   saltation and, 43—46
      concept of particle saltation, 44f
Environmental Protection Agency (EPA)
   regulations, 291
environmental sustainability, 120—121
equation of state, 33, 127

F

Fan, L. T., 88
fans, *See specific types*
feedstock, 265
Ferron, J. R., 52, 58, 174–177
field kilns, 1–2
fixed carbon recovery and burnout, 271–272
flame shape and character
  air-fuel ratio in, 113, 113f
  Craya-Curtet parameters, 38, 129
  droplet fuel size and, 125
  heat dissipation and, 107–108
  heat flux and, 38
  luminosity-emissions relation, 41–42, 164
flame temperature, adiabatic, 113–114
flame theory, 38
flotsam, 90–91, 93, *See also* percolation
flow behavior, transverse, 20–21
flow visualization modeling, 128–129
flue gases, 111, 218, 223t
FLUENT (CFD software), 285–286
fluid flow
  laminar and turbulent, 29–32, 32f
  in pipes, 28–32
  sources of, 33–35
forced draft fans, 46
4-nine destruction efficiencies, 114–115
freeboard aerodynamics
  in dust entrainment, 27–28, 33, 43–46
  fluid flow in pipes, effect on, 28–32
  introduction, 27–28
  mixing and
    confined jets, 37–39
    precessing jets, 40–42
    particle-laden jet, 42–43
    swirling jets, 39–40
    turbulent jet, development of, 35–37
  total flow inducement, 46
freeboard gas flow, 280, 284t–285t
freeboard phenomenon-bed phenomena synergy, 13–14, 27–28
freeboard(s)
  dimensionless parameters of combustion, 34
  heat transfer
    among exposed surfaces, 166–167
    coefficients for radiation, 165–166
    to exposed bed and wall, 2–3, 3f, 13, 167–170
    introduction, 145–146
    to the wall, 155–159
  in indirect fired kilns, purging volatiles, 9
  thermal model, 184–187
free jet-confined jet relation, 38
free jets, 35–36
  turbulent free jet, 35–37, 36f
friction velocity, 40
fuel, *See also specific fuel types*
  atomization of liquid, 123–126
  combusion and, 107–108
  energy calculations
    of fuel, 213, 218t
    of natural gas, 213, 219t
  types used in rotary kilns, 107–108, 114–115
fuel-air systems, 113–114
fuel efficiency, 41–42, 112, 114–115, 128, 207–208
fuel heating value, 107–108, 110–111
fuel NO_x, 133–134
fuel oil firing, 123–126

G

gas components, natural and exit gas, 209, 212t
gaseous fuels, 108
  expected flame geometry for, 114, 114f
gases
  H, U, S, G, 315, 323t–331t
  radiative effect of combustion, 164
  specific heats, 314
gas flow rate, 27–28
gas-phase combustion, 131–132
Gauthier, C., 70–71
Gibbs free energy and entropy, 132, 204–206
grain inertia regime, 52
grain temperature, 53–54
granular flow
  defined, 49
  equations of motion, 52–54
  kinetic theory application to, 52
  mechanisms of momentum transfer, 51
  modeling, 49–51
  observed, in a rotary drum, 17, 20–21, 55–58

Index 363

other flow types compared, 50, 50f
theories governing, 51
granular temperature, 53–54, 71–72, 74
Greek nomenclature, 311–312
green coke, 116, 265–266, 268–270
green petroleum coke (GPC), 265
greenhouse gas (GHG) sources, 304
Guruz, H. K., 165
Gyro-Therm, 40–41

H

Hagen-Poiseuille equation of laminar flow, 30
Hawthrone method, 128
Hayde, Stephen J., 2, 247–248
hazardous waste incineration, 114–115, 299–301
  risk assessment, 301
heat balance, 203
  cement kiln, 242–243, 243t
  summary and analysis, 218, 228t
heat content, units of conversion, 314
heat exchange, cross section, 13, 14f
heat flux, 38, 41–42, 146
heat of reaction/heat of formation, 203–204
heat recuperators, 6–7
heat transfer
  in the bed, 145, 174–175, 178–182
  bed to wall, 166–167, 175–177
  combustive gases, radiative effect, 164
  conduction, 145–151, 178–182
  conduction–convection problems, 152–154
  conduction in the wall, 155–159
  convection, 145–146, 151–152
  estimating, 31–32
  exposed surfaces
    bed and wall surfaces, 166–167
    freeboard surfaces, 167–170
  flame to bed, 164
  in the fluid, 34
  methods overview, 145–146
  packed beds, 178–181
  radiant, 41–42
  radiation, 41–42, 114–115, 145, 159
    blackbody, concept of, 159–161
  radiation exchange, 163
  radiation shape factors, 161–162
  refractory lining materials and, 155

rotating bed mode, 181–182
shell losses, 154–155
wall to bed, 166–167, 175–177
heat transfer, freeboard
  among exposed surfaces, 167–170
  coefficients for radiation, 165–166
  to exposed bed and wall, 166–167
  introduction, 145
  to the wall, 155–159
heat transfer model
  results and application, 192–195
  single particle for shale processing, 196–200
  solution procedure, 189–192
Henein, H., 55–57, 94–95, 99
higher heating value (HHV), 111
horizontal kilns, 3, 3t, 5
Hottel, H. C., 164

I

ilmenite, 245–247
indirect fired kilns, 9
induced draft (ID) fan, 27–28, 46
industrial kilns, temperature profiles of, 272
infiltration air, 27–28, 35, 46
instantaneous jet, 41–42
international standard atmosphere, 332t–333t
iron oxide
  magnetite (Fe_3O_4), 349t–350t
  wutsite (FeO), 348t–349t
irradiation, 161

J

Jenkins, B. G., 128, 173
jets, *See* flame shape and character; *specific types*
jetsam, 90–91, *See also* percolation
jetsam loading-number concentration, 103–104

K

kaolinite ($Al_2O_3 \cdot 2SiO_2 \cdot 2H_2O$), 337t
kerosene, 123
kidney, 89
kiln design, 25, 45–46, 112, *See also specific types of kilns*

kiln geometry
  determining capacity, 23
  plug-flow estimations from, 24–25
  residence time relation, 15, 19, 22–23
kiln loading, 17–19
kiln sizing, 280, 283t
kiln throughput estimation, 279–280, 281t–282t
kiln wall
  heat conduction, 155–159
  heat transfer, 159–161
    freeboard, 165–170
    refractory lining materials and, 155
    shell losses, 154–155
    wall to bed, 166–167, 175–177
Kocaefe, D., 272–274

L

laminar jets, 36–37
Lewis number, 125–126
lightweight aggregate (LWA), 2, 247–255
  feedstock mineralogy, 250–252
  raw material characterization, 250
  thermal history, 252–255
lime kilns
  about, 236–238
  active layer, 49
  air velocities, 35t
  Craya-Curtet parameter, 129, 238
  history of, 1–2
  modern, 8
lime manufacturing, 231–232, 238
lime products, 218, 223t
limestone, 209, 211t, 213, 221t–222t
  dissociation, 232–236, See also limestone calcination
limestone calcination, 209
  heat consumed by, 218, 226t
limestone calcination kilns, 39
liquid burnable materials (LBM), 107–108, 125, 126t
long dry kilns, 7–8
long kilns, history of, 1–2
lower heating value (LHV), 111
Luminis Pty. Ltd., 40–41
Lun, C. K. K., 53, 68, 81–82
Luxton, R. E., 40–42

M

Mach number, 34
magnesium carbonate ($MgCO_3$), 348t
magnesium metasilicate ($MgO \cdot SiO_2$), 352t–353t
magnesium orthosilicate ($2MgO \cdot SiO_2$), 353t–354t
magnesium oxide (MgO), 345t–346t
magnetite (Fe_3O_4), 349t–350t
mass
  balance, 269–270
  units of conversion, 313
mass and energy balance
  calcination calculation, 209, 217t
  chemical compositions, 208
  combustion, 213, 218t
  energy balance inputs, 209, 209f
  global heat and material, 206–207
  introduction, 203
  mass balance inputs, 208
  sensible energy, input and output, 213, 218, 221t–228t
  shell heat loss, 209, 214t–216t
  site survey, measured variables, 209
    ambient air to the cooler, 210t
    coal conveying air, 210t
    cooling air to the burner, 210t
    dust and coal composition, 212t
    kiln dimensions and shell temperatures, 213t
    limestone, 211t
    natural gas and exit gas components, 212t
  thermal module for chemically reactive system, 207–208
metallurgy, extractive, 243–244
methane combustion, 110–111
microexplosions, 125–126
Millen, Thomas, 1
minerals process applications
  carbothermic reduction processes, 243–244
    rotary kiln SL/RN process, 244–245
    titaniferous materials, roasting of, 245–247
  cement kilns, 238
  cement process chemistry, 238
    decomposition zone, 240

Index 365

sintering zone, 241–242
transition zone, 240–241
introduction, 231
lightweight aggregate, 2, 247–255, *See also* lightweight aggregate (LWA)
lime manufacturing, 231–232, 238
limestone dissociation (calcination), 232–236
ore reduction processes, 243–244
rotary kiln SL/RN process, 244–245
titaniferous materials, roasting of, 245–247
rotary lime kiln, 236–238
mixing
atomization requirements, 123–126
in carbothermic processes, 245
Hawthrone method, 128
introduction, 85
jet properties and
confined jets, 37–39
precessing jets, 40–42
particle-laden jet, 42–43
swirling jets, 39–40
turbulent jet, development of, 35–37
modeling within the bed, 87–89
purpose of, 20
turbulent, 31, 127–128
Moles, F. D., 128, 164, 173
momentum equation, 127
Mullinger, P. J., 128
mullite ($3Al_2O_3 \cdot 2SiO_2$), 337t
municipal solid waste (MSW) and power generation, 301–303
rotary kiln dual use, 302–303

N
Nathan, G. J., 41–42
natural gas, 114–115, 164, 209, 212t, 213, 219t, 221t–222t
Navier-Stokes, 45–46, 61, 130–132, 151–152, 51
Newby, M. P., 38
Nicholson, T., 21–23
nickel carbide (Ni_3C), 355t
nickel carbonate ($NiCO_3$), 356t
nickel monoxide (NiO), 356t–357t
nitric oxide (NOx) emissions, 41–42, 133–134
nomenclature, 309–311

notation, variables, unit, and description, 309–311
nozzle tip velocities, 35

O
Owen, P. R., 44
oxycombustion cement plant, 304–305
flow diagram, 305, 305f
oxy-hemoglobin formation, 110

P
Parker, D. J., 24–25
particle-laden jets, 42–43, 118–119
particulate flow behavior in rotary kilns, 54–55, 132–133
particulate flow model
active layer
analytical expression for thickness of the, 72–74
momentum equation solution, 68–69
velocity profile in the, 70–71
application, 77–78
density and granular temperature profiles, 71–72
description, 59–60
introduction, 59–74
momentum conservation
governing equations, 61–64
integral equation, 64–68
momentum equation
numerical solution scheme, 74
solution in the active layer, 68–69
results and validation, 74–76
simplifying assumptions, 60
particulate matter detection system (PMDS), 294
Peclet number, 34
Peray, K. E., 238
percolation
defined, 86
spontaneous, 87, 89, 93–94, 99
percolation velocity, 88–89, 91–93, 95
Pershin, V. F., 49
pet coke, 265–274, 273f, 285–286
direct-fired, 274–278
pet-coke calcination, 265, 274
kiln incinerator, 285–286, 286f
petroleum coke, 116–117, *See also* pet coke
Phillips, E. L., 251

pipes, fluid flow in, 27–32
PJ burner, 41–42
plug flow estimations, 24–25
Polak, S. L., 116
pollution control, 27–28, 206–207, 304, *See also specific pollutants*
Portland cement, 1–2
pot kilns, 1–2
Prandtl number, 34, 36–37
precalciners, 8
precessing jets, 40–42
　jet nozzle, 41f
preheaters, 8, 8f
pressure, units of conversion, 314
principal organic hazardous constituent (POHC), 295–298
　thermal stability index, 295–297, 297t
prompt NO, 135
pulverized fuel
　burner evaluation using CFD, 135–140
　combustion systems
　　confined jets, 27–28, 37–39
　　particle-laden jets, 42–43
　delivery and firing systems, 120–121
　reaction kinetics of carbon particles, 122–123
　sources of, 115–116, 115t
pyroscrubber, 285–289
　petroleum coke-calcining kiln incinerator, 285–286, 286f

Q
quartz (SiO_2), 350t

R
radiosity, 161
Ransome, F., 1
reacting flows, multicomponent, basic equations, 33–35
recirculation, 38
recirculation vortex, 39–40, 129
relaxation time, 43
residence time, 38, 123–124, 127–128
　dimensionless, 24–25, 25f
residence time-kiln geometry relation, 15, 19, 22–23
Reynolds number, 29–31, 34, 37, 151–152
Reynolds' principle, 29–31
Ricou, F. P., 36–37

rolling bed, 16–18, 18f
Rosin-Rammler relation, 118–119
rotary calcining integrated system, 266
rotary kilns, 266–268, 266f, *See also specific types of rotary kilns*
　basics, 2–3, 107
　competitive features, 2, 9, 249–250
　dams and tumblers, purpose of, 5–6, 8–9, 45
　design challenges, 46
　design structure, 278–280
　　comments, 280
　　mass balance scheme, 278–280, 279f
　efficiency, 41–42, 46
　environmental applications, 291
　evolution of, 1–5
　fuel types used in, 107–108, 114–115
　lifters, energy-savings with, 5–6
　other contact kilns compared, 3–5
　regulation of, 107–108, 117, 117t, 249–250
　sizing, 15–16
　types of, 5–7
　　coolers and dryers, 8–9
　　indirect fired kilns, 9
　　long dry kilns, 7–8
　　short dry kilns, 8
　　wet kilns, 7
rotary kilns, operation basics
　axial motion, 21–24
　bed phenomenon, 13–15
　dimensionless residence time, 24–25
　freeboard phenomenon, 13–14
　geometrical features and their transport effects, 15–16
　introduction, 13–14
　transverse bed motion, 16–19
　transverse flow behavior, 20–21
rotary reactor, 13, 15
round jets, 36–37
Ruhland, W., 128
rutile, 245–247

S
saltation, 43–46
　concept of particle saltation, 44f
　velocity, 121
Sarofim, A., 164
Savage, S. B., 49–50, 86, 92

Index 367

Schlichting, H., 28
Schneider, G. M., 41–42
scrubbing, 117
seals, 27–28
Seaman, W. C., 19, 21–24, 49–50
segregation
  mechanisms of, 85–87
  reducing, 87
segregation equation solutions
  mixing and segregation, 97, 104–106
  radial mixing, 96–97
  strongly segregating system, 95–96
segregation model
  application, 100–102
  within the bed, 89–91
  boundary conditions, 94–95
  governing equations, 91–94
  governing equations numerical solution, 97–99
  introduction, 87–89
  validation, 99
segregation rates, 85, 88–89
sensible heat calculations, 54
sensible heat defined, 203–204
serpentine ($3MgO \cdot 2SiO_2 \cdot 2H_2O$), 354t–355t
shaft-type kilns, 3, 3t
shale, 2, 247–255
  density versus temperature curve for, 252f–253f
  heat transfer modeling for, 196–200
  metal oxide composition of, 251t
shell heat loss, 154–155, 209, 214t–216t
ships, concrete, 247–248
short dry kilns, 8, 8f
sieving, inverse, 86
silicon dioxide (SiO_2)
  cristobalite, 351t–352t
  quartz, 350t
Singh, D. K., 49–50, 56–58, 174–177
skin friction coefficient, 32
slate, 248–253
  metal oxide composition of, 251t
slipping bed, 67
SL/RN kilns, 244–245
slumping bed, 16–17, 67
Spalding, D. B., 36–37
specific heat, 317t
squeeze expulsion, 89, 91

stack gases, 111
stationary kilns, 1
Stefan-Boltzmann constant, 179, 313
Strouhal number, 40–42
submerged jets, 35–36
sulfur dioxide (SO_2) emissions, 117
surface velocity, deep beds, 21
swirling jets, 39–40
swirl number, 39–40

T

Tackie, E. N., 45–46
Taylor, P., 295–297
temperature, formulas for calculating, 314,
  See also specific types of temperature
$4CaO \cdot 3TiO_2$, 343t
thermal balance and energy use, 270
thermal model
  description, 182–183
  one-dimensional for bed and freeboard, 184–187
  quasi-three-dimensional for the bed, 182–183
  of rotary kiln processes, 182
  two-dimensional for the bed, 187–188
thermal NO, 134–135
thermodynamics, chemical, 203–204
thermodynamic tables
  gases, 315
  inorganic materials, 316–349
Thring, M. W., 38
tire combustion, 117–118
titaniferous materials, 245–247
tongue, 89
toxicity equivalency factor (TEF), 292
toxicity equivalent quotient (TEQ), 292
trajectory segregation, 85–86
transport phenomenon
  history of, 1–2
  modeling, 13–14
trialumino disilicate ($3Al_2O_3 \cdot 2SiO_2$), 337t
$3CaO \cdot 2TiO_2$, 342t–343t
$3CaO \cdot V_2O_5$, 340t–341t
Tscheng, S. H., 177
turbulence modeling, 131
turbulent diffusion flames, 27–28, 34–35, 38
turbulent jets, 35–37, 127–128

turbulent kinetic energy (TKE), 127–128
turbulent variable swirl burners, 39–41

V
van der Hegge Zijnen, 36–37
vaporization, 111, 123–125, 126t
vapors
  H, U, S, G, 315, 323t–331t
  specific heats, 314
velocity, units of conversion, 313
Venkateswaran, V., 245
vertical kilns, 3–5, 3t
viscosity, 28–29, 81–82
volatile matter, 271
von Karman, T., 64
vortex shedding, 39–41, 129, 136–140

W
Wallouch, R. W., 272–274
Wang, R. H., 88
waste burning kilns, 291–299
  calculations, 293
waste fuels, 107–108, 114–115, 164, 206–207
Watkinson, A. P., 177
wet kilns, 7
wutsite (FeO), 348t–349t

Z
Zhang, Y., 57
zinc carbonate ($ZnCO_3$), 357t
zinc oxide (ZnO), 357t–358t